Advances in Rock and Mineral Materials

Advances in Rock and Mineral Materials

Guest Editors

Gorazd Žibret
Vilma Ducman
Lea Žibret

MDPI Basel • Beijing • Wuhan • Barcelona • Belgrade • Novi Sad • Cluj • Manchester

Guest Editors
Gorazd Žibret
Department of Mineral Resources and Geochemistry
Geological Survey of Slovenia
Ljubljana
Slovenia

Vilma Ducman
Laboratory for Cements, Mortars & Ceramics
Slovenian National Building and Civil Engineering Institute
Ljubljana
Slovenia

Lea Žibret
Laboratory for Cements, Mortars & Ceramics
Slovenian National Building and Civil Engineering Institute
Ljubljana
Slovenia

Editorial Office
MDPI AG
Grosspeteranlage 5
4052 Basel, Switzerland

This is a reprint of the Special Issue, published open access by the journal *Materials* (ISSN 1996-1944), freely accessible at: https://www.mdpi.com/journal/materials/special_issues/6N9G4OT7Y3.

For citation purposes, cite each article independently as indicated on the article page online and as indicated below:

Lastname, A.A.; Lastname, B.B. Article Title. *Journal Name* **Year**, *Volume Number*, Page Range.

ISBN 978-3-7258-6143-9 (Hbk)
ISBN 978-3-7258-6144-6 (PDF)
https://doi.org/10.3390/books978-3-7258-6144-6

Cover image courtesy of Gorazd Žibret

Contents

About the Editors

Gorazd Žibret

Gorazd Žibret is a senior researcher at the Department of Mineral Resources and Geochemistry, Geological Survey of Slovenia, and assistant professor at the Postgraduate School ZRC SAZU, Ljubljana. His main profession is geology, with more than 20 years of experience in geochemistry and economic geology. His research focuses on the environmental impacts of mining and ore processing activities, primary raw material exploration and extraction, secondary raw material assessments, and data mining techniques in geosciences.

Vilma Ducman

Vilma Ducman is a senior researcher and head of the Laboratory for Cements, Mortars & Ceramics at the Slovenian National Building and Civil Engineering Institute. Her research interests include studies of alkali-activated materials, geopolymers, artificial lightweight aggregates, and carbonation for up-cycling in the construction sector.

Lea Žibret

Lea Žibret is a research associate at the Laboratory for Cements, Mortars & Ceramics at the Slovenian National Building and Civil Engineering Institute. Her research interests include studies of alkali-activated materials, clay-based materials, secondary raw materials, geological studies and X-ray diffraction studies.

materials

MDPI

Article

Use of a Glaciogene Marine Clay (Ilulissat, Greenland) in a Pilot Production of Red Bricks

Louise J. Belmonte, Lisbeth M. Ottosen and Gunvor M. Kirkelund *

Department of Environmental and Resource Engineering, Technical University of Denmark (DTU), 2800 Lyngby, Denmark; limo@dtu.dk (L.M.O.)
* Correspondence: gunki@dtu.dk

Abstract: Uplifted occurrences of fine-grained glaciogene marine sediments are found throughout the northern hemisphere. These sediments could be used to produce local construction materials, to rely less on imported construction materials from southern regions. In this study, a representative occurrence from Ilulissat, West Greenland, was investigated as a potential resource for local brick production. The study comprised three parts: (1) raw material characterization based on grain size distribution, major element chemistry, including total carbon, sulfur, and chloride concentrations, mineralogy, morphology, and Atterberg limits; (2) the production of test bricks at a Danish brickwork; and (3) testing of the bricks based on total shrinkage, water absorption, hygroscopic adsorption, open porosity, bulk density, compression strength, and mineralogy. The bricks produced proved to have excellent compression strength, low open porosity, and low water absorption. The shrinkage could be reduced by adding 10% chamotte to the marine sediment. Based on the investigated properties, this indicates that this type of clay is highly suitable as a resource for bricks.

Keywords: Arctic; marine sediment; construction material; fired brick; brickwork; pilot

1. Introduction

The Arctic climate places high demands on the construction materials used, which are, at the same time, mostly imported from southern regions, in the climatic conditions in which they were intentionally developed. The high cost and logistical challenges of importing construction materials to the Arctic make it relevant to explore locally sourced materials. Large occurrences of fine-grained glaciogene marine sediments were deposited after the last glaciation and some later uplifted above sea level due to isostatic movements. Today, they can be found all over the formerly glaciated northern hemisphere regions, such as Canada, northern Scandinavia, and Greenland [1–6]. The marine sediments are dominated by a <60 μm fraction (clay and silt fractions) and contain only a minor amount of the 60–2000 μm fraction (sand fraction).

Studies of Greenlandic, Norwegian, and North American marine sediments have shown qualitatively similar mineralogy results often consisting of feldspars, quartz, amphibole, mica/illite, chlorite, expandable clay minerals (e.g., smectite, vermiculite, and mixed layer clays), and calcite, and their fine and colloid fractions contain a high percentage of primary minerals, such as quartz, feldspars, and amphibole [1–5,7–9]. The chemical composition of the studied samples of marine sediments from this region is shown in Table 1, compared to marine sediments from other areas and Danish clay, both used for brick production.

Table 1. Major element composition of marine sediments and brick clays.

	Glaciogene Marine Sediments					
Chemical Oxide (wt%)	Greenland (North Atlantic Craton) [9]	Greenland (Nagssugtoqidian Orogen) [10–12]	Norway [5,8]	Canada [2,3]	Marine Sediments Used for Clay Bricks [13–16]	Typical Clay for Danish Red Bricks [17]
SiO_2	53.99	57.5–64.2	45.8–53.6	46.8–62.6	47.13–63	55.46–76.97
TiO_2	0.71	0.6–0.8	0.8–1.1	0.0–1.0	0.63–0.81	
Al_2O_3	16.44	14.5–16.2	14.3–21.5	15.4–23.4	7.69–17.23	9.59–14.41
Fe_2O_3T	8.55	6.2–6.6	8.2–15.2	4.4–8.8	4.13–21.55	3.76–6.96
MnO	0.12	0.1	0.0–0.2	0.1	0.03–0.31	
MgO	4.94	2.4–3.8	3.9–5.1	2.1–4.9	0.05–3.6	0.87–1.71
CaO	3.72	2.4–4.4	1.0–4.1	2.9–5.4	0.68–7.49	0.68–2.48
Na_2O	3.84	3.3–4.5	1.2–2.4	1.3–4.1	0.61–1.9	0.48–1.19
K_2O	3.5	2.7–2.9	4.7–6.0	2.4–3.7	1.36–2.13	2.49–3.0
P_2O_5	0.12	0.1–0.2		0.1–0.3	0.15–0.32	0.06–0.12
LOI	3.72	2.4–5.1	3.5–10.3	1.6–11.9	10–19.35	

In West Greenland, quaternary glaciogene marine sediment deposits are observed throughout the inhabited coastal regions and are, in most cases, easily accessible from towns and settlements. In addition, the Greenlandic ice sheet is continuously transporting sediments toward the coastal areas for deposition [18], which make sediments an abundant resource in Greenland. The geology of West Greenland is dominated by Precambrian crystalline basement rocks (approximately 3.8–1.7 Ga) consisting of gneisses (associated with the greenschist to granulite facies metamorphism) and some mafic, ultramafic, pegmatitic, and granitic occurrences [19]. Larger occurrences of quaternary marine sediments have also been mapped in these areas [20], and the chemical composition of the sediments will vary to some extent based on their provenance area. The Greenlandic marine sediments in Table 1 include samples from the Nagssugtoqidian orogen [21] (Ilulissat, Kangerlussuaq, and Qasigiannguit) as well as the North Atlantic Craton (Ilulialik). The Greenlandic sediments generally fall within the same range of chemical composition as Canadian sediments. However, in comparison to Norwegian sediments, Greenlandic sediments contain slightly more SiO_2 and Na_2O and slightly less Fe_2O_3T (i.e., total Fe_2O_3), MgO, and K_2O. This reflects regional differences in sediment provenance.

Research has established that marine sediments are suitable as primary and secondary resources in the production of clay ceramics [8–11]. The chemical composition of Greenlandic marine sediments is overall comparable to the chemical composition of marine sediments [13–16] and Danish clay [17] that have previously been used for bricks (Table 1). Today, bricks are neither commonly used nor produced in Greenland, where construction materials, such as wood and concrete, are dominant [22], and the total import of materials for building and construction is about 25% of the total import to the country [23]. Depending on the raw materials and production methods, bricks have several advantages that make them an interesting building material in Arctic climate regions, such as Greenland. They are durable and require almost no maintenance compared to, e.g., wood. Furthermore, they are fire-resistant and have good thermal and acoustic properties [24]. Previous studies of small clay discs (d = 20 mm, h = 3 mm) produced using Greenlandic marine sediment, combined with local particulate waste materials, such as waste incineration ashes and mine tailings, have shown physical and chemical properties that would make the marine sediment suitable for brick production [10,25]. Brick discs with Greenlandic marine clay from the town of Sisimiut even showed less shrinkage during firing, as well as lower porosity and water adsorption and a higher density than brick discs made from commercially used Danish clay material [10,25]. Low shrinkage and high density indicate that Greenlandic marine sediments will be stable during brick production and result in strong bricks, whereas low porosity and water absorption can indicate a more durable brick. Therefore, there could be potential in producing bricks from Greenlandic clay materials, and these promising laboratory-scale results are to be verified by real-size bricks produced in industrial settings. In this study, marine sediment from the town of Ilulissat, West Greenland, was characterized and tested for use in industrial brick manufacturing. Furthermore, the technical suitability of the produced bricks as a local building material was evaluated by parameters such as compression strength and water absorption.

2. Materials and Methods

2.1. Materials and Sampling

The investigated raw material was a fine-grained glaciogene marine sediment (MS) from the town of Ilulissat, West Greenland (sampling point red arrow (22 W 0495779 7678250) in Figure 1). Approximately 150 kg of MS was sampled from a depth interval of 20–100 cm below surface. The material was roughly homogenized with a shovel and distributed into 10 L plastic buckets, which were sealed and stored at room temperature until use. Furthermore, two additional raw materials (CR and CH) were used to produce the bricks. CR is a fine-grained residue produced from crushing granitic rocks for aggregate production, which is produced at Betoncentralen, Nuuk, Greenland. CH—chamotte—is a ceramic product with a high alumina and silicon content. CH is currently not produced in Greenland, but can be manufactured by crushing and milling discarded bricks from a future production.

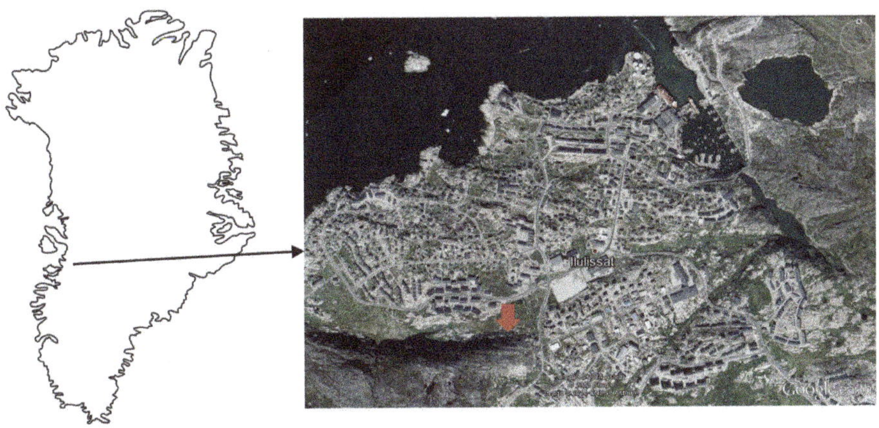

Figure 1. Location of sample extraction (marked by the red arrow): Ilulissat, Greenland.

2.2. Characterization of Raw Material

The particle size distribution of MS, CR, and CH was determined by laser diffractometry. The samples were pre-dispersed in approximately 20 mL 0.005 M $Na_4P_2O_7$ using ultrasonics, and then analyzed by wet dispersion in deionized water on a Malvern Mastersizer 2000 laser diffractometer (Malvern Instruments, Malvern, UK). The Mie theory was applied for calculating the particle size distribution using a real refractive index of 1.5 and an imaginary refractive index of 0.1.

Majorbelement analyses were determined using the X-ray fluorescence (XRF) technique on pressed powder tablets prepared from crushed bulk material. The tablets were analyzed on a PanAlytical MagiX PRO (PanAlytical, Almelo, The Netherlands) equipped with a Rh-tube and a gas flow detector with a resolution (Mn-Kα) of 35 eV. Excitation voltage and current were 60 kV and 53 mA, respectively. The loss on ignition (LOI) was based on weight loss between 105 °C and 1000 °C and was measured after heating at 1000 °C for 2 h.

The mineralogy was analyzed by X-ray diffraction (XRD) on a X'Pert PRO diffractometer (PanAlytical, Almelo, The Netherlands), using Cu Kα radiation generated at 45 kV and 40 mA. In addition to bulk analyses, the <2 μm fraction of MS was separated by sedimentation and analyzed as orientated samples after treatments of air-drying, ethylene glycolation (vapor) at room temperature for at least 48 h, and heating at 350 °C and 550 °C for a minimum of 2 h. The mineral phases in the <2 μm fraction were identified according to [26]. The morphology of the materials (mounted on carbon tape) was investigated by scanning electron microscopy (SEM) on an FEI Quanta 200. Furthermore, total carbon (TC), sulfur (S), chloride (Cl^-), and Atterberg limits were measured for MS. TC and S were measured using the combustion infrared detection method on LECO CS-200 (LECO Nordic, Täby, Sweden). The Cl^- concentration of the pore water: 4 g of dried MS was dispersed in 50 mL of distilled H_2O for 24 h and filtrated through a 0.45 μm filter by use of vacuum pumping. Cl^- (mg/L) was analyzed by ion chromatography (IC). The Atterberg limits (liquid and plastic limits) were determined in accordance with DS/CEN ISO/TS 17892-12 [27]. A 0.425 mm sieve was used for the determination of the liquid limit. An additional determination of the liquid limit was carried out using the Casagrande method [28].

2.3. Production of Fired Specimens

To evaluate the technical properties, a small-scale test production of fired specimens was carried out at the Danish brickworks, Petersen Tegl. The raw materials were untreated prior to production; however, larger particles (>5 mm) were removed by hand if encountered during mixing and molding. Three main compositions consisting of MS (in pure form or with 10% substitutions of either CR or CH) were prepared (Table 2).

Table 2. Specifications for the test specimens.

Specimen Name	MS (wt%)	CR (wt%)	CH (wt%)	No. of Specimens	Coal Furnace	Electrical Furnace	Firing Temperature (°C)
1A	100			2	X		1045–1055
1B	100			3		X	1070–1080
2A	90	10		3	X		1045–1055
2B	90	10		2		X	1030–1040
3A	90		10	3	X		1045–1055
3B	90		10	3		X	1030–1040

The compositions were mixed thoroughly in an industrial mixer and water was added to obtain a workable mass. The specimens were prepared by hand using the soft-mud method [29] in a water lubricated wooden mold measuring 18.6 cm × 12 cm × 6 cm (L × W × H). A wooden piston was used to aid the removal of the mold from the formed specimens. After forming, the specimens were included in the normal production routine at Petersen Tegl. They were dried at a maximum of 70 °C for 3–5 days. From each composition, half of the specimens were fired in a fully automated coal-fired tunnel kiln (reaching a maximum temperature of approximately 1045–1055 °C) and the other half were fired in an electrical furnace (reaching a maximum temperature of either 1030–1040 °C or 1070–1080 °C). In both furnaces, the heating and cooling rates were approximately 20 °C/hour, using a lowered rate of 4–10 °C/h in the interval of 520–620 °C. The specimens were maintained (soaked) at the maximum temperature for approximately 4 h.

2.4. Characterization of Fired Specimens

The total shrinkage of the specimens after firing was calculated as a percentage of the initial molded length before drying. Water absorption due to capillary action was determined after DS/EN 772-11 [30] at a temperature of 20 °C (±2 °C). The samples were placed in a large tray filled with distilled water to cover 5 mm of the total brick height. The weight of the samples was measured on dry-wiped samples at certain time intervals until the weight was stable for 24 h. Thereafter, the initial rate of water absorption (IRA) and capillary coefficient (k-value) were determined. For the remaining tests, the specimens were cut (parallel to the width of the specimen) into smaller pieces of approximately 4 × 10 × 5 cm (length × width × height). The open porosity and bulk density were determined according to DS/EN 772-4 [31] using a desiccator. The samples were first placed under vacuum (2 kPa) for 24 h, then under vacuum with added distilled water for 24 h, before the vacuum was released and the bricks were submerged in water for 24 h. The samples were thereafter weighed under water and immediately after being wiped dry. The compression strength was measured on an Amsler 60 tonne hydraulic press on three pieces from each specimen, which were dried at 105 °C for a minimum of three days. For comparison, two Danish red bricks, WT131 (produced by Wienerberger, Helsinge, Denmark) and D46 (produced by Petersen Tegl, Broager, Denmark), were included in the compression tests. The hygroscopic adsorption properties were determined according to ISO 12571 [32] using the desiccator method, where desiccators with different salts (LiCl, $MgCl_2$, NaBr, NaCl, KCl, KNO_3, and K_2SO_4) were used to keep different percentages of constant relative humidities. For this test, small samples of approximately 1 cm × 1 cm × 2 cm (L × W × H) and a weight of 3–11 g were cut from a piece from each specimen. Only one square prism from each sample was placed in the individual desiccators. The adsorption curves were established at a temperature of 20 °C (±2 °C). The bulk mineralogy of the fired specimens was investigated by XRD using the same equipment and operating conditions as described in Section 2.2.

3. Results

3.1. *Raw Materials*

The particle size distributions are presented in Figure 2. The d50 values were approximately 0.02 mm, 0.6 mm, and 0.05 mm for MS, CR, and CH, respectively. MS was dominated by the <60 µm fraction (clay and silt fractions).

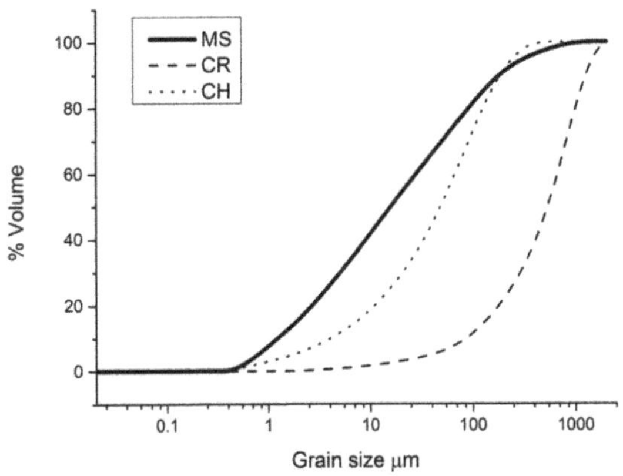

Figure 2. The particle size distributions of MS, CR, and CH.

Raw material properties are presented in Figure 3 and Table 3. SiO_2, Al_2O_3, and Fe_2O_3T were the dominating oxides for all raw materials; however, CR had a high percentage of MgO (9.8 wt%), which made MgO more dominant than Fe_2O_3 for this raw material. The total amounts of alkali and alkaline earth elements (K_2O, Na_2O, MgO, and CaO) were 12 wt%, 18.9 wt%, and 4.3 wt% for MS, CR, and CH, respectively. Additional analyses of TC, S, Cl^-, and Atterberg limits measured for MS are also shown in Table 3.

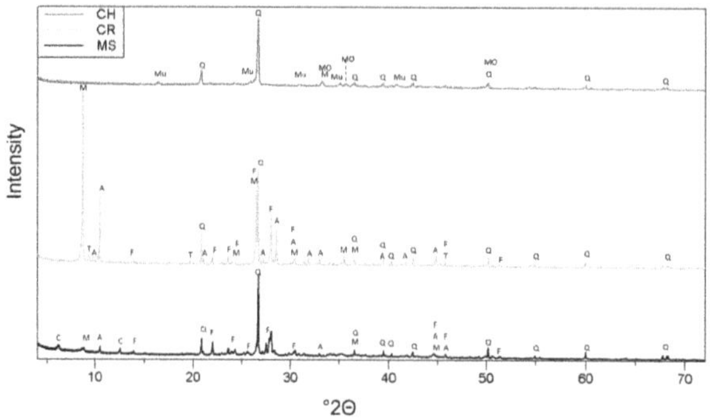

(**a**)

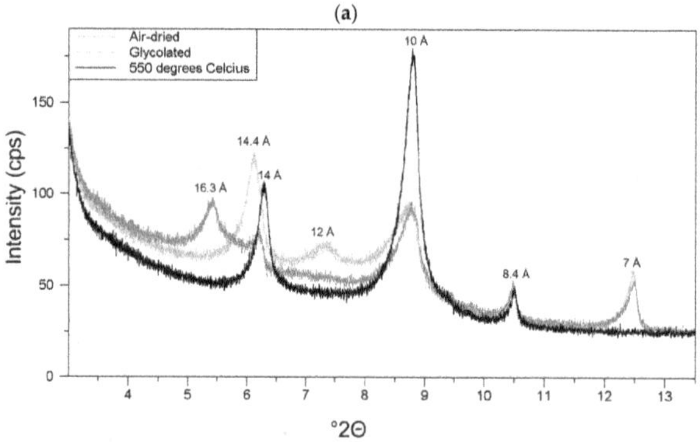

(**b**)

Figure 3. (**a**) The bulk mineralogical composition of MS, CR, and CH. A = amphibole, C = chlorite, F = feldspar, M = mica, MO = metal oxide (most likely an Fecompound, e.g., hematite), Mu = Mullite, Q = quartz, and T = talc. (**b**) The clay mineral composition of MS.

Table 3. Characteristics of raw materials. N.M. = not measured.

		MS	CR	CH
Major elements (XRF)	SiO_2 (wt%)	61.12	61.77	66.94
	TiO_2 (wt%)	0.59	0.39	1.28
	Al_2O_3 (wt%)	14.52	10.63	19.22
	Fe_2O_3T (wt%)	6.48	6.45	7.80
	MnO (wt%)	0.08	0.11	0.06
	MgO (wt%)	3.52	9.67	0.99
	CaO (wt%)	2.36	3.64	0.33
	Na_2O (wt%)	3.31	3.11	0.43
	K_2O (wt%)	2.81	2.51	2.59
	P_2O_5 (wt%)	0.12	0.04	0.06
	$LOI_{1000\,^{\circ}C}$ (wt%)	5.09	1.68	0.30
	TC (wt%)	1.41 ± 0.18	N.M.	N.M.
	S (wt%)	0.10 ± 0.08	N.M.	N.M.
	Cl^- (mg/L)	3.29 ± 0.04	N.M.	N.M.
Atterberg limits	Water content (%)	23.5 ± 1.7	N.M.	N.M.
	Liquid Limit (%)	30.2 ± 0.2	N.M.	N.M.
	Plastic Limit (%)	20.5	N.M.	N.M.
	Plasticity Index	9.7	N.M.	N.M.
	Activity	0.23	N.M.	N.M.

3.2. Fired Brick Specimens

The total linear shrinkages, bulk (apparent) densities, and open porosities of the six different compositions are presented in Table 4. Composition 2B had the lowest linear shrinkage but also the highest open porosity and lowest bulk density; the opposite was true for composition 1B. The initial rate of water absorption (IRA) and capillary coefficients (k-values) are also presented in Table 4. They generally correlated well, with the IRA values being about a factor-ten higher than the k-values. The compression strengths of the six compositions and of the WT131 and D46 bricks are plotted in Figure 4, and the Greenlandic bricks showed remarkable strength, up to eight-times higher than the values obtained for the regular bricks. The adsorption curves are shown in Figure 5. The moisture content (U) is defined according to ISO 12571 [32] as $(m1 - m0)/(m0) \cdot 100\%$, where m0 is the initial dry mass and m1 is the mass of the same prism when equilibrium with the environment is achieved. All samples adsorbed moisture at very high relative humidity levels and only up to a few percent. As these determinations were based on the measurements of a single square prism, only the general trend can be outlined from the investigations. For compositions 2B and 3B, the adsorption initiated at a low relative humidity level, whereas it was only apparent at >80% relative humidity for compositions 1A, 1B, 2A, and 3A.

Table 4. Properties of brick specimens.

	Total Shrinkage %	Bulk (Apparent) Density kg/m^3	Open Porosity %	IRA kg/(m$^2\times$ min)	k-Value kg/(m$^2\times$ s^{½})
1A	13.8 ± 0.5	2350	4.9	0.170 ± 0.060	0.018 ± 0.005
1B	14.9 ± 0.4	2400	2.9	0.080 ± 0.050	0.008 ± 0.004
2A	11.0 ± 0.5	2260	16.8	0.300 ± 0.010	0.035 ± 0.003
2B	7.7 ± 0.8	1960	38.2	1.630 ± 0.060	0.183 ± 0.010
3A	13.8 ± 0.2	2360	9.3	0.080 ± 0.020	0.009 ± 0.003
3B	11.0 ± 0.4	2090	27.2	0.890 ± 0.180	0.094 ± 0.019

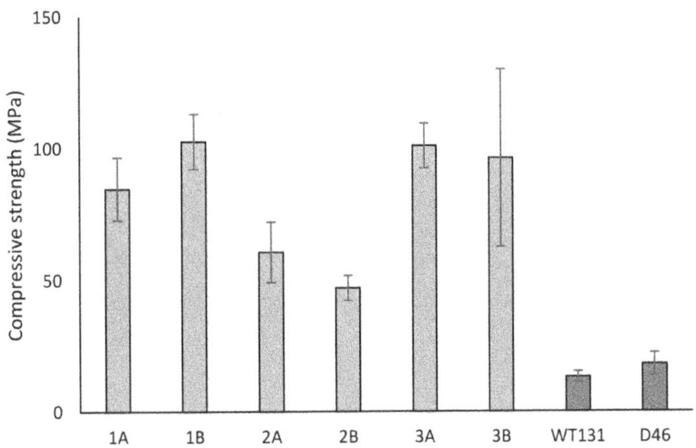

Figure 4. Compressive strength of bricks.

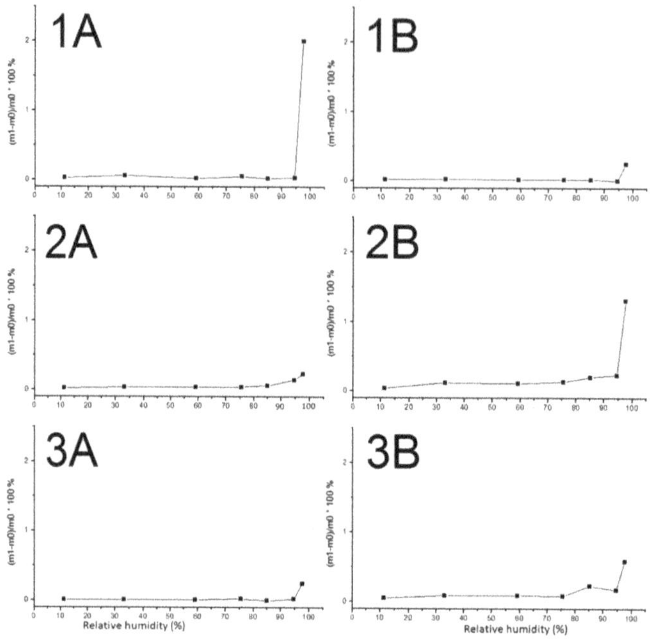

Figure 5. The adsorption curves of the six tested compositions (Table 2) measured at 20 ± 2 °C.

The results from the bulk mineralogy analyses of all the fired compositions are shown in Figure 6. All patterns had inherited features from MS (compared with Figure 3a). Most of the fired specimens lacked phases, such as chlorite, mica, and amphibole, in comparison to MS, but gained metal-oxide (hematite) and showed indications of the high-temperature phase, cristobalite. Peak shifts were observed for some of the feldspar peaks and were likely explained by high-temperature phase transitions, e.g., microcline to sanidine. The SEM micrographs (Figure 7) show the morphology of the raw materials and fired specimens.

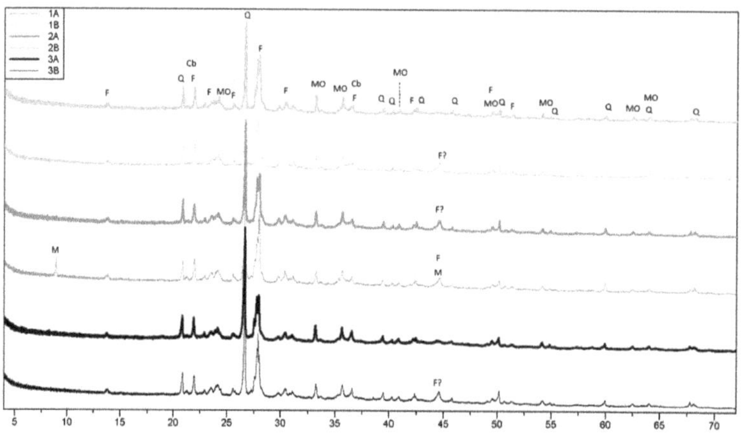

Figure 6. X-ray diffractograms of the fired specimens: cristobalite (Cb), feldspar (F), mica (M), metal oxide (MO), and quartz (Q).

(a)

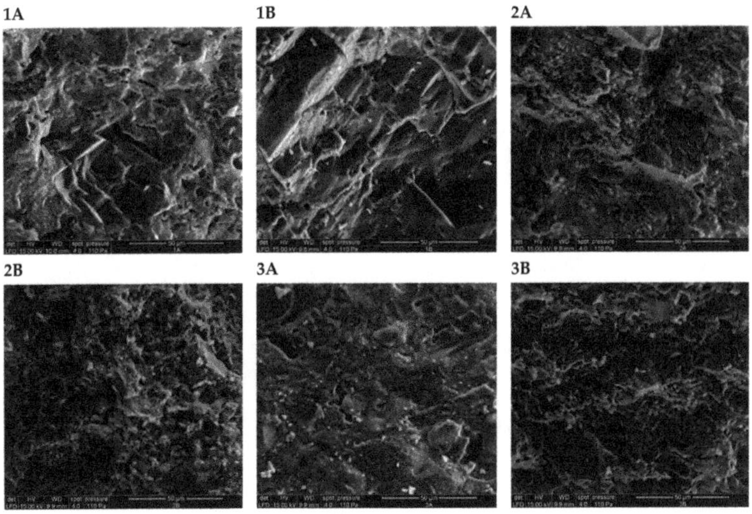

(b)

Figure 7. Micrographs of (**a**) raw materials (MS—marine sediment, CR—crushed rock, CH—chamotte) and (**b**) fired specimens (Table 2). The magnification of all images (bar showing 50 μm) is the same, except for CR (300 μm).

4. Discussion

4.1. Raw Material Characterization

The particle size distribution, mineralogy, major element distribution, carbon content, and Atterberg limits showed that MS is very similar to other known marine clays from Greenland, Canada, and Norway [2–5,7,8]. The results from the bulk mineralogy analyses of MS, CR, and CH (Figure 3a) show the mica in MS to be dominantly trioctahedral (biotite-type), based on a (060) reflection of around 1.54 Å and very weak peaks in the 1.50 Å region [22]. For CR, the octahedral coordination was less clear, but appeared to be more trioctahedral based on stronger reflections around 1.54 Å than around 1.50 Å. The high MgO wt% observed for CR (Table 3) corresponds well with the abundant mica (biotite) found from the XRD investigations. For CR, the presence of pyroxene was also indicated; however, due to peak overlapping, this could not be clearly resolved. For CH, cristobalite was observed, but is not displayed in the figure due to the low intensity of the peaks. The result of the clay mineral analyses of MS is presented in Figure 3b. Expandable (smectitic) clay was identified based on the glycolated peak around 16.3 Å. Mixed-layer clay was identified based on the air-dried peak around 12 Å, which disappeared after glycolation and chlorite was identified based on the presence and increase in the 14 Å peak and the disappearance of the second-order (7 Å) reflection at 550 °C. The shift from 14.4 Å to 14 Å from air-dried to 550 °C could indicate the presence of vermiculite. Illite/biotite was identified from the 10 Å peak (before heating). The 8.4 Å peak (Figure 3b) was assigned to amphibole, which together with quartz and feldspar peaks from the bulk analysis (Figure 3a) confirmed the presence of primary minerals in the <2 μm fraction.

The low TC (Table 3) was consistent with the low content of organic carbon observed for other marine clays in Greenland, Canada, and Norway (e.g., [1]). Furthermore, the low TC also indicated a low content of $CaCO_3$, which, combined with the relatively high amount of Fe_2O_3T, contributed to the red coloration of the bricks after firing. The low chloride concentration of MS indicates that, in spite of its marine origin, the original salt content in the pore water has since been leached out. The morphology of the three raw materials as studied by SEM revealed that CR and CH generally contained grains of euhedral textures, whereas the grains in MS had a more subhedral texture. Furthermore, the surfaces of larger grains in MS were generally covered by smaller grains with a flaky or powdery appearance, which were likely to be weathering products, e.g., clay minerals. This observation was less pronounced for CR and CH.

The relatively low plasticity index of MS (Table 3) presents a challenge when producing bricks by the soft-mud method. The extrusion (stiff mud) or dry-pressing methods [29] would therefore possibly be better for industrial production using this type of clay.

4.2. *Influence of Production Settings*

To observe the effect of the different temperatures on the technical properties of the produced bricks, the pure MS compositions, 1A and 1B, were fired initially at 1045–1055 °C and 1070–1080 °C, respectively. The higher firing temperature of 1B led to an increase in total shrinkage, bulk density, and compression strength, and a decrease in water absorption and open porosity. As no obvious benefits could justify the higher temperature and therefore energy consumption, it was decided to use lower firing temperatures of 1030–1055 °C for compositions 2A, 2B, 3A, and 3B. Furthermore, the relatively high total shrinkage of both 1A and 1B gave rise to the idea of introducing the two additives, CR and CH, both of which had a coarser grain size distribution and a lower LOI than MS and therefore were less likely to contribute to shrinkage.

Compared to small brick discs produced at the laboratory scale, which were formed and pressed at 47 MPa, the open porosity (39.5%), density (1691 kg/m^3), and water absorption (23.3%) [15] were significantly different from the bricks made at the brickwork in this study, even when using the same marine sediment. Additionally, the firing curve in the laboratory was faster and performed at a lower temperature (maximum 1000 °C for 1 h [10]). These findings emphasis the importance of developing laboratory production methods that closely correlate with industrial production, both in terms of forming pressure and firing curve, to enable direct comparisons of brick properties between laboratory and pilot scales.

4.3. *Shrinkage*

As expected, the total shrinkage reduced for the CR-containing composition (2A) compared to the pure MS composition (1A). However, despite a coarser grain size distribution of CH compared to MS (Figure 2), there was no observable difference in total shrinkage between compositions 3A and 1A. Lowering the firing temperature to 1030–1040 °C for compositions 2B and 3B also reduced the total shrinkage. According to the Brick Industry Association [29], the total size shrinkage after drying and firing can vary for different clays, but will usually be around 4.5–8%. For the compositions tested here, only 2B falls within the higher end of this range. However, although the total shrinkage is generally in the range of 11–15% for the compositions in this study, it does not appear to be detrimental, as only very few cracks were observed on the surfaces of the specimens after firing. Most of these cracks were correlated to underlying larger rock fragments and could possibly have been avoided by crushing or by particle size sorting of the material before production, which is standard practice in industrial brick manufacturing.

4.4. Initial Rate of Water Absorption and Hygroscopic Behavior

According to ASTM C216-14 [33], the initial rate of water absorption (IRA) for facing bricks should be up to 1.5 kg/(m^2·min). Bricks with higher IRA values require pre-wetting, as they will otherwise prevent the complete hydration of the cement, due to the absorption of water from mortar. Bricks with a low IRA should be covered on the construction site in order to prevent wetting, as this might prevent or delay the formation of bonds between the brick and the cement. The compositions 1A, 1B, 2A, and 3A, which were fired at temperatures of 1045–1080 °C, all have low IRA values, whereas compositions 2B and 3B, which were fired at 1030–1040 °C, have IRA values closer to 1.5 (Table 4). All compositions had very low hygroscopic adsorption (Figure 5). However, the adsorption of water at lower relative moisture contents for compositions 2B and 3B compared to the other compositions indicates that 2B and 3B have larger available inner surface areas where water can adsorb. Furthermore, the combined studies on hygroscopic adsorption and capillary uptake (Table 4) indicate that the higher firing temperatures of compositions 1A, 1B, 2A, and 3A reduced the amount of pores capable of capillary uptake compared to the lower-temperature compositions 2B and 3B.

4.5. Degree of Sintering

Vieira and Monteiro [34] investigated a fine-grained granitic waste material for incorporation in ceramics, which had a d50 of approximately 10 μm. They concluded that the fine particle size distribution favored sintering and reduced the porosity in the produced ceramics. Furthermore, the material was shown to have a considerable fluxing potential at temperatures around or slightly above 1000 °C due to a high amount of alkaline oxides present in the form of biotite and Na,K-feldspar, which contributed to the formation of a liquid phase. As both the grain size distribution and qualitative mineralogy of the material characterized by Vieira and Monteiro [34] are comparable to MS, it is likely that similar reactions would occur here. In addition, the very fine particle sizes observed on the surfaces of the grains in MS (Figure 7) are also likely to enhance sintering. When comparing the XRD results of the raw materials to the XRD results of the fired compositions, it is evident that phases such as mica (biotite), amphibole, and chlorite disappeared after firing, which could be explained by the breakdown/melting of these phases at or below the temperatures investigated. Interestingly, composition 2B, which was fired at 1030–1040 °C, still contains some mica (peak at approximately 8.9°2Θ), which is not evident in composition 2A fired at 1045–1055 °C. This indicates that at least some of the mica encountered in CR is stable at temperatures above the range of 1030–1040 °C, but will breakdown at or below temperatures of 1045–1055 °C. The mica in MS appears to breakdown at temperatures below the range of 1030–1040 °C as no mica peaks are apparent in composition 3B, which was fired at this temperature.

The coarser particle size distribution of CR, which has a d50 of approximately 600 μm, is less likely to enhance sintering and reduce porosity, which is consistent with the reduced strength (Figure 4) and higher open porosity values found for composition 2A compared to 1A (Table 4). In general, the strength, density, and shrinkage of most of the tested compositions indicate a high degree of sintering.

5. Conclusions

The results demonstrate that the marine sediment from Ilulissat (MS) is suitable as brick clay and that it is possible to produce high-quality bricks. The produced specimens were fired at temperatures in the range of 1030–1080 °C. The highest temperatures did not contribute to further enhancements of the technical properties, and the temperature range of 1030–1055 °C was therefore sufficient for the marine sediment. The specimens fired at 1045–1055 °C generally had very high compression strengths, low porosities, and low water absorptions. The introduction of 10% of the crushing rock residue gave rise to a reduction in strength, higher porosity, and water absorption. In addition, the specimens fired at 1030–1040 °C had the lowest strength, highest porosity, and water absorption. On the contrary, the use of 10% chamotte with marine sediment did not impair the bricks' properties. The bricks showed greater shrinkage compared to previous, small laboratory samples, and show the importance of conducting large-scale tests under industrial forming pressures and firing conditions.

Author Contributions: Conceptualization, L.J.B., L.M.O. and G.M.K.; investigation, L.J.B.; writing—original draft preparation, L.J.B.; writing—review and editing, L.M.O. and G.M.K.; supervision, L.M.O. and G.M.K. All authors have read and agreed to the published version of the manuscript.

Funding: This research received no external funding.

Institutional Review Board Statement: Not applicable.

Informed Consent Statement: Not applicable.

Data Availability Statement: The raw data supporting the conclusions of this article will be made available by the authors on request.

Acknowledgments: The authors wish to thank Petersen Tegl for collaborating on the production of the brick specimens.

Conflicts of Interest: The authors declare no conflicts of interest.

References

1. Foged, N. Ingeniørgeologiske Undersøgelser af Kvartære Marine Leraflejringer på Vestgrønland. Ph.D. Thesis, Technical University of Denmark, Lyngby, Denmark, 1979. (In Danish)
2. Gillott, J.E. Fabric, composition and properties of sensitive soils from Canada, Alaska and Norway. *Eng. Geol.* **1979**, *14*, 149–172. [CrossRef]
3. Locat, J.; Lefebvre, G.; Ballivy, G. Mineralogy, chemistry, and physical properties interrelationships of some sensitive clays from Eastern Canada. *Can. Geotech. J.* **1984**, *21*, 530–540. [CrossRef]
4. Ramesh, R.; d'Anglejan, B. Mineralogy, Chemistry and particle size interrelationships in some Post-Glacial Marine Deposits of the St. Lawrence Lowlands. *J. Coast. Res.* **1995**, *11*, 1167–1179.
5. Roaldset, E. Mineralogy and geochemistry of Quaternary clays in the Numedal area, southern Norway. *Nor. Geol. Tidsskr.* **1972**, *52*, 335–369.
6. Rosenqvist, I.T. Origin and mineralogy glacial and interglacial clays of southern Norway. *Clays Clay Miner.* **1975**, *23*, 153–159. [CrossRef]
7. Bentley, S.P.; Smalley, I.J. Mineralogy of sensitive clays from Quebec. *Can. Mineral.* **1978**, *16*, 103–112.
8. Pederstad, K.; Jørgensen, P. Weathering in a marine clay during postglacial time. *Clay Miner.* **1985**, *20*, 477–491. [CrossRef]
9. Gunnarsen, K.C.; Jensen, L.S.; Rosing, M.T.; Dietzen, C. Greenlandic glacial rock flour improces crop yield in organic agricultural production. *Nutr. Cycl. Agroecosystems* **2023**, *126*, 51–66. [CrossRef]
10. Belmonte, L.J.; Ottosen, L.M.; Kirkelund, G.M.; Jensen, P.E.; Vestbø, A.P. Screening of heavy metal containing waste types for use as raw material in Arctic clay-based bricks. *Environ. Sci. Pollut. Res.* **2018**, *25*, 32831–32843. [CrossRef] [PubMed]
11. Belmonte, L.J.; Villumsen, A.; Ottosen, L.M.; Kirkelund, G.M. Use of clay from Kangerlussuaq in the Greenlandic construction industry. In Proceedings of the Rock Mechanics in the Nordic Countries, Kongsberg, Norway, 9–12 June 2010.
12. Kalvig, P. Industrimineraler i Grønland—En Vurdering af Nogle Udnyttelsesmuligheder. Ph.D. Thesis, Technical University of Denmark, Lyngby, Denmark, 1990. (In Danish)
13. Hamer, K.; Karius, V. Brick production with dredged harbour sediments. An industrial-scale experiment. *Waste Manag.* **2002**, *22*, 521–530. [PubMed]
14. Mezencevova, A.; Yeboah, N.N.; Burns, S.E.; Kahn, L.F.; Kurtis, K.E. Utilization of Savannah Harbor river sediments as the primary raw material in production of bricks. *J. Environ. Manag.* **2012**, *113*, 128–136. [CrossRef] [PubMed]
15. Romero, M.; Andrés, A.; Alonso, R.; Viguri, J.; Rincón, J.M. Sintering behaviour of ceramic bodies from contaminated marine sediments. *Ceram. Int.* **2008**, *34*, 1917–1924. [CrossRef]
16. Salim, W.S.W.; Sadikon, S.F.; Salleh, S.M.; Noor, N.A.M.; Arshad, M.F.; Wahid, N. Assessment of physical properties and chemical compositions of Kuala Perlis dredged marine sediment as a potential brick material. In Proceedings of the IEEE Symposium on Business, Engineering and Industrial Applications, Bandung, Indonesia, 23–26 September 2012; pp. 509–512.
17. Holmboe, T. *Teglværksler i Danmark—Sammensætningen af Dansk Teglværksler, Hårdtbrændende Ler, Alternative Lertyper og Kortlægning Med Stang Slingram*; Danmarks og Grønlands geologiske undersøgelse rapport 92; GEUS: Copenhagen, Denmark, 2001. (In Danish) [CrossRef]
18. Overeem, I.; Hudson, B.; Syvitski, J.P.M.; Mikkelsen, A.B.; Hasholt, B.; Broeke, M.R.; Noël, B.P.Y.; Morlighem, M. Substantial export of suspended sediment to the global oceans from glacial erosion in Greenland. *Nat. Geosci.* **2017**, *10*, 859–863. [CrossRef]

19. Stendahl, H.; Stensgaard, B.M.; Secher, K. Geological environments favourable for future mining. *Geol. Ore* **2009**, *16*, 1–12. [CrossRef]
20. MacGregor, J.A.; Colgan, W.T.; Paxman, G.J.G.; Tinto, K.J.; Csathó, B.; Darbyshire, F.A.; Fahnestock, M.A.; Kokfelt, T.F.; MacKie, E.J.; Morlighem, M.; et al. Geologic provinces beneath the Greenland Ice Sheet constrained by geophysical data synthesis. *Geophys. Res. Lett.* **2024**, *51*, e2023GL107357. [CrossRef]
21. Hollis, J.A.; Keiding, M.; Møller Stensgaard, B.; van Gool, J.A.; Garde, A.A. Evolution of Neoarchaean supracrustal belts at the northern margin of the North Atlantic Craton, West Greenland. *GEUS Bull.* **2006**, *11*, 9–32. [CrossRef]
22. Bjarløv, S.P.; Vladykova, P. The potential need for energy saving in standard family detached and semi-detached wooden houses in arctic Greenland. *Build. Environ.* **2011**, *46*, 1525–1536. [CrossRef]
23. Statbank Greenland. Available online: https://bank.stat.gl/pxweb/en/Greenland/Greenland__IE__IE10/IEXANV.px/ (accessed on 26 June 2024).
24. Hornbostel, C. *Construction Materials: Types, Uses and Applications*, 2nd ed.; John Wiley and Sons, Inc.: Hoboken, NJ, USA, 1991.
25. Kirkelund, G.M.; Skevi, L.; Ottosen, L.M. Electrodialytically treated MSWI fly ash use in clay bricks. *Constr. Build. Mater.* **2020**, *254*, 119286. [CrossRef]
26. Brown, G.; Brindley, G.W. X-ray diffraction procedures for clay mineral identification. In *Crystal Structures of Clay Minerals and Their X-ray Identification*; Brindley, G.W., Brown, G., Eds.; Mineralogical Society: London, UK, 1980; pp. 305–359.
27. *DS/CEN ISO/TS 17892-12*; Geotechnical Investigation and Testing—Laboratory Testing of Soil—Part 12: Determination of Liquid and Plastic Limits. ISO: Geneva, Switzerland, 2018.
28. *DIN 18.122-1*; Baugrund, Untersuchung von Bodenproben—Zustandsgrenzen (Konsistenzgrenzen)—Teil 1: Bestimmung der Fließ- und Ausrollgrenze. Deutsches Institut für Normung: Berlin, Germany, 1997.
29. Brick Industry Association. *Manufacturing of Brick*; Technical notes on brick construction, 9; Brick Industry Association: Reston, VA, USA, 2006.
30. *DS/EN 772-11*; Methods of Test for Masonry Units—Part 11: Determination of Water Absorption of Aggregate Concrete, Autoclaved Aerated Concrete, Manufactured Stone and Natural Stone Masonry Units Due to Capillary Action and the Initial Rate of Water Absorption of Clay Masonry Units. Dansk Standard: Nordhavn, Denmark, 2011.
31. *DS/EN 772-4*; Methods of Test for Masonry Units—Part 4: Determination of Real and Bulk Density and of Total and Open Porosity for Natural Stone Masonry Units. Dansk Standard: Nordhavn, Denmark, 1998.
32. *ISO 12571*; Hygrothermal Performance of Building Materials and Products—Determination of Hygroscopic Sorption Properties. ISO: Geneva, Switzerland, 2021.
33. *ASTM C216-23*; Standard Specification for Facing Brick (Solid Masonry Units Made from Clay or Shale). ASTM: West Conshohocken, PA, USA, 2023.
34. Vieira, C.M.F.; Monteiro, S.N. Characterization of granite waste for incorporation in red ceramic. *Mater. Sci. Forum* **2005**, *498–499*, 728–733. [CrossRef]

materials

MDPI

Article

Exploring the Structure–Activity Relationship of Bentonites for Enhanced Refinement of Recycled Vegetable Oil

Alberto Mannu [1,*], Simona Castia [2], Giacomo Luigi Petretto [3], Sebastiano Garroni [2], Franca Castiglione [1] and Andrea Mele [1]

[1] Department of Chemistry, Materials and Chemical Engineering "G. Natta", Politecnico di Milano, Piazza L. da Vinci 32, 20133 Milano, Italy; franca.castiglione@polimi.it (F.C.); andrea.mele@polimi.it (A.M.)
[2] Department of Chemical, Physics, Mathematics and Natural Science, INSTM, University of Sassari, Via Vienna 2, 07100 Sassari, Italy; s.castia@studenti.uniss.it (S.C.); sgarroni@uniss.it (S.G.)
[3] Department of Medicine, Surgery and Farmacy, University of Sassari, 07100 Sassari, Italy; gpetretto@uniss.it
* Correspondence: alberto.mannu@polimi.it

Abstract

The use of bentonite for recycling vegetable oils presents challenges, as even minor variations in the clay composition and structure can lead to significant differences in its ability to retain various chemical groups. This study investigates the structure–activity relationship of four bentonites—two hydrophilic and two hydrophobic (both in commercial and ground forms)—to better understand these effects. Solid-state NMR spectroscopy revealed subtle differences between hydrophobic and hydrophilic materials, as well as distinctions between ground and unground hydrophilic clays, through ^{29}Si and ^{27}Al experiments. These structural variations directly influenced the bentonites' capacity to retain specific chemical groups, which in turn affected the pour point and volatile profile of the processed oils. A simplex lattice design of experiments, combined with multivariate analysis, facilitated the development of a predictive model to optimize process efficiency. Remarkably, this model achieved an improvement in pour point of up to 14.5 °C (from −2 °C to −16.5 °C) for oils treated with hydrophilic unground bentonite. This research underscores the critical role of bentonite morphology in enhancing the efficiency of vegetable oil recycling.

Keywords: solid-state NMR; bentonite; vegetable oil; pour point; design of experiments

1. Introduction

Recycling waste cooking oils (WCOs) have emerged as an important topic in recent years [1,2]. Within the available methods and technologies for WCO refining for non-biodiesel applications, water treatment under controlled conditions has proven able to provide a high-purity recycled vegetable oil [3,4]. Nevertheless, while the purity assessment conducted through ^{1}H NMR spectroscopy indicates a high level of triglyceride purity [5] upon water processing, minor constituents remain detectable, as revealed by UV-VIS [6] or SPME-GC [7] analyses. These impurities arise from several sources and are associated with the lifecycle of the oil prior to recycling.

WCOs can be contaminated by metal traces, aldehydes, ketones, heterocycles and other organic molecules originating from cooking tools, the Maillard process [8,9] and food leaching [10]. These minor components can also be removed by treatment with adsorbent materials such as natural clays or alternative powders [11]. Preliminary analysis of the volatile fraction prior to and after bentonite treatment of WCO samples revealed a complex mechanism underlying the action of bentonite. The results indicated that even slight variations in clay composition led to the preferential removal of different chemical groups [12]. This structure–activity relationship can also be observed by UV-VIS spectroscopy [6].

In this context, any knowledge regarding the mechanisms behind the absorption capacity of clay can contribute to the development of specific protocols aimed at vegetable oil recycling processes. Recently, a study has been conducted on the recycling of used vegetable oils with hydrophilic or hydrophobic bentonites, revealing the pivotal role of hydrophobicity in the removal of polar compounds from waste [13]. In that study, the presence of a small amount of water was crucial in determining the effectiveness of hydrophobic bentonites when used with recycled vegetables oils. Looking at recent studies, the topic of bentonite processing of WCOs has been thoroughly addressed [14,15]. Nevertheless, recent sustainability estimations [5] have indicated that water processing is more feasible than clay adsorption. In this context, two relevant aspects were not addressed: the use of bentonite clays to treat already recycled used vegetable oils, and the reason for the different performances within hydrophilic and hydrophobic bentonites sold with the same CAS number and apparently with the same physical characteristics.

Thus, four bentonites were used to capture the volatiles present in a recycled mixture of triglycerides, considering the variation in the pour point of the treated oil, which served as an indicator to highlight different outcomes related to different bentonites. Using a design of experiments (DOE) approach, the potential to prepare mixtures of different bentonites to be exploited as hybrid media in vegetable oil recycling has been assessed. A detailed characterization of the bentonites was carried out by solid-state magic angle spinning (MAS) NMR (^{29}Si and ^{27}Al), providing valuable insights into the morphology of the materials in relation to their properties, allowing us to relate the coordination mode of Si and Al atoms with the physical properties of the material.

2. Materials and Methods

2.1. Chemicals

Hydrophilic S001 (MKCR3493) and hydrophobic S002 (MKCN7873) bentonites were purchased from Merk Europe, Darmstadt, Germany (CAS number 1302-78-9).

2.2. Ball Milling

A total of 8 g of bentonite (S001 or S002) was ball-milled for 20 min in an SPEX 8000 M MIXER/MILL (Metuchen, NJ, USA) shaker mill at 875 rpm. The as-obtained S001M1 and S002M1 samples were then used for oil refinement.

2.3. Vegetable Oil Processing

For vegetable oil processing, 100 g of vegetable oil and 15 g of clay (15 wt%) were placed in a beaker, and the mixture was stirred for 30 min and then filtered under vacuum. The processed oil was then subjected to pour point analysis.

2.4. Pour Point Determination

The ASTM D97 method was adapted to our purposes and employed for pour point determination [16]. Oil samples (20 g) were placed in a falcon and slowly cooled in an ice/salt bath. The temperature of the oil was monitored by an internal thermometer every minute, until visual confirmation of the change from liquid to solid.

2.5. Headspace Solid-Phase Microextraction (HS-SPME)

A 100 μm PDMS/DVB/CAR (Polydimethylsiloxane/Divinylbenzene/Carboxen) coated fiber 50/30 Stableflex (Supelco, Sigma Aldrich, St. Louis, MO, USA) was preconditioned at 270 °C for 1 h. A total of 5 g of sample was added to a 20 mL SPME vial (75.5 × 22.5 mm) equipped with a septum. The system was conditioned for 5 min at 60 °C. Thus, the fiber was exposed to the headspace for 30 min. Once the extraction was completed, the fiber was desorbed for 2 min into the injector at a temperature of 250 °C in a splitless injection mode.

2.6. *Gas Chromatography–Mass Spectrometry (GC-MS) Analysis*

GC-MS measurements were carried out using an Agilent 6890 GC (Palo Alto, CA, USA), coupled with an Agilent 5973 MSD detector (Palo Alto, CA, USA). A ZB-Wax 30 m × 0.25 mm i.d., 0.25 μm film thickness column was used for the chromatographic separation. The following temperature ramp was used: held at 40 °C for 4 min, increased to 150 °C at a rate of 3.0 °C/min, held for 3 min and then increased to 220 °C at a rate of 10 °C/min. Finally, the temperature was held for 10 min. The constant flow of helium (carrier gas) was at a rate of 1 mL/min. The data were analyzed using a MassHunter Workstation B.06.00 SP1 (Agilent, Palo Alto, CA, USA). Identification of the analytes (Section 3.3) was achieved by comparing them with co-injected standards and matching their MS fragmentation patterns and retention indexes against built-in libraries, literature data or commercial mass spectral libraries (NIST/EPA/NIH 2008; HP1607 from Agilent Technologies, Palo Alto, CA, USA).

Retention Indexes

A hydrocarbon mixture of n-alkanes (C9–C22) was analyzed separately under the same chromatographic conditions to calculate the retention indexes with the generalized equation by Van del Dool and Kartz [17]:

$$Ix = 100[(t_x - t_n)/(t_{n+1} - t_n) + n] \quad (1)$$

where t is the retention time, x is the analyte, n is the number of carbons of alkane that elute before the analyte and n + 1 is the number of carbons of alkane that elute after the analyte.

2.7. *NMR Analysis*

Solid-state NMR spectra were acquired with a Bruker NEO 500 spectrometer, equipped with a MAS iprobe, operating at 500.13 MHz for ^{1}H, 99.35 MHz for ^{29}Si and 130.31 MHz for ^{27}Al. Solid samples were packed in a 4 mm zirconium oxide rotor. ^{29}Si direct excitation (DE-MAS) spectra were acquired with a recycle delay of 20 s, spinning speed of 10 kHz and 560 scans. ^{27}Al spectra were obtained with recycle delay of 5 s, spinning speed of 12 kHz and 256 scans. All the experiments were performed at 298 K. Deconvolutions were performed with origin 2024 using a Lorentzian fitting model. The chemical shifts are reported relative to tetramethylsilane (TMS) for ^{29}Si using octakis-(trimethylsiloxy)silsesquioxane (Q8M8) and for ^{27}Al using $Al(NO_3)_3 \cdot 9H_2O$ as a secondary standard, respectively [18].

2.8. *Karl Fischer Titration*

Water content was determined through Karl Fischer titration with a coulometric MKV-710B titrator (Alkimia SRL, Milan, Italy).

2.9. *FT-IR Spectroscopy*

For the FT-IR measurements, a FT-IR Jasco 480 Plus spectrophotometer (Jasco International Co., Ltd., Tokyo, Japan) was employed. The powders were analyzed in the form of tablets prepared by mixing the bentonite with anhydrous KBr and pressing them into disks using a hydraulic press.

2.10. Design of Experiments (DoE)

A simplex lattice design was used to study the effects of four components in four runs. The design was run in a single block and the order of the experiments was randomized to minimize the effects of lurking variables. Statgraphics 18 software was used for the statistical multivariate analysis.

3. Results and Discussion

3.1. FT-IR Spectroscopy Analysis

Bentonite is a natural material containing different minerals which can influence its properties [19]. According to the provider, the commercial bentonites herein considered are composed of montmorillonite, $Al_2O_3 \times 4SiO_2 \times H_2O$ (molecular weight: 180.1 g/mol), with a particle size lower than 74 μm and a mean particle size of 5.45 μm [20]. Nevertheless, no additional data are available regarding the differences between the two products. Previous SEM and TEM analyses on the same batches of bentonites revealed, respectively, different interlayer distances and particle shapes, while powder X-ray analysis did not highlight any relevant difference [5]. Nevertheless, the behavior of the two clays, prior to and after grinding, was different, confirming the different properties of the apparently very similar materials. After being subjected to a ball-milling process, significant differences, in terms of microstructure, were observed, and interlayer distances were modified with respect to the pristine materials, probably due to the partial release of trapped water upon grinding [13]. To detect variations in the chemical groups, FT-IR analysis was conducted on both bentonites (Figure 1).

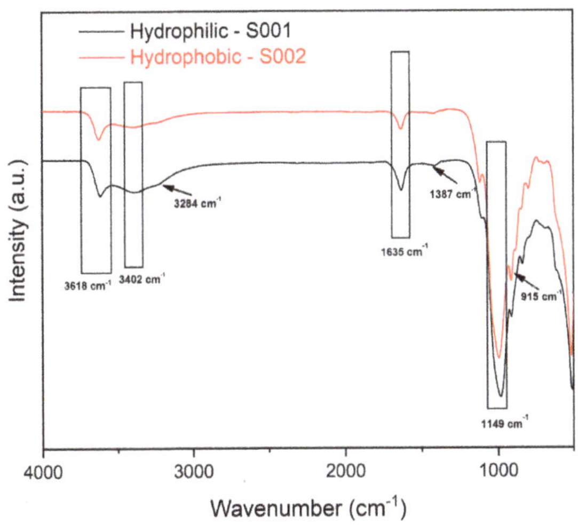

Figure 1. FT-IR spectra of hydrophilic S001 (black line, lower spectrum) and hydrophobic S002 (red line, upper spectrum) bentonites.

Looking at the spectra shown in Figure 1, it is possible to clearly see the different IR profiles of S001 and S002 bentonites. It is known that hydrophobic bentonites are obtained by chemical treatment of hydrophilic ones with the consequent removal of minor components, resulting in a less complex mixture of chemical groups in the FT-IR spectrum [21]. Regarding the specific composition observed, both samples show typical bands relative to the Si-O stretching (1149 cm^{-1}), the bending of the adsorbed water (1635 cm^{-1}) and the stretching of the adsorbed water (3402 cm^{-1}), as well as the OH-Al stretching band (3618 cm^{-1}). For the hydrophilic S001 bentonite, relevant bands associated with the

-OH vibrations were detected at 3284 cm^{-1}, and CO_3 stretching of calcite was detected at 1387 cm^{-1}. The band at 915 cm^{-1}, visible in the hydrophobic S002 sample, can be attributed to Al-Al-OH bending [21].

In addition to the characterization of the starting materials, both the bentonites were monitored during the ball-milling procedure, specifically at 30, 60 and 120 min (Figure 2).

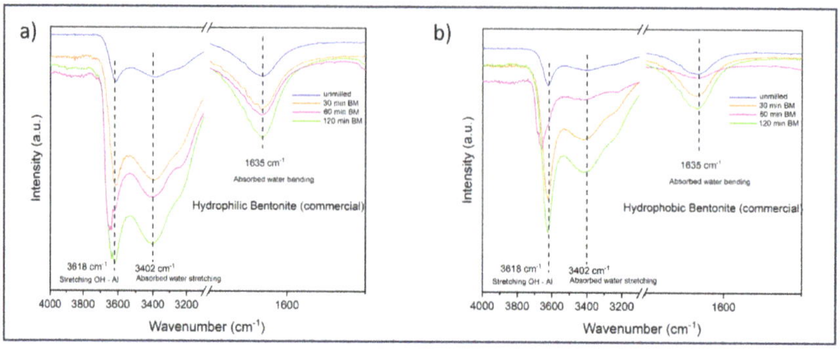

Figure 2. FT-IR spectra of hydrophilic (**a**) and hydrophobic (**b**) bentonite subjected to a ball-milling process (0, 30, 60 and 120 min).

As already reported for similar samples subjected to the same milling treatment [13], it is possible to notice, for both bentonites, some slight changes in the FT-IR spectra after 60 min of ball-milling (shifting of the stretching OH-Al band and consistent attenuation of the adsorbed water bending band), followed by a further change, which apparently results from the restoration of the original chemical structure at 120 min. For this reason, the milled bentonites, which were further employed to treat the triglyceride mixture, were ground for 60 min.

3.2. NMR Morphological Analysis

To overcome the difficulties that occurred with the powder X-ray diffraction characterization and to add further information to the FT-IR spectroscopy data, thus providing a satisfying description of the differences between the bentonites studied, multinuclear solid-state NMR spectroscopy was performed. In fact, the limited selectivity of X-ray diffraction, commonly used for the characterization of such minerals, due to the tiny changes in scattering factor values can be overcome by looking at the solid-state NMR profile of the main heteroatoms [22]. In fact, Si and Al exhibit different gyromagnetic ratios γ (6.976×10^7 rad T^{-1} s^{-1} for ^{27}Al, and -5.3188×10^7 rad T^{-1} $s^{-1}\gamma$ for ^{29}Si), allowing us to highlight even small differences in the chemical environment of the analyzed nuclei.

Thus, ^{29}Si MAS NMR analysis was performed on hydrophilic and hydrophobic bentonites as received and after grinding (Figure 3).

All spectra show a sharp peak at −93.4 ppm and a rather broad signal covering the range of −105–115 ppm, indicating local disorder of the Si environment in the 3D structure. According to the classic nomenclature and peak assignment proposed by Lippmaa et al. [23], the ^{29}Si SS NMR signals in Figure 3 are labeled as Q3 and Q4, respectively. Q3 refers to "chain branching sites", $Si(OAl)(OSi)_2OH$ in our case, whereas Q4 is associated with the "three-dimensional cross-linked framework" of the fully condensed $Si(OAl)(OSi)_3$ frame, usually present in the tetrahedral layers of the clay [24]. After deconvolution of the spectral region containing the Q3 and Q4 signals, the peak's width, as well as the absolute area ratios (Q4/Q3), were estimated and are reported in Table 1.

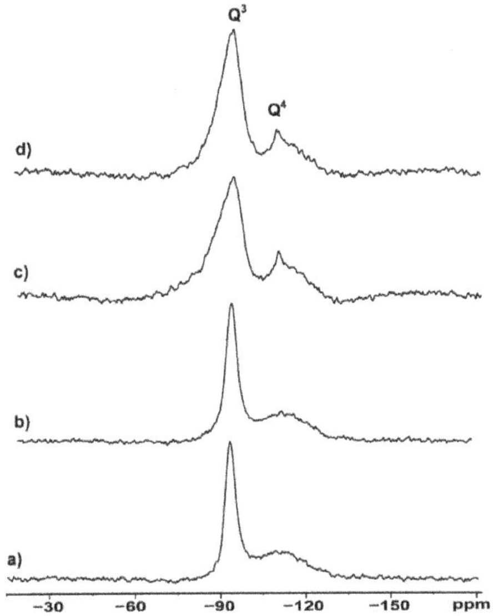

Figure 3. ^{29}Si MAS NMR spectra of (**a**) hydrophilic bentonite S001, (**b**) ball-milled hydrophilic S001M1, (**c**) hydrophobic bentonite S002 and (**d**) ball-milled hydrophobic bentonite S002M1.

Table 1. ^{29}Si SS NMR spectral data for samples of hydrophilic and hydrophobic bentonite (prior to and after milling). LW indicates the NMR linewidth (Hz) measured at half-height. Q4/Q3 indicates the peak integral ratio as obtained from peak deconvolution.

Spectrum	Sample	LW Q3 (Hz)	LW Q4 (Hz)	Q4/Q3 Ratio
A	hydrophilic (S001)	491	345	0.137
B	hydrophilic milled (S001M1)	489	354	0.150
C	hydrophobic (S002)	834	476	0.190
D	hydrophobic milled (S002M2)	841	531	0.186

Hydrophilic bentonite shows a sharper Q3 signal compared to the hydrophobic one (491 Hz vs. 834 Hz) and a lower Q4/Q3 ratio (0.137 vs. 0.190). The higher Q4/Q3 ratio observed in the hydrophobic samples compared to their hydrophilic counterparts can be related to their lower Brönsted acidity compared to the hydrophilic samples. The rationale behind this is that Q3 contains the contribution of the Si-OH, which are responsible for the acidity. The effect of the ball-milling procedure is evaluated by comparing the suitable spectra in Figure 3a vs. Figure 3b for hydrophilic bentonites, and Figure 3c vs. Figure 3d for hydrophobic bentonites. The data highlight that grinding does not affect the chemical structure—as no chemical shift changes were observed—but rather it increases, in a selective manner, the structural disorder of the materials, as pointed out by the linewidth (LW) values. The variation, appreciable in Q4 especially, can be expressed as D_{LW} (Q4) = $LW(Q4)_{milled} - LW(Q4)_{not\ milled}$ for both the hydrophilic and hydrophobic materials. The experimental data lead to values of D_{LW} $(Q4)_{hydrophilic}$ = 9 Hz and D_{LW} $(Q4)_{hydrophobic}$ = 55 Hz, providing evidence of more effective morphological changes in hydrophobic samples upon mechanical stress.

To provide an exhaustive overview of the chemical structure of the clays, ^{27}Al MAS spectra were also acquired (Figure 4).

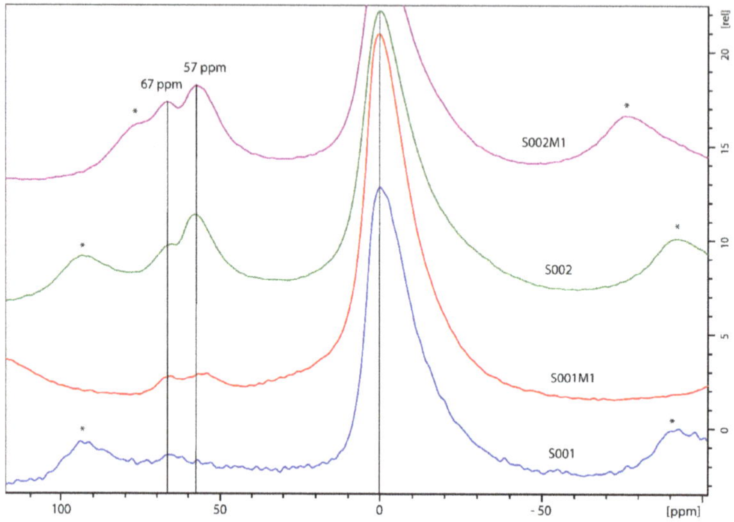

Figure 4. ^{27}Al MAS NMR spectra of hydrophilic bentonite S001, ball-milled hydrophilic bentonite S001M1, hydrophobic bentonite S002 and ball-milled hydrophobic bentonite S002M1. Spinning sidebands are marked with an *.

The experimental spectra show the presence of peaks in two regions, from 70 to 50 ppm and from −25 to +25 ppm, corresponding, respectively, to Al in its tetrahedral (Al^{IV}) and octahedral (Al^{VIII}) forms (Figure 4 and Table 2) [25]. It is known that the presence of Al in tetrahedral frameworks increases the Brönsted acidity of the material [26]. Also, two different peaks in the Al tetrahedral region can be observed, which correspond, respectively, to Al sites in layered bentonite (peak at around 70 ppm) and to the Al characteristics of fully condensed sites (55–59 ppm) [27]. The relative percentages of the areas of each peak are reported, along with the corresponding chemical shifts, in Table 2.

Table 2. ^{27}Al SS NMR spectral data for samples of hydrophilic and hydrophobic bentonite (prior to and after milling).

Bentonite	Octahedral	d (ppm)	Al Sites Fully Condensed	d (ppm)	Al Sites in Layered Bentonite	d (ppm)
S001	91.63%	0.25	0%	-	8.36%	69.32
S001M1	91.61%	0.87	4.22%	55.70	4.15%	55.97
S002	74.50%	0.10	13.76%	58.45	11.70%	68.54
S002M1	74.50%	0.10	13.76%	58.45	11.70%	68.54

Looking at the data relative to the four bentonite samples analyzed, it is possible to draw the following conclusions. Hydrophilic bentonite (commercial) just shows octahedral Al sites (>91%) with about 8% tetrahedral Al layered sites. After ball-milling, about half of the layered sites are converted into fully condensed ones. The chemical shift, in this case for the fully compacted sites (about 56 ppm), is slightly anomalous as it should be close to 70 ppm. Thus, it cannot be excluded that this new peak corresponds to layered sites (not fully condensed) arranged in a different way than the original ones. However, a clear effect, concentrated on the tetrahedral Al sites, of the milling procedure is observed for hydrophilic bentonite. On the other hand, hydrophobic bentonites show all three configurations, with a distribution between octahedral, tetrahedral layered and tetrahedral fully condensed figurations, which seems to not be affected by the ball-milling procedure. Independently of the milling procedure, it is possible to understand the different behavior of the two types of bentonites (hydrophilic and hydrophobic) by considering their octahedral/tetrahedral site ratios, which are, respectively, about 92/8 and 75/25. As reported by Vajglova and coworkers [25], the presence of Al in the tetrahedral frameworks increases the Brönsted acidity of the material; thus, the bentonite sold under the label hydrophobic differs from the hydrophilic one as it has a consistent increased number of Bronsted acidic sites.

3.3. Design of Experiments (DoE)

Once the relevant structural differences between the considered bentonites were highlighted by solid-state NMR spectroscopy, a simplex lattice design of experiments (DoE) [26–29] was used to build a viable statistical model describing the effect of the specific bentonite used on the pour point of the processed oil. According to the DoE, four experiments were conducted, and for each one, the volatile content (VC) and pour point (PP) were determined (Table 3).

Table 3. List of experiments conducted and relative details.

Experiment	Bentonite	Response 1	Response 2
1	S002M1	PP *	VC **
2	S002	PP *	VC **
3	S001M1	PP *	VC **
4	S001	PP *	VC **

* PP stands for pour point; ** VC stands for volatile content.

The volatile content was determined by SPME GC-MS analysis, which provided a semiquantitative assessment of several chemicals in different relative amounts (Table 4).

Table 4. Semiquantitative volatile profiles for the triglyceride mixtures prior to bentonite treatment (TQ), and oil samples treated, respectively, with bentonites S001 (hydrophilic), S001M1 (hydrophilic milled), S002 (hydrophobic) and S002M1 (hydrophobic milled).

Compound	Retention Times (RTs) *	TQ	S001	S001M1	S002	S002M1
Pentanal	3.87	1.09	0.87	1.27	0.30	0.75
Hexanal	6.63	12.02	9.32	17.82	9.22	6.55
2-Hexenal, (E)-	10.68	0.43	0.33	0.51	0.06	0.26
Heptanal	9.69	1.21	0.85	1.80	0.04	0.58
Octanal	12.81	3.10	2.48	3.59	1.15	1.14
2-Heptenal, (Z)-	13.82	9.54	8.37	10.00	5.94	5.94
2-Octenal, (E)-	16.68	4.13	3.69	3.82	2.85	2.83
Nonanal	15.77	8.37	8.05	7.17	5.60	4.30
2-Nonenal, (E)-	19.36	1.08	1.25	0.93	1.33	1.24
2-Decenal, (Z)-	22.00	5.73	9.22	3.82	8.59	8.92
2,4-Decadienal, (E,E)-	25.74	3.23	7.92	2.90	8.69	10.14
2-Undecenal	24.44	3.23	5.42	2.05	6.41	6.97
2-Heptanone	9.61	1.36	0.53	2.00	0.12	0.33
2-Octanone	12.68	1.10	0.65	1.35	0.37	0.25
Total ketones		2.46	1.19	3.35	0.48	0.59
Acetic acid	17.83	4.00	7.31	6.34	9.67	10.59
Butanoic acid	22.34	2.12	1.41	2.09	2.61	1.94
Pentanoic acid	24.79	4.38	5.19	3.58	5.37	5.42
Hexanoic acid	27.12	15.25	11.35	11.30	15.45	16.01

Table 4. *Cont.*

Compound	Retention Times (RTs) *	TQ	S001	S001M1	S002	S002M1
Heptanoic acid	30.06	1.07	0.98	0.55	1.71	2.15
Octanoic acid	32.21	0.60	0.95	0.28	1.61	1.70
Nonanoic acid	33.81	0.32	0.59	0.13	1.24	0.94
1-Pentanol	11.86	1.17	0.83	1.32	0.84	0.61
1-Hexanol	14.80	4.12	2.03	5.73	1.89	1.73
1-Octen-3-ol	17.40	3.17	2.06	2.88	0.96	0.91
1-Heptanol	17.52	1.28	0.85	1.58	0.60	0.72
2-Hepten-1-ol, (E)-	18.94	1.03	0.60	1.10	0.68	0.48
1-Octanol	20.06	3.27	4.92	1.17	5.72	6.28
Dodecane	10.02	0.99	0.67	0.65	0.19	0.14

* Retention times are expressed in minutes.

The general chemical compositions of the volatile fraction of the oil prior to and after treatment with bentonite are in agreement with previously reported HS-SPME GC/MS data relative to similar matrixes [7,30]. Looking at the main organic volatile compounds (VOCs) detected and reported in Table 4, it is possible to explain the presence of many organic compounds, such as aldehydes and acids, as resulting from food leaching [31], the Maillard reaction, oxidation promoted by temperature and the hydrolysis of triglycerides [32]. Furthermore, regarding the aldehyde profiles, the autoxidation of linoleic acid has been reported as a source of dienals (2,4-heptadienal or 2,4-decadienal) and alkenals (2-undecenal, 2-decenal and 2-octenal) [33].

To better compare the outcomes of the different treatments in terms of variations in their volatile profiles, the detected compounds have been grouped into chemical families, as shown in Table 5, which also integrates the PP values of each sample.

Table 5. Semiquantitative assessment of the main composition of samples TQ (prior to bentonite treatment), S001 (hydrophilic bentonite), S001M1 (milled hydrophilic bentonite), S002 (hydrophobic bentonite) and S002M1 (milled hydrophobic bentonite) and their PP values.

		1	2	3	4	5
		TQ	S001	S001M1	S002	S002M1
Volatiles	Aldehydes	53.15	57.78	55.71	50.17	49.62
	Ketones	2.46	1.19	3.35	0.48	0.59
	Acids	27.73	27.77	24.26	37.64	38.76
	Alcohols	14.04	11.30	13.79	10.68	10.74
	Alkanes	0.99	0.67	0.65	0.19	0.14
Pour Point (PP, °C)		−2 °C	−16.5	−9.5	−10	−10

Looking at the data reported in Table 5, the most evident difference between the starting oil, TQ, and the treated ones is related to the PP, which consistently improves from −2 °C to about −10 °C for samples treated with bentonites S002, S002M1 and S001M1, reaching −16.5 °C after processing with bentonite S001. The 6.5 °C drop in the treated samples of hydrophobic bentonite (S002 series) compared to S001 is indicative of the specific ability of the hydrophilic bentonite to remove some chemicals. It is known that water molecules in waste vegetable oil samples are better removed by hydrophilic bentonites, and increasing the specific surface area can enhance this phenomenon [12]. Nevertheless, herein, we observe an inverse trend, as the non-milled hydrophilic bentonite is much more efficient that the milled one (S001 vs. S001M1). The latter conclusion seems to contrast with the published data [4]; regardless, it should be considered that when recycled vegetable oil is treated, a low tenor of water (0.4 wt% from Karl Fischer titration) is present than in raw used vegetable oil. In such an environment, the action of the bentonite is related to the trapping of organic molecules that have not been saturated by water, as happens in waste cooking oils [5]. It is also known that the absorption process of porous materials involves a combination of pore inclusion (determined by pore size) and the interactions between polar/non-polar groups. Regarding the differences in morphology, the ^{29}Si MAS NMR spectra showed sharper Q3 signals in hydrophilic bentonites than hydrophobic ones, meaning that hydropilic bentonites show less contribution of $Si(OAl)(OSi)_2OH$ phases in their structures. Also, the enlargement of the peaks in the milled samples suggests a decrease in crystallinity due to the ball-milling process. On the other hand, the ^{27}Al MAS spectra revealed that milled bentonites have a high-density packed structure (increasing the Al^{IV} peak). Thus, it seems that the combination between the reduced number of O-H groups and the more tiny and regular structure are responsible for the better performances of S001 than the other clays in removing organic contaminants.

Looking at Table 2, it is possible to gain some insights about the specific chemical groups retained during the four processes. By comparing columns 1 and 2 (hydrophilic bentonites) with 3 and 4 (hydrophobic bentonites), a reduction in the relative amounts of ketones, acids, alcohols and alkanes is observed after processing with hydrophobic bentonites. Considering that hydrophobic bentonites have fewer -OH groups, it is possible to attribute this behavior to the physical retention of the low-molecular-weight contaminants.

Looking for differences between milled and non-milled hydrophilic bentonites, the most relevant difference lies in the relative amounts of ketones and alcohols, which are reduced in S001-treated samples (non-milled). It is possible to relate this outcome to the reduction in Al^{IV} signal intensity in S001-processed oils (^{27}Al MAS spectra). Again, there is an indication that the structural organization of the clay has a relevant impact on specific chemical groups' retention ability.

3.4. Optimization by Statistical Multivariate Analysis

In accordance with the simple lattice DoE [34], four experiments were conducted, and the variation in the PP, depending on the composition of the bentonite employed, was expressed in a surface plot [35], as reported in Figure 5.

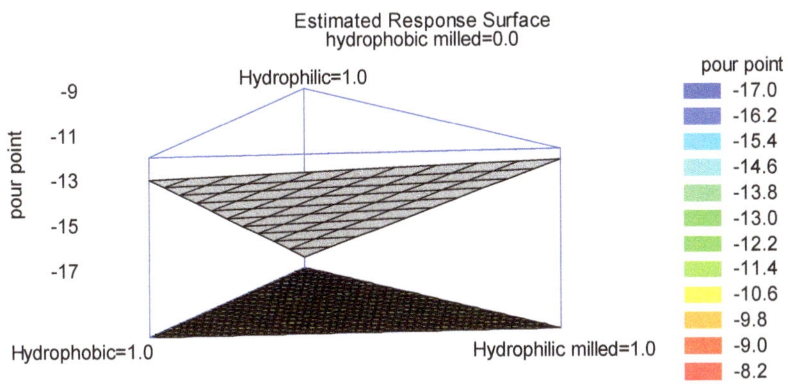

Figure 5. Contours of the estimated response surface.

Looking at the plot in Figure 5, it is possible to observe the variation in the PP in the oils treated with different bentonites. If a lower PP is desirable, hydrophilic non-milled bentonite should be used. By looking at the red and dark-orange areas on the top of the triangle, it is possible to estimate that even a mixture of hydrophilic non-milled and hydrophobic (milled or not) bentonites should guarantee a PP within the range of −15 and −17 °C.

In the end, it was highlighted how even small differences in the structure of the clay can result in a sensible variation in the ability of the material to effectively remove undesired chemicals from recycled vegetable oils. This experimental evidence is open to a wide range of customization possibilities if mixtures of different clays are considered.

4. Conclusions

In this study, we investigated the effectiveness of four bentonites—two commercial types (hydrophilic and hydrophobic) and two modified via 60 min of ball-milling—in processing a refined mixture of triglycerides from waste sources. The chemical structures of these bentonite powders were thoroughly characterized using FT-IR and solid-state ^{29}Si and ^{27}Al NMR spectroscopies.

The ^{29}Si solid-state NMR results provided additional insights, particularly in the form of an increased Q4/Q3 ratio, highlighting the structural variations between the hydrophilic (S001) and hydrophobic (S002) bentonites. Furthermore, the ^{27}Al solid-state NMR spectra revealed significantly lower intensity in the Al^{IV} signal in hydrophilic bentonite after milling (when comparing S001 with the milled sample, S001M1). These findings fill a critical gap left by prior studies using X-ray diffraction, SEM and TEM, which could not capture consistent differences in the chemical structures of the materials.

Equipped with this comprehensive structural characterization, the bentonites were employed as adsorbents in the treatment of recycled vegetable oil. Gas chromatography of the volatile fractions showed a higher affinity of the hydrophobic bentonite S001 for trapping ketones, along with a superior capacity to retain ketones and alcohols compared to its milled counterpart, S001M1. Additionally, the pour point (PP) was used as a further indicator of the bentonites' ability to retain impurities during oil treatment. A simplex lattice design of experiments, combined with multivariate analysis, was applied to develop a predictive model capable of estimating the PP of treated vegetable oil based on the specific bentonite employed.

Overall, this study underscores the critical role of bentonite morphology and chemical structure, as characterized by solid-state NMR techniques, in determining their efficiency in trapping volatile organic compounds, which is related to the acidic characteristics of the material. Solid-state NMR, in particular, proved invaluable in detecting subtle yet significant differences in the structural features of bentonites, offering insights that could guide the optimization of clays for industrial applications involving VOC adsorption.

Supplementary Materials: The following supporting information can be downloaded at: https://www.mdpi.com/article/10.3390/ma18051059/s1, Figure S1: Deconvolutions of ^{27}Al MAS NMR spectrum for hydrophilic bentonite; Figure S2: Deconvolutions of ^{27}Al MAS NMR spectrum for hydrophilic milled bentonite; Figure S3: Deconvolutions of ^{27}Al MAS NMR spectrum for hydrophobic bentonite; Figure S4: Deconvolutions of ^{27}Al MAS NMR spectrum for hydrophobic milled bentonite; Figure S5: ^{29}Si MAS NMR spectra of (a) hydrophilic bentonite S001, (b) ball milled hydrophilic S001M1, (c) hydrophobic bentonite S002, (d) ball milled hydrophobic bentonite S002M1; Figure S6: ^{27}Al MAS NMR spectra of (a) hydrophilic bentonite S001, (b) ball milled hydrophilic bentonite S001M1, (c) hydrophobic bentonite S002, (d) ball milled hydrophobic bentonite S002M1. Spinning sidebands are marked.

Author Contributions: Conceptualization, G.L.P. and A.M. (Alberto Mannu); methodology, G.L.P., A.M. (Alberto Mannu) and F.C.; validation, A.M. (Alberto Mannu) and S.G.; investigation, S.C. and F.C.; resources, S.G. and A.M. (Andrea Mele); data curation, A.M. (Alberto Mannu) and A.M. (Andrea Mele); writing—original draft preparation, G.L.P., A.M. (Alberto Mannu), A.M. (Andrea Mele) and F.C.; writing—review and editing, A.M. (Alberto Mannu) and A.M. (Andrea Mele);

supervision, G.L.P. and S.G.; funding acquisition, A.M. (Andrea Mele) and S.G. All authors have read and agreed to the published version of the manuscript.

Funding: The present work was funded by the WORLD Project-RISE, a project that has received funding from the European Union's Horizon 2020 research and innovation program, under the Marie Skłodowska-Curie, Grant Agreement No. 873005.

Institutional Review Board Statement: Not applicable.

Informed Consent Statement: Not applicable.

Data Availability Statement: The original contributions presented in the study are included in the article/Supplementary Material; further inquiries can be directed to the corresponding author.

Conflicts of Interest: The authors declare no conflicts of interest.

References

1. Cárdenas, J.; Orjuela, A.; Sanchez, D.L.; Narváez, P.C.; Katryniok, B.; Clark, J. Pre-treatment of used cooking oils for the production of green chemicals: A review. *J. Clean. Prod.* **2021**, *289*, 125129. [CrossRef]
2. Foo, W.H.; Koay, S.S.N.; Chia, S.R.; Chia, W.Y.; Tang, D.Y.Y.; Nomanbhay, S.; Chew, K.W. Recent advances in the conversion of waste cooking oil into value-added products: A review. *Fuel* **2022**, *324*, 124539. [CrossRef]
3. Mannu, A.; Garroni, S.; Ibanez Porras, J.; Mele, A. Available technologies and materials for waste cooking oil recycling. *Processes* **2020**, *8*, 366. [CrossRef]
4. Gharby, S. Refining Vegetable Oils: Chemical and Physical Refining. *Sci. World J.* **2022**, *2022*, 6627013. [CrossRef]
5. Mannu, A.; Almendras Flores, P.; Briatico, F.; Di Pietro, M.E.; Mele, A. Sustainable Production of Raw Materials from Waste Cooking Oils. *RSC Sustain.* 2025; *Advance article.*
6. Mannu, A.; Poddighe, M.; Garroni, S. Application of IR and UV–VIS spectroscopies and multivariate analysis for the classification of waste vegetable oils. *Resour. Conserv. Recycl.* **2022**, *178*, 106088. [CrossRef]
7. Tu, D.; Li, H.; Wu, Z.; Zhao, B.; Li, Y. Application of Headspace Solid-Phase Microextraction and Multivariate Analysis for the Differentiation Between Edible Oils and Waste Cooking Oil. *Food Anal. Methods* **2014**, *7*, 1263–1270. [CrossRef]
8. Shi, B.; Guo, X.; Liu, H.; Jiang, K.; Liu, L.; Yan, N.; Farag, M.A.; Liu, L. Dissecting Maillard reaction production in fried foods: Formation mechanisms, sensory characteristic attribution, control strategy, and gut homeostasis regulation. *Food Chem.* **2024**, *438*, 137994. [CrossRef]
9. Bai, S.; You, L.; Ji, C.; Zhang, T.; Wang, Y.; Geng, D.; Gao, S.; Bi, Y.; Luo, R. Formation of volatile flavor compounds, maillard reaction products and potentially hazard substance in China stir-frying beef sao zi. *Food Res. Int.* **2022**, *159*, 111545. [CrossRef]
10. Lee Kuek, S.; Tarmizi, A.H.A.; Razak, R.A.A.; Jinap, S.; Norliza, S.; Sanny, M. Contribution of lipid towards acrylamide formation during intermittent frying of French fries. *Food Control.* **2020**, *118*, 107430. [CrossRef]
11. Mannu, A.; Di Pietro, M.E.; Petretto, G.L.; Taleb, Z.; Serouri, A.; Taleb, S.; Sacchetti, A.; Mele, A. Recycling of used vegetable oils by powder adsorption. *Waste Manag. Res.* **2023**, *41*, 839–847. [CrossRef] [PubMed]
12. Mistry, N.; Patel, V.; Parekh, P.; Naik, P.; Kumar, N.S.; Vekariya, R.; Khimani, M. *Adsorbent Materials, Materials from Natural Sources*, 1st ed.; CRC Press: Boca Raton, FL, USA, 2024; 26p, ISBN 9781032636368.
13. Mannu, A.; Poddighe, M.; Mureddu, M.; Castia, S.; Mulas, G.; Murgia, F.; Di Pietro, M.E.; Mele, A.; Garroni, S. Impact of morphology of hydrophilic and hydrophobic bentonites on improving the pour point in the recycling of waste cooking oils. *Appl. Clay Sci.* **2024**, *262*, 107607. [CrossRef]
14. Meirawaty, M.; Palit, C.; Setyorini, D.A.; Jambak, M.A. Bentonite applications in simple purification of bulk cooking oil as alternative solutions for household cost efficiency. *J. Community Based Environ. Eng. Manag.* **2021**, *5*, 63–72. [CrossRef]
15. Aritonang, B.; Ritonga, A.H.; Harefa, K.; Wiratma, D.Y.; Herlina. Purification of used Cooking Oil using a Combination of Activated Carbon and Bentonite Adsorbents. *J. Farm. (JFM)* **2024**, *7*, 31–40. [CrossRef]
16. *ASTM D97*; Standard Test Method for Pour Point of Petroleum Products. ASTM International: West Conshohocken, PA, USA, 2022.
17. Van Del Dool, H.; Kartz, P.D. A generalization of the retention index system including linear temperature programmed gas-liquid partition chromatography. *J. Chrom. A* **1963**, *11*, 463–471. [CrossRef]
18. Harris, R.K.; Becker, E.D.; de Menezes, S.M.C.; Goodfellow, R.; Granger, P. NMR Nomenclature. Nuclear Spin Properties and Conventions for Chemical Shifts—(IUPAC Recommendations 2001). *Pure Appl. Chem.* **2001**, *73*, 1795–1818. [CrossRef]
19. Murray, H.H. Chapter 6 Bentonite Applications. *Dev. Clay Sci.* **2006**, *2*, 111–130.
20. Available online: https://www.sigmaaldrich.com/IT/it/product/sigald/285234 (accessed on 12 December 2024).

21. Hayati-Ashtiani, M. Use of FTIR Spectroscopy in the Characterization of Natural and Treated Nanostructured Bentonites (Montmorillonites). *Part. Sci. Technol. Int. J.* **2012**, *30*, 553–564. [CrossRef]
22. Pavón, E.; Alba, M.D. Swelling layered minerals applications: A solid state NMR overview. *Prog. Nucl. Magn. Reson. Spectrosc.* **2021**, *124–125*, 99–128. [CrossRef]
23. Lippmaa, E.; Magi, M.; Samoson, A.; Engelhardt, G.; Grimmer, A.-R. Structural studies of silicates by solid-state high-resolution 29Si NMR. *J. Am. Chem. Soc.* **1980**, *102*, 4889–4893. [CrossRef]
24. Rhee, C.H.; Kim, H.K.; Chang, H.; Lee, J.S. Nafion/sulfonated montmorillonite composite: A new concept electrolyte membrane for direct methanol fuel cells. *Chem. Mater.* **2005**, *17*, 1691–1697. [CrossRef]
25. Okada, K.; Arimitsu, N.; Kameshima, Y.; Nakajima, A.; MacKenzie, K.J.D. Solid acidity of 2:1 type clay minerals activated by selective leaching. *Appl. Clay Sci.* **2006**, *31*, 185–193. [CrossRef]
26. Vajglová, Z.; Kumar, N.; Peurla, M.; Peltonen, J.; Heinmaad, I.; Yu Murzin, D. Synthesis and physicochemical characterization of beta zeolite–bentonite composite materials for shaped catalysts. *Catal. Sci. Technol.* **2018**, *8*, 6150–6162. [CrossRef]
27. Danner, T.; Norden, G.; Justnes, H. Characterisation of calcined raw clays suitable as supplementary cementitious materials. *Appl. Clay Sci.* **2018**, *162*, 391–402. [CrossRef]
28. Oloyede, C.T.; Jekayinfa, S.O.; Alade, A.O.; Ogunkunle, O.; Laseinde, O.T.; Adebayo, A.O.; Abdulkareem, A.I.; Smaisim, G.F.; Fattah, I.M.R. Synthesis of Biobased Composite Heterogeneous Catalyst for Biodiesel Production Using Simplex Lattice Design Mixture: Optimization Process by Taguchi Method. *Energies* **2023**, *16*, 2197. [CrossRef]
29. Monton, C.; Wunnakup, T.; Suksaeree, J.; Charoenchai, L.; Chankana, N. Investigation of the Interaction of Herbal Ingredients Contained in Triphala Recipe Using Simplex Lattice Design: Chemical Analysis Point of View. *Int. J. Food Sci.* **2020**, *2020*, 1–14. [CrossRef] [PubMed]
30. Purcaro, G.; Szafnauer, R. Tackling Food Fraud: Solid-Phase Microextraction Using Multi-Step Enrichment to Enhance Aroma Profiling in Olive Oil. *Column* **2021**, *17*, 20–25.
31. Choe, E.; Min, D.B. Chemistry of Deep-Fat Frying Oils. *J. Food Sci.* **2007**, *72*, R77–R86. [CrossRef]
32. Zhang, Q.; Liu, C.; Sun, Z.; Hu, X.; Shen, Q.; Wu, J. Authentication of edible vegetable oils adulterated with used frying oil by Fourier Transform Infrared Spectroscopy. *Food Chem.* **2012**, *132*, 1607–1613. [CrossRef]
33. Wu, C.-M.; Chen, S.-Y. Volatile compounds in oils after deep frying or stir frying and subsequent storage. *J. Am. Oil Chem. Soc.* **1992**, *69*, 858–865. [CrossRef]
34. Duquenne, V. What can lattices do for experimental designs? *Math. Soc. Sci.* **1986**, *11*, 243–281. [CrossRef]
35. Jankovic, A.; Chaudhary, G.; Goia, F. Designing the design of experiments (DOE)—An investigation on the influence of different factorial designs on the characterization of complex systems. *Energy Build.* **2021**, *250*, 111298. [CrossRef]

materials MDPI

Review

Advances in the Applications of Clinoptilolite-Rich Tuffs

Jelena Pavlović [1], Jasna Hrenović [2], Dragan Povrenović [3] and Nevenka Rajić [4,*]

[1] Institute of Soil Science, Teodora Drajzera 7, 11000 Belgrade, Serbia; soils.pavlovic@gmail.com
[2] Faculty of Science, Department of Biology, University of Zagreb, 10000 Zagreb, Croatia; jasna.hrenovic@biol.pmf.hr
[3] Faculty of Technology and Metallurgy, University of Belgrade, Karnegijeva 4, 11120 Belgrade, Serbia; povrenovic@tmf.bg.ac.rs
[4] Faculty of Ecology and Environmental Protection, University "Union—Nikola Tesla", Cara Dušana 62–64, 11158 Belgrade, Serbia
* Correspondence: nrajic@unionnikolatesla.edu.rs

Abstract: Adsorptive, catalytic, and antibacterial properties of clinoptilolite-rich tuffs (ZT) are presented here. ZT transformed into Fe-containing ZT (Fe-ZT) removes various organic and inorganic anions from water. Fe-ZT, which contains selenium, is beneficial for growing *Pleurotus ostreatus* mushrooms. The fungi convert inorganic Se from Fe-ZT into a more useful organically bonded form. ZT and Fe-ZT as supplements retain nitrogen and potassium in sandy, silty loam and silty clay soils. ZT shows an affinity toward toxic metal cations, which are essential for cleaning contaminated water. The adsorption of atenolol, acetylsalicylic, and salicylic acid onto M-ZT (M–Cu^{2+}, Mn^{2+}, Ni^{2+}, or Zn^{2+}) from water solutions suggests that both the natures of M and pharmaceuticals have a significant impact on the adsorption mechanism and determine the adsorption capability of the ZT. ZT is an excellent carrier for ultrafine (2–5 nm) nano oxide particles, which have been shown to have catalytic activity in different chemical processes and photodegradation reactions of organic pollutants. ZT can also be transformed into SO_4-SnO_2-ZT, which is catalytically active as a solid acid. M-ZT is an effective carrier of valuable bacteria. Ag-ZT possesses beneficial bactericidal activity in disinfecting water and soil remediation.

Keywords: natural zeolite; adsorption; catalysis; antimicrobial activity; nano oxides; biomass

1. Introduction

Zeolites are crystalline porous aluminosilicates with a three-dimensional lattice structure built from a network of corner-sharing tetrahedra, TO_4 (T= Si, Al), which contain Si and Al atoms in the center of tetrahedra. Only species with the proper size may diffuse through the lattice owing to the crystal structure's geometrically precise structured channels and cages. As a result, zeolites can be employed as molecular sieves and in ion-exchange, -adsorption, and -separation processes. Zeolites can also be converted into strong solid acids and used as heterogeneous catalysts [1–3].

The most common type of natural zeolite is clinoptilolite. Eight- and ten-member ring channels with diameters of up to 0.7 nm reveal an easily accessible open reticular structure [4,5]. Two channels run parallel to the c-axis. The A channels comprise tightly compressed ten-membered rings with an aperture of 0.31 × 0.75 nm, and the B channels comprise eight-membered rings with an aperture of 0.36 × 0.46 nm. The C channels have an aperture of 0.28 × 0.47 nm and are formed by eight-membered rings parallel to the a-axis. Water molecules and exchangeable alkali and alkali earth cations occupy crystallographically specific extra-framework sites. Clinoptilolite samples from different regions have different ion-exchange abilities due to variances in the Si/Al molar ratio and the types of extra-framework cations.

Clinoptilolite-rich deposits are primarily found in regions that have experienced hydrothermal processes and volcanic activity. Europe's principal deposits of clinoptilolite-rich tuffs are in Slovakia, Ukraine, Turkey, Italy, and Romania. Typically, the zeolite minerals coexist in the specific rock. Certain minor and accessory minerals, including feldspar, the mica group, chlorite, cristobalite, zircon, iron, sulfur, and clay minerals, are linked to zeolite mineralization. The components of zeolite mineralization are regarded as undesirable impurities [6].

In recent decades, the adsorption, catalytic, and antibacterial characteristics of clinoptilolite-rich tuffs (ZT) from two deposits, Vranjska Banja and Slanci (Serbia), containing up to 80 wt.% clinoptilolite, with quartz and feldspar as the primary mineral impurities, have been extensively studied [7–38]. This review compares these results with ZTs from other regions. It is important to note that the high degree of grain size, porosity, cation-exchange capacity, clinoptilolite chemical composition, and experimental conditions all affect the properties under study. As a result, comparisons are frequently challenging to make and may produce conclusions that are not sufficiently trustworthy.

The clinoptilolite content of the ZT from deposits in Serbia is similar to that of the clinoptilolite-rich tuff from Ukraine, including roughly 70 wt.% [6], the tuff from Slovakia containing approximately 85 wt.% clinoptilolite [39], the tuff from Iran containing 82 wt.%, and the tuff from Turkey containing 76 wt.%. Clinoptilolite from Serbia has a Cation-Exchange Capacity (CEC) of 170–180 mmol $M^+/100$ g, while clinoptilolite from Ukraine has a CEC of 64 mmol $M^+/100$ g [39], clinoptilolite from has a CEC of 130 mmol $M^+/100$ g, and clinoptilolite from Iran has a CEC of 126 mmol $M^+/100$ g [16]. The variation in the Si/Al molar ratio and the type of ion-exchangeable cations can be attributed to the variation in CEC value.

2. Conversion of ZT

2.1. Conversion of ZT to Fe-Containing ZTs

Because Fe-oxyhydroxides, which are available commercially, are usually used to remove toxic As from subterranean water, which is not cost-effective, upgrading ZT to Fe-ZT or Fe_3O_4-ZT makes it an economically viable adsorbent [40–45]. A straightforward two-step process has typically been employed. Treatment of ZT with a water solution of $Fe(NO_3)_3$ in an acetate buffer (pH = 3.6) followed by adding NaOH until pH = 7 and heating of the product at 80 °C to a constant mass gives Fe-ZT. EDS analysis of Fe-ZT revealed that the content of exchangeable cations significantly decreased relative to ZT and that the content of Fe was significantly higher than the decrease in the ZT-exchangeable cations. This indicates that the conversion of ZT to Fe-ZT includes at least two processes: (1) ion exchange and (2) precipitation of the Fe(III) species at the ZT surface. The conversion of ZT to Fe_3O_4-ZT can be performed in a water solution of $FeCl_3 \cdot 6H_2O$ and $FeSO_4$ in a molar ratio of 2:1 under N_2 [17,46–49].

The treatments led to an increase in Fe content from 0.21 (ZT) to 18.1 wt.% for Fe-ZT in the clinoptilolite from deposits in Serbia [19], whereas, from Slovakia, the Fe content was lower [50] increased from 1.03 (ZT) to 5.82 wt.% (Fe-ZT) under similar conditions. For Fe_3O_4-ZT, the increase in Fe was 5.63 wt.%. It was proposed that Fe species precipitation would occur at the ZT because the EDS analysis revealed that the Fe content is higher than predicted based on the CEC value. TEM analyses verified it (Figure 1).

Accumulating the Fe-containing precipitates increases the specific surface area of ZT from 28.6 to 140.3 $m^2\ g^{-1}$ for Fe-ZT and 45.2 $m^2\ g^{-1}$ for Fe_3O_4-ZT. Amorphous precipitate in Fe-ZT significantly increases the specific surface area of ZT, whereas in Fe_3O_4-ZT, the well's crystalline magnetite nanoparticles in the range of 5–30 nm cover the surface of ZT [17,19]. Similarly, Fe-ZT obtained from ZT from the deposit Donje Jesenje (Croatia) was found to have a specific surface area around three times larger than that of parent ZT [40]. Furthermore, a marginally greater rise in specific surface area has been noted for Fe_3O_4-ZT, synthesized from ZT originating from a deposit from Turkey [47].

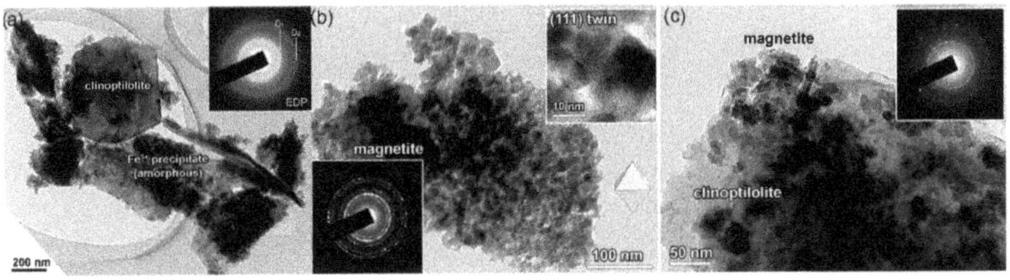

Figure 1. TEM images of (**a**) Fe-ZT: bright-field image of clinoptilolite sheets coated with a precipitate rich in Fe(III). The amorphous nature of the Fe(III) precipitate is confirmed by its electron diffraction pattern (EDP) in the upper right corner. (**b**) Nano oxide particles of magnetite: EDP is shown in the lower left corner. The upper right corner shows characteristic magnetite octahedral morphology with many crystals connected by the {111} spinel-twin law that forms through self-assembly. (**c**) Fe_3O_4-ZT: the EDP of magnetite layers on clinoptilolite sheets is given in the upper right corner [19].

It seems likely that the increase in the specific surface of Fe-ZT can be attributed to forming a second porous system at the ZT surface. The modification of ZT to Fe_3O_4–ZT increased the specific surface area to a lower extent but had two additional effects on ZT: (a) a partial dealumination (the Si/Al ratio increased from 5.0 to 6.1) and (b) magnetic properties. Powder XRD analysis showed that the partial dealumination did not significantly affect the crystallinity of ZT [17]. Most importantly, magnetite on the ZT surface introduced magnetism to ZT. In magnetic fields lower than 1.5 T, pure magnetite and Fe_3O_4-ZT exhibit comparable magnetic behavior with saturation magnetization values of 49.57 and 8.93 emu g^{-1}, respectively. The lower saturation magnetization value of Fe_3O_4-ZT was ascribed to the precipitation of magnetite on the ZT surface. The important notice is the fact that Fe_3O_4-ZT maintains its magnetic characteristics during the adsorption process. This lets the spent adsorbent be magnetically separated from water media [17].

Recently, a report was published on the green synthesis of clinoptilolite Fe_3O_4-zeolite nanocomposite using leaf extract from *Laurus nobilis* L. [51]. The composite displays characteristics of superparamagnetic properties. A novel solid-phase microextraction technique was created to extract and determine Rhodamine B utilizing high-performance liquid chromatography and clinoptilolite zeolite-coated magnetic nanocomposite.

2.2. Formation of Ultrafine Nano Oxide Particles inside ZT

Many nano oxide particles are effective catalysts in various chemical processes. However, most often, they must be supported by a suitable material that offers exposed surface area and mechanical strength because of their small size. With both of the desired qualities, clinoptilolite might be a promising option. Thus, the treatment of ZT with MCl_2 aqueous solutions (M = Ni, Cu, Mn, Zn) gave M-containing products (Ni-ZT, Cu-ZT, Mn-ZT, and Zn-ZT), which, after drying to a constant mass and calcination at 550 °C, produced nano oxide particles at the clinoptilolite surface [9,12]. EDS analyses showed that Na^+ ions are replaced in the Na-ZT by Mn(II), Ni(II), Cu(II), and Zn(II). The degree of ion exchange is different and depends on the chemical nature of the cations in aqueous solution: Ni(II) has the least amount of ion exchange (8%); Mn(II) and Zn(II) replace roughly 21% of the Na^+ ions, while for Cu(II), the exchange degree is the highest at 38%. According to a recent report, cations prefer four crystallographic positions in the clinoptilolite channels depending on their chemical nature [52]. Mn^{2+} prefers the cation site inside channel B (the site M2), Ni^{2+} is located primarily at two sites in the A channel (M1, M3: M1 is positioned in the center, and M3 is at the entrance to channel C), and Zn^{2+} occupies all three different sites inside three channels [53]. The differences in cation's accessibility to the extra-framework ion-exchange sites explain the differences observed during the calcination and dehydration of M-ZT. According to SEAD, the dehydrated Mn-ZT has no extra oxide phase compared to Ni-ZT, Cu-ZT, and Zn-ZT. As shown in Figure 2, NiO, ZnO, and Cu_2O oxide nanoparticles are randomly distributed in the clinoptilolite matrix. The last three solids contain crystal clusters that are populated with spherical nanocrystalline particles to varying degrees, which are oxides of Ni(II), Zn(II), and Cu(I) [9].

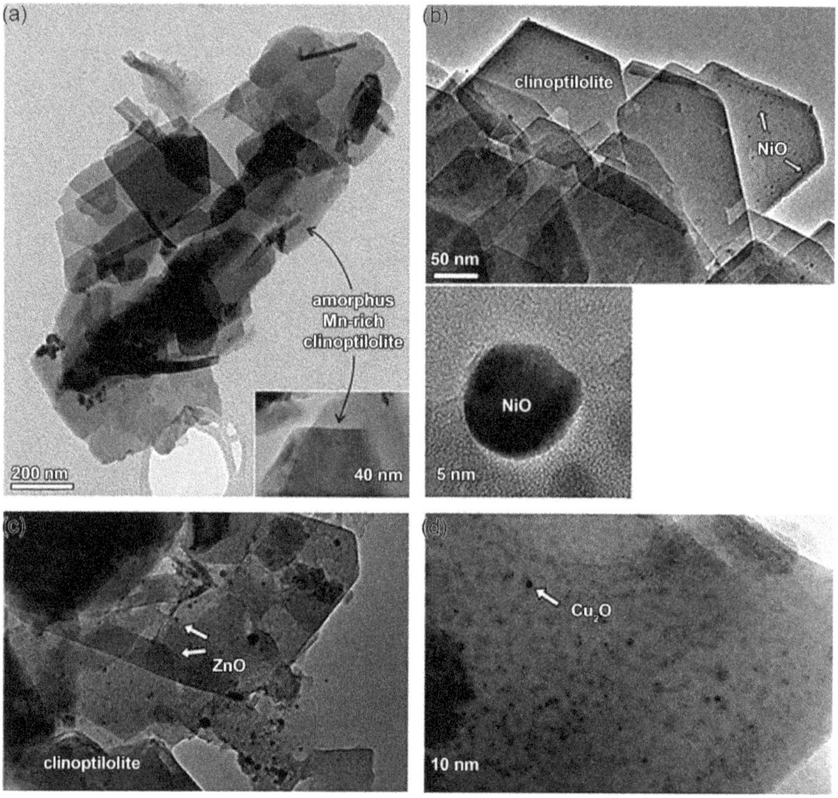

Figure 2. TEM image of the calcined M-ZT: (**a**) high-magnification image (bottom right) shows a uniform surface of an amorphous clinoptilolite grain without any visible oxide nanoparticles; (**b**) Ni-ZT grains with NiO nanoparticles. A single NiO particle and SAED pattern recorded across the NiO particles are shown below: (**c**) Zn-ZT grains with ZnO nanoparticles and (**d**) the surface of a Cu-ZT grain with multiple Cu_2O nanoparticles [9].

The average particle size of the NiO crystallites is approximately 5 nm in diameter, and their sizes range from 2 to 7 nm. SEAD pattern revealed that they belong to the cubic NiO structure (JCPDF # 78-0643). It is implied by the lack of preferential faceting of the cubic NiO crystals that a rapid crystallization occurred during the calcination. Similarly, the SAED identified a polycrystalline ZnO with a wurtzite structure in the calcined Zn-ZT (JCPDS 00-003-0888). The Cu_2O (cuprite) nanoparticles crystallized due to the calcination of the Cu-ZT. In this case, the reduction of Cu(II) to Cu(I) occurred most probably due to the higher lattice enthalpy of the Cu_2O. The estimated average size of cuprite particles was only 2 nm.

Nano-sized MnxOy-containing clinoptilolite has been synthesized using a hydrothermal method in a temperature range of 80–180 °C. Manganese found in three oxidation states, Mn^{2+} (37.8%), Mn^{3+} (14.2%), and Mn^{4+} (48%), explained the high catalytic activity of the composite [54].

The small pore size of the clinoptilolite lattice, the particular channel arrangement, and the crystallographic sites occupied by particular cations all appear to impact the crystallization of oxide phases during the calcination process. The steric restriction imposed by the clinoptilolite lattice enables the uniform distribution of the ultrafine nanoparticles, and the clinoptilolite lattice itself becomes a very good carrier of nanoparticles.

2.3. Conversion of ZT into SnO_2-ZT and SO_4-SnO_2-ZT

The main benefits of solid acids and super acids over ordinary mineral acids or Lewis acids are that they do not need as many toxic or corrosive reagents, nor do they have handling, post-reaction separation, recovery, recycling, or contamination issues. Some metal oxides, such as titanium, zirconia, and SnO_2, have become highly effective solid acid catalysts because of their super-acidity, high activity, and selectivity. Sulfated SnO_2 has recently been shown to have greater acid strength and catalytic activity than other sulfated metal oxides [55,56]. With this in mind, the immobilization of SO_4-SnO_2 onto ZT was studied to prevent possible sulfate leaching reported for SO_4-SnO_2 [18].

ZT was converted into SO_4-SnO_2-ZT using a three-step procedure: (1) conversion of ZT to H-ZT by treating ZT with 1 mol dm^{-3} HCl at 100 °C and then with 0.2 mol dm^{-3} NH_4OH at 65 °C; (2) preparation of the SnO_2-containing ZT (SnO_2-ZT) using an ethanolic solution of $SnCl_2$ (C_0 = 2 g dm^{-3}) and NH_4OH solution followed by calcination at 400 °C; and (3) sulfation of SnO_2-ZT using $(NH_4)_2SO_4$ followed by calcination at 400 °C to obtain SO_4-SnO_2-ZT [18].

In the first step of the procedure, ZT was converted into its hydrogen form, H-ZT. The treatment does not affect clinoptilolite's crystallinity, but the lattice underwent a partial dealumination, as evidenced by the increase in the Si/Al molar ratio from 4.9 to 7. The content of Na drastically decreased, and K, Ca, and Mg were not found, which confirms that ZT is essentially transformed into H-ZT. Subsequent transformation of H-ZT into SnO_2-ZT includes the treatment of H-ZT in a Sn(II) solution in an alkaline medium, resulting in a minor reduction of crystallinity, which is not affected by further transformation of SnO_2-ZT into SO_4-SnO_2-ZT. The solid-state Al27 NMR evidences a partial dealumination of ZT. Besides the peak characteristic for Al atoms located in tetrahedral positions of the zeolite framework (AlO_4 structural units) at 55 pm, maxima corresponding to octahedrally coordinated extra-framework Al species (0 ppm), as well as to the extra-framework five-coordinated Al (30 ppm), appear in the spectra. Moreover, the ^{29}Si MAS NMR spectra display peaks connected to various Si environments and are assigned to the Si atoms with H atoms in their vicinity.

The presence of SnO_2 in SnO_2-ZT was verified by X-ray photoelectron spectroscopy (XPS). According to the XPS depth profiles, the Sn concentration decreases from the top to the bottom of the sample. There is a slight accumulation of Sn at the surface, which is more noticeable in the samples with a higher Sn content (1.6–2.4 at.%). The unexpected result was the observation of Sn within the clinoptilolite lattice, indicating that the Sn species are present both inside the lattice and on the zeolite's surface. It is proposed that holes are formed in the lattice via a dealumination modification. Moreover, covalent Si–O–Sn bonds are suggested at the Sn(IV) species in the holes.

The last modification step is the sulfation of SnO_2-ZT to SO_4-SnO_2-ZT, including treatment with $(NH_4)_2SO_4$. This brings about a partial pore blockage of the clinoptilolite lattice, evidenced by a 20% decrease in the specific surface area for the sample with the highest Sn amount.

The purpose of this conversion of SnO_2-ZT to SO_4-SnO_2-ZT was to give the ZT an acidic quality that would be potent enough to catalyze the esterification of levulinic acid and thus replace the highly corrosive H_2SO_4 that is typically used in this process. An FTIR spectra analysis of adsorbed pyridine was used to investigate the acidity of SnO_2-ZT and SO_4-SnO_2-ZT. The SnO_2-ZT spectra contain both Lewis and Brönsted acid sites. There are significantly more Lewis acid sites than Brönsted sites due to the contribution of coordinated unsaturated Sn ions from the Sn(IV) species inside the lattice. The role of Sn(IV) in the creation of Lewis acid sites is supported by the fact that the number of Brönsted acid sites does not change significantly with the Sn content while the number of Lewis acid sites increases. Finally, the amounts of Brönsted and Lewis acid sites are significantly higher for sulfated samples, indicating that sulfate species increase the Brönsted acid strength of the hydroxyl groups on the surface of SnO_2 particles and function as Lewis acid sites.

3. Adsorption Studies

3.1. M(II) Adsorption (M–Mg, Mn, Ni, Cu, Zn, Pb) onto Na-ZT

Water and wastewater have been treated with ZTs, and this method continues to show promise for environmental cleaning processes. Over the past few decades, the application of ZTs has focused on eliminating ammonium and heavy metals due to ZT's pronounced ion-exchange capabilities [57–64]. Here, ZT's affinity (from deposits in Serbia) is presented for some metal cations. The experiments were conducted at 25–55 °C with an initial M(II) concentration of 1.5–6.0 mmol M dm^{-3}, using a batch method, and using the (solution volume)/(solid weight) ratio 100 cm^3: Na-ZT (ZT enriched with Na^+ by ion-exchange pretreatment) removes the studied cations from water media via an endothermic ion-exchange reaction. Specifically, metal ions from an aqueous solution that come into contact with the ZT replace sodium ions from the clinoptilolite lattice. It is not found that intra-particle diffusion is the rate-limiting step in the ion-exchange process, even though all of the metal ions under study have hydrated radii that are noticeably larger than the clinoptilolite lattice aperture. It can be proposed that the coordination sphere of the cations alters during the reaction. Table 1 gives the removal efficiency of the Na-ZT.

Table 1. Removal efficiency (%) of Na-ZT toward M(II) at the initial concentration of 1.5 mmol M(II) dm^{-3} at 25 °C.

M(II)	Mg	Mn	Ni	Cu	Zn	Pb
Removal, %	60	47	15	84	50	100

For all studied cations, the removal efficiency rises with temperature. As a result, the removal efficiency toward Cu(II) increased by roughly 14% when the temperature increased from 25 to 45 °C. Moreover, the pseudo-second-order rate model gives the most accurate description of adsorption kinetics for each of the cations. However, the rate of adsorption increases with temperature only for Mn^{2+} and Ni^{2+}; the increase is significant only for Mn^{2+}. This can be due to the influence of the hydrolysis of metal ions on the adsorption kinetics [10]. The outcomes of the detailed crystal-structure analyses of Pb-exchanged zeolites provide a potential explanation. Research revealed that a change in the coordination sphere of Pb(II) ions occurs concurrently with the ion exchange of Pb(II) in zeolite lattices [65]. The last modification makes the current Pb(II) ion species smaller, facilitating a simple exchange of Na^+ for Pb^{2+} within the zeolite channels. Because of the increased hydrolysis at higher temperatures, the formation of polynuclear metal species is likely. The large polynuclear hydrolysis products are less able to fit through the zeolite pore system due to their larger size, which explains the irregularities in the rate change by temperature. At 25 °C, the rate increases in the series $Mn^{2+} < Zn^{2+} < Pb^{2+} < Ni^{2+} < Cu^{2+} \approx Mg^{2+}$, whereas at 55 °C, the rate of adsorption is the highest for Mn^{2+} [8,10,11,20–23].

3.2. Adsorption of Nitrate and Phosphate onto Fe-ZT

Both nitrogen and phosphorus are essential elements found in a variety of sectors, including agriculture and various industries. But phosphate and nitrate ions are also contaminants. The primary cause of eutrophication in lakes, reservoirs, and rivers is an overabundance of these anions [66–68]. Globally, elevated concentrations of these ions during the last ten years have sparked grave worries. As a result, effective techniques for removing them from water are required to protect water supplies.

The negatively charged zeolite lattice reduces phosphate and nitrate adsorption efficiency by clinoptilolite; therefore, zeolite surface modification, such as coating with cationic surfactants [68–71], is applied to enable zeolite removal efficiency. Recent research, however, highlights the toxicity of cationic surfactants and advises using them with care [72,73].

Considering this, modified zeolite forms that employ non-compromising materials for modification are investigated. Fe-ZT was shown to be a suitable adsorbent for oxoanionic species [40–45,74].

Solutions of KNO_3 and KH_2PO_4 (1–6 mmol NO_3^-/PO_4^{3-} dm^{-3}) were used for the nitrate and phosphate adsorption by Fe-ZT [19,23]. The equilibrium adsorption data for both anions agree with the Langmuir isotherm, giving the Langmuir constant (R_L) values in the range of 0–1, which is characteristic of favorable adsorption.

The adsorption mechanism is explained by electrostatic interactions between negatively charged nitrate and phosphate anions and positively charged M–OH_2^+ groups on the metal-hydroxy-containing Fe-ZT. At lower pH, the M–OH_2^+ groups arise due to the reaction between the hydronium ions from the solution and the surface hydroxyl groups present on Fe-ZT. However, nitrate ions may also be partially bound through ion exchange, whereby they replace hydroxyl ions [24].

The highest affinity toward phosphate ions Fe-ZT was shown at pH = 6.5. Removal efficiency increases with temperature, and the adsorption is described well by the Langmuir isotherm. Adsorption kinetics follows the Lagergren pseudo-second-order model. However, the ^{31}P NMR study showed that the adsorption mechanism of phosphate is complex, including electrostatic interaction and more prominent covalent bonding between phosphate ion and Fe(III). The phosphate is predominantly bonded as a bidentate ligand [19].

3.3. Adsorption of Ciprofloxacin onto Fe_3O_4-ZT

Adsorption of ciprofloxacin (CIP) was studied for the initial CIP concentrations in the range of 0.04 to 0.2 mmol dm^{-3} [17]. The adsorption experiments were conducted at pH ~5, where CIP is present as the CIP$^+$ ion. Fe_3O_4-ZT was an excellent adsorbent in the 15 to 30 °C temperature range. The adsorption is very fast, so more than 80% of the maximum adsorption capacity is reached within the first 10 min. Adsorption kinetics follows the Lagergren pseudo-second-order equation, and the Langmuir isotherm model describes the equilibrium adsorption data. Because the adsorbed CIP could not be removed by different treatments, including the acid-base treatment and ion exchange, it seems likely that the adsorption mechanism involves very strong electrostatic interactions between the negatively charged Fe_3O_4-ZT surface and the cationic form of CIP.

The presence of nano-magnetite particles on ZT brings magnetism to ZT, which allows the easy separation of the saturated adsorbent by simple magnetic separation. Moreover, it is worth noticing that the CIP-containing Fe_3O_4-ZT exhibits strong antibacterial activity toward pathogens (*E. coli and S. aureus*), suggesting its possible application in water disinfection [17].

The regeneration of the CIP-containing Fe_3O_4-ZT is successfully achieved by a plasma treatment operating at atmospheric pressure with air as the working gas [24]. Non-thermal plasma does not affect the zeolite crystal structure nor its textural properties, suggesting that the treatment can be a convenient method for regenerating the mineral adsorbent.

Rouhani et al. [49] have also verified the adsorption efficiency of Fe_3O_4-ZT toward antibiotics. Tetracycline was removed with 98.6% efficiency under optimal reaction conditions, with pH = 7–8 yielding the maximum removal efficiency. The physical adsorption mechanisms included tetracycline polar molecules' van der Waals forces and hydrogen bonds with Fe_3O_4-ZT functional groups.

3.4. Adsorption of Atenolol, Acetylsalicylic Acid, and Salicylic Acid onto M-ZT

ZT and M-ZT (M–Mn(II), Ni(II), Cu(II), Zn) were tested for their adsorption capacities toward three pharmaceuticals that are known to be emerging water contaminants: atenolol (ATL), a beta-blocker drug, acetylsalicylic acid (ASA), a non-steroidal anti-inflammatory drug commonly used in organic synthesis, and salicylic acid (SA), widely used in organic synthesis [26].

The studies were carried out microcalorimetrically. The profiles of the heat flow signals are completed after 30 min, showing that the adsorption reached equilibrium. Adsorption isotherms differ for the same type of pharmaceutical, suggesting that the chemical nature of M influences the adsorption process. Also, different shapes of adsorption isotherms and different adsorption capacities are obtained for the different M-ZTs.

ATL adsorption capacities are found: Ni-ZT displays the highest adsorption capacity with multilayer adsorption. The final surface concentrations of ATL are from 40 to 115 μmol g^{-1}. Cu-ZT displayed the lowest adsorption abilities. A decrease in the differential heat (Q_{diff}) indicates that ATL molecules interact with heterogeneous solid surfaces: a drop from 40 kJ mol^{-1} to 10 kJ mol^{-1} is evident for Cu- and Ni-ZT and from 35 to 10 kJ mol^{-1} for Zn- and Mn-ZT. Initial Q_{diff} values suggest that a fraction of ATL molecules strongly interact with active sites, whereas most ATL molecules are physisorbed on M-ZTs.

In the case of SA, different adsorbed amounts and equilibrium concentrations were found for ZT and M-ZTs. The isotherm shapes differ significantly: the isotherms for Ni-Z and Mn-Z indicate multilayer adsorption, whereas the isotherms of SA on ZT, Cu-Z, and Zn-Z show saturation. The final surface concentrations of SA (amounts adsorbed per gram of solid adsorbent) are from 5 to 15 μmol g^{-1}. The values obtained for the SA adsorption show a strong interaction with M-ZT for at least a fraction of the salicylic acid molecules, as they are significantly higher than those found for ATL. For the Ni-, Cu-, Zn-, and Mn-ZT, the corresponding initial differential heat values for the SA adsorption are 110, 109, 105, and 55 kJ mol^{-1}. The initial Q_{diff} for SA on ZT is significantly lower (21 kJ mol^{-1}), indicating that ZT has a considerably lower affinity toward SA than M-ZTs. The heat of interaction could not be measured for ASA because of the ASA decomposition on the adsorbent surface.

The heats of adsorption obtained by microcalorimetry studies confirm that the adsorption of ATL and SA happens on energetically heterogeneous surfaces. This can be explained by the fact that the studied pharmaceuticals possess electron-donor groups (SA and ASA have O-donor groups (–OH and –OOH), and ATL has both –NH and –OH) and can interact with M^{2+} cations (which are electron acceptors) in different ways depending on the nature of M as well as on the steric requirements imposed by their crystallographic positions in the zeolite lattice [26].

4. Use of ZT in Agriculture

4.1. Use of ZT and Fe-ZT as Soil Supplements

Natural zeolites have good adsorptive and ion-exchange qualities, making them suitable for agricultural use. They have been viewed as a promising option for a soil supplement that can enhance soil's chemical and physical characteristics, including their ability to retain water and acidity. This quality is especially crucial for sandy soils [75]. One of the primary advantages of adding ZT to soils is its capacity to extend plant nutrient retention, which prevents nutrient leaching and helps preserve water [76–78]. Furthermore, ZT has successfully remediated heavy-metal-polluted soils [79–81].

ZT and Fe-ZT were tested as soil supplements for preserving plant nutrients (nitrogen and potassium) in different soil types: sandy, silty loam, and silty clay soils [13]. This was performed by leaching experiments performed in laboratory conditions using column systems at room temperature (Figure 3).

KNO_3 was used as a mineral fertilizer. Adding KNO_3 corresponded to 10 mg N (100 g soil)$^{-1}$ and 28 mg K (100 g soil)$^{-1}$, which amounts to 200 kg N and 550 kg K ha^{-1} soil.

Both ZT and Fe-ZT were active in soils, retaining the nutrients. Their retention efficiency depends on the soil type. Nitrate ions readily leach out, irrespective of the soil type. The high leaching was detected at the beginning of the experiment, and then the leaching proceeded slowly. This is ascribed to the high solubility of nitrate. Silty clay and sandy soils treated with chabazite (Italy) likewise exhibited a high nitrate leaching rate at the start of the leaching experiments [82]. Adding ZT and Fe-ZT increases nitrate retention in silty loam and clay soils. The retention effect is more pronounced for Fe-ZT (42% for

Fe-ZT and 26% for ZT) in silty loam soil in contrast to silty clay, where Fe-ZT retains only 14% of N (ZT efficiency was 4%).

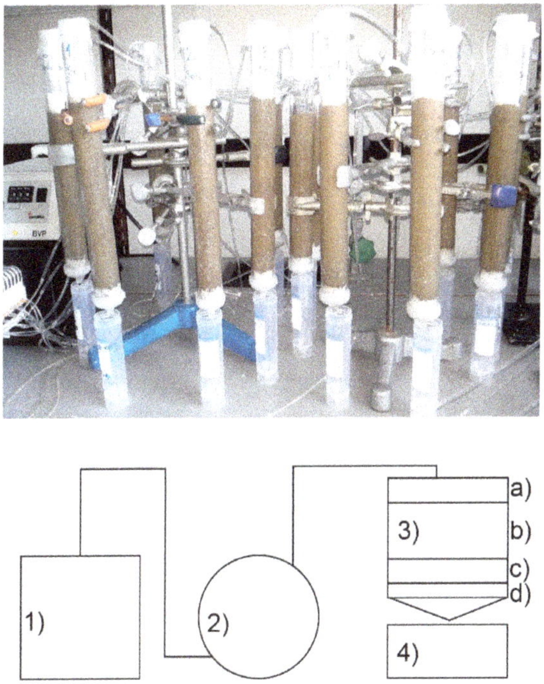

Figure 3. (a) Experimental setup used in the leaching experiments and (b) schematic presentation of an individual column: (1) water tank; (2) peristaltic pump; (3) plexiglass column; (a) PVC balls; (b) soil or soil/Fe-ZT mixture; (c) PVC balls; (d) PVC filter; (4) sample collector.

Their physico-chemical properties can explain the difference in retention activity in different soil types. Silty clay soil is alkaline, and it seems likely that there is a competition between the nitrate and hydroxyl ions for the adsorption sites on the zeolite, causing leaching of the nitrate.

Adding both ZT and Fe-ZT improves the retention of potassium in the studied soils in the following order: silty loam < silty clay << sandy soil. The highest retention was evident for sandy soil, which is important since sandy soils are poor in clay content and have limited ability to bind potassium. This effect is minor for the silty loam and clay soils since both can retain potassium [13].

The results confirm the applicability of ZT and Fe-ZT as soil supplements for retaining plant nutrients (nitrogen and potassium) in different soil types. On sandy soils, the addition has the most noticeable effect. Similarly, research demonstrates that K^+ leaching can be considerably reduced even in sandy soil supplemented with municipal compost [83]. It has also been reported that K-enriched ZT can be utilized to stop the loss of K^+ from soil when applied to sandy soil and sandy soil that has been supplemented with chemical fertilizers [84].

4.2. Use of the Se-Containing Fe-ZT for the Growth of Pleurotus Ostreatus

To obtain SeO_3-Fe-ZT and SeO_4-Fe-ZT, Fe-ZT was treated with water solutions of Na_2SeO_3 (4.5 mmol dm^{-3}) and Na_2SeO_4 (3.0 mmol dm^{-3}) at pH = 8 and pH = 3, respectively [14]. The Se XANES spectra of SeO_3-Fe-ZT confirmed the presence of selenite (37%) and selenate (63%), which were found in SeO_4-Fe-ZT. The Se K-edge EXAFS analysis provides insight into the local structure around the Se atoms in both Se-containing products. In the case of SeO_3-Fe-ZT, it is suggested that the selenite adsorption involves the formation of both Se–O–Fe and Se–O–Si bonds, while in the case of SeO_4-Fe-ZT, adsorption primarily occurs through the formation of Se–O–Fe bonds.

Both products containing selenium were utilized to cultivate *P. ostreatus* mycelia. The Se content in mushrooms cultivated on the substrate with SeO_3-Fe-ZT was about 211 $\mu g\ g^{-1}$ and lower than in mushrooms grown on the substrate with SeO_4-Fe-ZT (~254 $\mu g\ g^{-1}$). The presence of Se in mushrooms shows that *P. ostreatus* converts inorganic Se into an organically bound form by adsorbing it. The amount of Se in the utilized substrate is determined by its oxidation number. The cultivation of *P. ostreatus* onto a substrate mixed with the Se-containing ZT is a promising method for obtaining Se-enriched dietary supplements since the latter have been known to possess antioxidant activity [15].

4.3. Catalytic Use of ZT

One of the main concerns in the engineering of the atmospheric environment is the removal of NOx. NOx gases are typically produced when fuels burn at high temperatures. One of the methods for reducing NOx that has received the greatest attention is the selective catalytic reduction of NOx (SCR-NOx). According to Moreno-Tost et al. [85], clinoptilolite has demonstrated excellent performance as a catalyst for the selective catalytic reduction of NOx by ammonia (NH_3-SCR). The propane-SCR-NOx process has studied the catalytic efficacy of clinoptilolite exchanged with transition metal ions (Zn^{2+}, Fe^{2+}, Cu^{2+}, and Mn^{2+}). The Cu- and Mn-ZT performed the best conversion rate. The concentration of strong acidic sites, redox centers, highly specific surface areas, and additional framework species were found to be responsible for the conversion activity [86].

At ZT, wet impregnation and calcination at 500 °C were used to deposit oxides of cobalt, manganese, and mixed cobalt-manganese oxides. Using the XPS technique, Mn^{3+} and Mn^{4+} ions were found. Co_3O_4 is produced in both bi-component Co-Mn-ZT and Mn-ZT Mn-ZT, whereas MnO_2 is produced at Mn-ZT. On the ZT surface, the active phases were uniformly distributed and highly dispersed, leading to complete oxidation of the n-hexane. The sample containing the highest concentration of Co^{3+} ions exhibited the highest catalytic activity [87].

ZT was modified with hydrochloric acid, iron, copper, and cobalt salts to obtain catalytic properties for converting dihydroxyacetone (DHA) based biomass. During the conversion of DHA, lactic acid, formic acid, pyruvic acid, acetic acid, and levulinic acid were obtained. The highest lactic acid yield (66.2%) was achieved with Co-ZT, formic acid (93.6%) with Cu-H-ZT, and 87.4% acetic acid with Fe-ZT. The catalytic activity was proposed for a partial dealumination of clinoptilolite lattice and reduction of the Fe and Cu species [88].

A partial dealuminated ZT from deposit Kučin (Slovakia) was tested in the liquid phase isomerization of α-pinene. The results indicate that the catalytic activity of ZT is an entangled function of chemical composition, crystallinity, overall acidity, and substrate access to the active sites [5].

ZT purchased from Sepifeed (Turkey) was catalytically active in the isomerization of geraniol under mild conditions. Thumbergol, used in cancer treatment, was obtained at 47 mol.% [89].

The catalytic performance of NiO-ZT, Cu_2O-ZT, and Zn-ZT was studied in the pyrolysis of hardwood lignin using a bench-scale fixed-bed reactor [12]. The catalysts were mixed with lignin and then treated under an N_2 atmosphere at 500 °C. GC and GC/MS were used to analyze the resulting pyrolysis gas and liquid products.

Synergistic interaction between the ZT lattice and nano oxide particles explains the bio-oil production with high phenol content. The type of nano oxide present in ZT impacts the yield of phenols, with NiO-ZT exhibiting the highest value (approx. 54%). In the presence of NiO-ZT, the yield of unwanted oxygenated compounds (esters, carbonyls, and organic acids) decreases. Still, harmful polycyclic aromatic hydrocarbons (PAHs) do not rise significantly. This indicates that the clinoptilolite lattice's microporous constrictions do not favor the side-effect catalytic reactions that produce PAHs.

The results lead to the conclusion that the pyrolysis of hardwood lignin cannot be related to the acidity of the catalysts. A significant number of Lewis acid sites were found for ZnO-ZT, but the phenol content decreased compared to Na-ZT. In contrast, Cu_2O-ZT possesses many Lewis acid sites, increasing the bio-oil phenol content [12].

The results show that the clinoptilolite-based catalysts can have a significant role in the catalytic pyrolysis of lignin to bio-oils.

TiO_2-ZT was obtained using $TiCl_4$ and partially dealuminated ZT at a high temperature. The oxide form of Ti-immobilized on dealuminated clinoptilolite was formed in the anatase phase, with the zeolite structure remaining intact, according to powder XRD and EDS analyses. This catalyst's activity was tested in the esterification of 1-octanol with acetic acid. It was suggested that firstly, acetic acid interacts with the surface hydroxyl groups of the catalyst with intermolecular forces (H-bonding), and then the alcohol reacts with the adsorbed acid molecule to form the alkyl acetate product with condensation [90].

Levulinic acid (LA) is one of the platform chemicals obtained from biomass, and its esterification to levulinate esters is an industrially important process. The esterification is an acid-catalyzed reaction for which strong, harmful sulfuric acid is used. Many efforts have been made to substitute for sulfuric acid. The catalytic activity of SnO_2-ZT and SO_4-SnO_2-ZT was tested in the LA esterification with ethanol (EOL) and octanol (OOL) [17].

The obtained catalytic results show that both SnO_2-ZT and SO_4-SnO_2-ZT are catalytically active in the LA esterification with both alcohols. SnO_2-ZT exhibits high activity (around 55%) in converting LA to octyl levulinate (OLE). The catalytic activity is not affected by the Sn content. That can be due to hindered access of the long chain of octanol to all acid sites, especially those inside the pores of the clinoptilolite lattice. Significantly lower catalytic activity is achieved in converting LA to ethyl levulinate (ELE). The conversion rate increases with the Sn content (up to 22%), indicating that the catalytic active sites on the catalysts become more accessible to smaller ethanol molecules.

A total conversion of LA to both OLE and ELE was accomplished by SO_4-SnO_2-ZT, which is explained by the acidity of the samples, i.e., the presence of high amounts of both the Brönsted and Lewis acid sites (Figure 4). It is concluded that sulfate groups significantly influence the esterification reaction, supported by reported data [91,92].

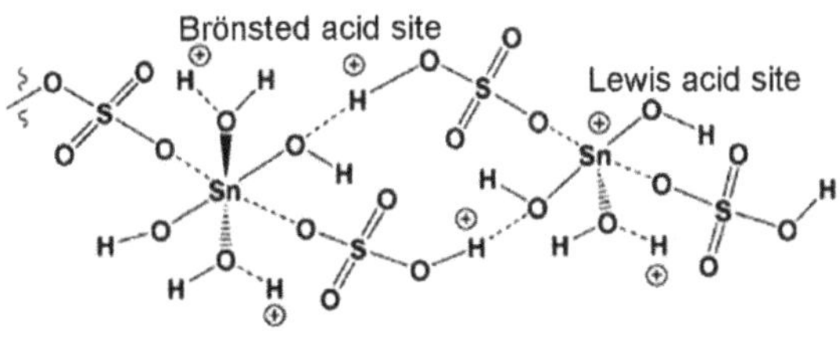

Figure 4. Schematic presentation of Lewis and Brönsted acid sites onto SO_4-SnO_2-ZT.

Reusability tests revealed that the process by which LA is converted to esters differs for EOL and OOL. Because of the length of the chain, it appears that the esterification reaction with OOL mostly happens at the catalyst's exterior surface. On the other hand, because EOL molecules can penetrate the catalyst's pores more deeply, there is less diffusion out of the intermediate products and ELE, which increases coke formation and reduces activity. Additionally, reusability tests revealed that (a) the first partial leaching of sulfate groups from the catalysts under test reduces the extent of the LA conversion and (b) the extent of the LA conversion is stabilized following the second cycle, indicating that structural characteristics of the clinoptilolite lattice prevent further leaching [18].

When all the data are considered, it seems that the clinoptilolite lattice can limit the extent of the sulfate leaching, providing reusable catalysts and preventing the aggregation of catalytically active SO_4-SnO_2 particles. This increases dispersion and improves accessibility for the reactants.

Photocatalytic Activity

Heterogeneous photocatalysis has emerged as the most widely studied advanced oxidation process (AOP). Suspensions of powdered TiO_2 in the treated solution have been used in most studies on photocatalytic degradation of organic pollutants [93]. However, several characteristics make it difficult to fully utilize TiO_2's photo efficiency, including poor adsorption, low surface area, the absorption of only a small portion of sunlight, rapid recombination of electron–hole pairs, and difficulty separating from solution. Several attempts have been made to circumvent this limitation to improve the efficiency of photocatalysts using suitable supports. Zeolites have been widely used among various supports due to their unique structural properties.

ZT belongs to the Semnan region (Iran) was used to prepare NiO-containing zeolite [94]. NiO-ZT has an essential role in the photodegradation of aqueous cefalexin (degradation efficiency was 73.5%), while pure NiO and zeolite did not have significant photodegradation efficiency [94]. Moreover, photocatalytic degradation of a 4-nitrophenol aqueous solution was investigated using ZnO-nano-ZT under UV irradiation. The photocatalyst was prepared by ion-exchanging nano-ZT in a zinc nitrate aqueous solution for 24 h, followed by calcination at 450 °C for 12 h. The results indicated that the photodegradation rate was affected by the initial 4-nitrophenol concentration, the pH, and the amount of catalyst [95].

CuO-ZT prepared via wet impregnation of parent zeolite with $CuSO_4$ aqueous solution and calcination was reported as an efficient photocatalyst in the degradation of p-aminophenol under sunlight irradiation. As for ZnO-ZT, it was concluded that the ZT host was important in the photodegradation process, so pure CuO and natural zeolite did not have significant photodegradation efficiency [96].

NiS-ZT can effectively degrade furfural exposed to UV light. Selecting the optimal parameters to accelerate the rate of degradation is crucial. Since it is outside the zeolite framework and does not exhibit a significant degradation efficiency, the zeolite lattice plays a crucial role in the degradation process, indicating that the active centers within the zeolite structure are NiS. One significant benefit of photodegradation is that very little photocatalyst (330 mg dm^{-3}) is used, which conserves photocatalyst, conserves photons because of the decreased scattering, and ultimately reduces environmental contamination [97].

ZT is suitable for photocatalytically active metal oxide particles such as SnO_2, TiO_2, ZnO, NiO, and CuO [98–102]. An enrichment in their catalytic activity was attributed to a synergistic effect between the particles of metal oxides and lattice of ZT. ZT prevents the aggregation of metal oxide particles by fixing them onto ion-exchange sites and enables electron–hole recombination.

Photocatalytic efficacy of SnO_2-ZT and ZT was evaluated in the degradation of methylene blue (MB), a representative cationic dye. Besides ZT, several zeolitic tuffs from deposits in different regions (Turkey, Iran, Romania, and Slovakia) were tested for comparison [14,16]. Using a batch reactor system, photocatalytic tests were carried out at room temperature and atmospheric pressure and under visible light irradiation.

SnO_2-ZT is photocatalytically active, increasing both the adsorption capacity and photocatalytic performance, which can be attributed to a high surface area and a partial increase in the negative potential of the surface. The degradation of MB is affected by the Sn content. Increasing the Sn content above an optimal amount decreases its photocatalytic activity under visible light illumination. This is attributed to the SnO_2 aggregation, a decrease in the effective surface area, and the collision of SnO_2 particles with free MB molecules.

It can be proposed how ZT functions in photocatalytic systems as present in this study. The synergistic effect of the clinoptilolite lattice and SnO_2 particles is responsible for the photolytic activity. Because of ZT's adsorption affinity for cationic organic dyes, more molecules are drawn to the catalyst surface, where the produced hydroxyl radicals create many active sites for the adsorption of intermediates. Additionally, the lattice inhibits their aggregation by anchoring the SnO_2 particles to specific crystallographic sites while simultaneously permitting electron–hole recombination.

ZT also exhibits exceptional photocatalytic activity in the MB degradation process. pH impacts the overall degradation of MB, peaking at pH = 6 (70% for C_0 = 10 mg MB dm^{-3}, 0.2 g ZT dm^{-3}, for 300 min). The photodegradation follows the kinetic model of Langmuir–Hinshelwood. The Fe species, typically found as impurities in zeolitic tuffs, have a combined effect on the initial adsorption and degradation of MB upon exposure to visible light, which is responsible for the entire degradation of the dye. Zeolitic tuffs from other regions give similar results. The activity rises with the tuffs' Fe content. Additionally, pure ZT from a deposit in Ukraine showed photocatalytic activity in the rhodamine B degradation due to iron impurities in tuff [103].

4.4. Interactions of Bacteria and ZT

Bacteria's interaction with zeolite depends on its chemical characteristics (native or modified ZT). The characteristics determine the type of zeolite-bacterium interactions, which can then influence the target species of bacteria (e.g., useful or pathogenic bacteria). As a model, the useful bacterium *Acinetobacter junii*, which is used in tertiary-stage wastewater treatment, is chosen. Besides the widely tested *Escherichia coli* and *Staphylococcus aureus*, an emerging hospital pathogen, *Acinetobacter baumannii*, is also chosen as a model pathogenic bacterium.

In a water column, bacteria have a native tendency to attach to the solid surface, where they continue to grow in the form of biofilm. The bacterial population in the biofilm, opposite to the planktonic population, will be protected from different environmental biotic and abiotic stresses [27]. The inorganic surface should be inert, nontoxic, of porous structure, relatively cheap, easily available, and environmentally friendly and provide a rough, irregular surface for bacterial colonization [104]. ZT meets all the mentioned characteristics to serve as a carrier of immobilized bacteria.

To elucidate the crucial factors determining the extent of bacterial immobilization onto different ZTs of the same particle size, ZTs of the particle size ~0.125 mm originating from Croatia, Turkey, and Serbia were examined [27]. After 24 h of contact, immobilized *A. junii* was in the order of a few billion bacteria per gram of dry ZT. The extent of bacterial immobilization could not be correlated with the main ZT features, such as the clinoptilolite content, cation-exchange capacity, or zeta potential. The extent of bacterial immobilization on a single ZT cannot be predicted by mineralogical and chemical analysis of ZT. The results conclude that each ZT is a candidate to be used as a carrier of bacteria, with the only limitation that ZT must not have a high concentration of toxic heavy metals.

In contact with bacteria suspended in water with ZT, the attachment of bacteria on the surface of the ZT will appear within one hour. The intensive immobilization of ZT (billions of bacteria per gram of ZT) is achieved after 12 h of contact, and prolonged contact until 24 h will ensure stable bacterial biofilm [27].

The rate of bacterial immobilization depends more on the features of the material and, to a lesser extent, on the features of the bacterium. The structure of the bacterial cell wall, widely classified as Gram-positive or Gram-negative, will not determine the immobilization. For the attachment to solid surfaces, bacteria use the structures on the cell surface that are pili or capsules. A hydrophilic (having pili at the surface) and hydrophobic (having capsule on the surface) isolate of *A. baumannii* was used to elucidate the effect of the hydrophobicity level of the bacterial surface on the intensity of immobilization onto the hydrophilic surface of ZT [28]. Both *A. baumannii*, with hydrophilic or hydrophobic cell surfaces, were immobilized onto ZT (5.2 and 6.9 log CFU (Colony Forming Units)/g, respectively) in the form of biofilm within 24 h of contact.

The nutrient availability for heterotrophic bacteria influences bacterial survival and multiplication in water. The immobilization of *A. baumannii* onto ZT in nutrient-poor and nutrient-rich water was followed for 72 h [28]. Planktonic cells of *A. baumannii* present in water became quickly attached to the surface of ZT particles, regardless of the nutrient concentration in water. Immobilized cells excrete the extracellular polymers and form stable biofilm on the particles within 24 h (Figure 5). No further incorporation of planktonic cells occurs in the formed biofilm.

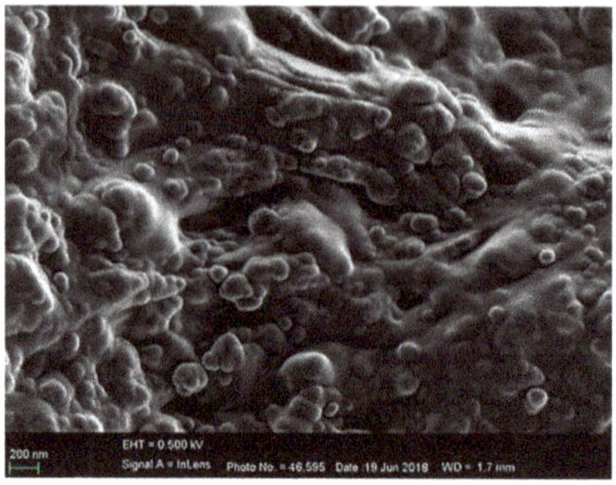

Figure 5. Cells of *A. baumannii* immobilized onto the ZT particles covered with extracellular polymer.

In nutrient-poor water, the shortage of nutrients prevents the multiplication of bacteria previously incorporated in the biofilm; consequently, the biofilm stays conserved. In nutrient-rich water, the availability of nutrients enables the multiplication of bacteria inside the initially formed biofilm, which increases the number of bacteria and further matures the biofilm. According to the behavior of bacteria in nutrient-poor and nutrient-rich water media, the mode of *A. baumannii* immobilization onto ZT is proposed, whereby this mode can be applied to the other bacterial species (Figure 6).

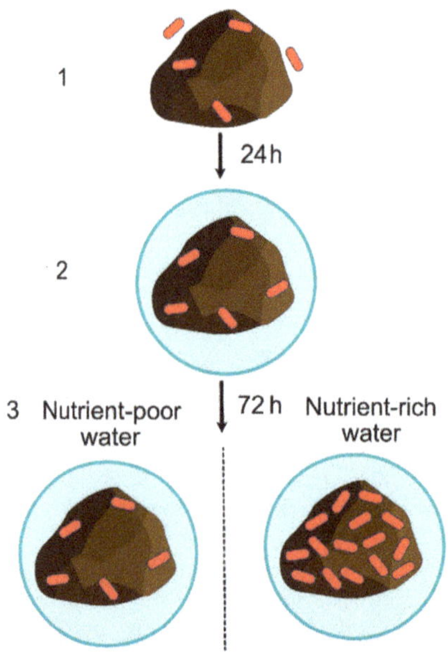

Figure 6. Mode of bacterial immobilization onto ZT. (1) Planktonic cells are quickly attached to the surface of ZT particles; (2) immobilized cells excrete the extracellular polymers and form biofilm; (3) in the nutrient-poor water, biofilm stays conserved, while in the nutrient-rich water, the number of bacteria in biofilm increases.

Another parameter that influences the extent of bacterial immobilization is the particle size of ZT. The extent of bacterial immobilization increases with the decrease in particle size of ZT [30]. Under the same conditions, immobilized *A. junii* decreased from 9.7 to 8.5 CFU/g when aparticle size < 0.125 mm or 0.5–1.0 mm was used, respectively.

4.4.1. Possible Applications of Bacterial Immobilization onto ZT

As bacteria can be easily immobilized onto ZT particles, ZT can be used as a carrier of beneficial bacteria. Bioparticles consisting of phosphate-accumulating bacteria *A. junii* immobilized onto ZT incorporate very well into activated sludge biomass [31]. The bioaugmentation of activated sludge with 5 g dm^{-3} of bioparticles significantly improved the phosphorus removal from fresh municipal wastewater. Better phosphorus removal is a function of increased biomass of phosphate-accumulating bacteria *A. junii*.

Bacteria immobilized onto ZT particles are protected from the grazing of ciliates and rotifers in the activated sludge and protected from being washed away by large amounts of wastewater [27]. Applying ZT in the tertiary stage of wastewater treatment can be used as a low-cost, efficient, and energy-saving technique to improve phosphorus removal from wastewater.

4.4.2. Interactions of Bacteria and M–ZT

The ZT containing heavy metals or bactericidal organic substances was studied to eliminate unwanted pathogenic bacteria [30]. The antibacterial effect of such modified ZT depends on the type and concentration of the modified ZT used, the chemical composition of the water medium, the concentration and species of bacteria, and contact time with bacteria. The modified ZTs act bactericidally by the direct contact of particles with bacteria or via the leaching of biologically toxic ions.

The antibacterial activity of Cu-ZT, Zn-ZT, or Ni-ZT at the concentration of 1 wt.% was tested against Gram-negative bacterium *E. coli* and Gram-positive bacterium *S. aureus* at the initial bacterial concentration of 7 log CFU/mL in different water media after short (1 h) and long-term (24 h) exposure [32]. The Ni-ZT showed weak antibacterial activity as compared to Cu-ZT and Zn-ZT. Antibacterial activities of the modified ZTs in a nutrient-rich medium were significantly lower than those in the nutrient-depleted media. The increased antibacterial activity of modified ZTs was evident by prolonging the contact time with bacteria. There was no remarkable difference in the antibacterial activity against Gram-negative and Gram-positive bacteria. The antibacterial activity of modified ZTs depends more on the features of the material and, to a lesser extent, on those of the bacterium.

The antibacterial activity of the Cu_2O, ZnO, or NiO nanoparticles supported on ZT was investigated in the secondary effluent wastewater [33]. The Cu_2O- and ZnO-containing ZT are more effective than the NiO-containing ZT, comparable to Cu-ZT, and Zn-ZT compared to Ni-ZT. After 24 h of contact, the Cu_2O and ZnO-containing ZT reduced the numbers of *E. coli* and *S. aureus* by four to six orders of magnitude. In the real wastewater, the bactericidal effect (meaning 100% reduction) against native *E. coli* was obtained after 1 h of contact. An increase in the NiO-ZT concentration from 1 to 5 wt.% did not enhance the antibacterial activity against *E. coli*. This fact suggests that the antibacterial activity of nano oxide particles supported on ZT depends on the chemical nature of the toxic metal and that an increase in the ZT content could not increase the antibacterial activity.

The Ag-ZT, Cu-ZT, and benzalkonium-containing ZT (BC-ZT) showed bactericidal activity against clinical isolates of *A. baumannii* in a physiological solution, depending on the type and concentration of ZTs used and the contact time [34]. Differences between the bacterial isolates only slightly influenced the bactericidal activity. Minimum bactericidal concentrations (MBC) after 24 h of contact varied from 31 to 250 mg dm^{-3}, 125 to 250 mg dm^{-3}, and 250 to 500 mg dm^{-3} for Ag-ZT, Cu-ZT, and BC-ZT, respectively.

Ag-ZT showed a superior bactericidal effect compared to Cu-ZT or Zn-ZT against isolates of *E. coli* [35]. Ag-ZT showed a bactericidal effect within 1 h of contact in a nutrient-rich medium, while Cu-ZT or Zn-ZT were bactericidal only in nutrient-poor water media.

Except in water media, Ag-ZT and Cu-ZT were tested for the elimination of *A. baumannii* from artificially contaminated natural soils [36]. Adding 1wt.% of Cu-ZT shortened the survival of *A. baumannii* from seven to three days in slightly acidic terra rossa and from four months to 14 days in slightly alkaline red palaeosol.

Adding 0.1 wt.% of Ag-ZT to slightly acidic or alkaline soil completely removed the *A. baumannii* within 1 h of contact, with a negligible impact on native soil bacteria.

Salycilate-containing BC-ZT at a concentration of 0.5 wt.% [37] immediately killed all *E. coli* and *S. aureus* in a nutrient-rich medium. The bactericidal effect was ascribed to the benzalkonium cations, while the salicylate anions had a negligible effect on the bacteria. It was shown that the arrangement of the BC layers onto ZT is crucial in the antibacterial activity of BC-ZT [3]. BC-ZT with the BC bilayer or patchy BC bilayer coverage shows a bactericidal effect. In the case of monolayer coverage, an antibacterial effect is expected, while in partial monolayer coverage, bacteria are unaffected.

Forming a biofilm on abiotic surfaces is one important virulence factor of *A. baumannii*. Novel composites with anti-biofouling activity were developed to prevent biofilm formation on the tubes for medical applications [38]. Ag-ZT was added to a polyvinyl chloride matrix, followed by coating with D-Tyrosine. Composites containing 1 wt.% of Ag-ZT showed a bactericidal effect against the clinical isolate of *A. baumannii* in the phosphate-buffered saline within 24 h of contact.

4.4.3. Other Possible Applications of M-ZT

The antibacterial activity of metal-containing ZT can be used to disinfect water containing low concentrations of nutrients, such as drinking water or secondary effluent wastewater. An M-ZT with low leaching of metal cations, such as Cu- or Cu_2O-containing ZT [32,33], is recommended for disinfection. The bactericidal activity of ZT with Cu(II), Ag(I), or BC can be applied in the disinfection of water or wet surfaces [34] but also in the remediation of soils contaminated with human pathogens [36].

ZTs can be used in medicine except for environmental applications. The BC-ZT and salicylate-containing BC-ZT [37] can be an alternative drug with simultaneous antibacterial and anti-inflammatory effects. Composite material containing the Ag-ZT is promising in preventing the biofilm formation of *A. baumannii* on tubes used for medical applications [38].

5. Conclusions

This review summarizes the results of extensive studies on tuffs rich in clinoptilolite from the deposits in Serbia and other countries. Because clinoptilolite is present there in high content, the tuffs can be recommended in wastewater treatments to adsorb and remove a wide range of inorganic (such as heavy metal cations, oxyanions, nitrates, and phosphates) and organic (such as organic micropollutants) pollutants from water.

Clinoptilolite's thermal and chemical stability makes it useful for creating ultrafine, catalytically active oxides or as a carrier of catalytically active species. Its catalytic activity has been confirmed in several catalytic processes.

Additionally, clinoptilolite is a beneficial addition to the soil since it improves the soil's nutritional qualities and can supply certain elements that it lacks, like selenium.

In conclusion, clinoptilolite is a potential antibacterial agent and a carrier of beneficial bacteria for the biotechnological treatment of activated sludge and wastewater.

This suggests that natural clinoptilolite has considerable potential for use as an environmentally friendly material in various applications, including synthetic zeolites that have long been regarded as superior.

Author Contributions: Conceptualization, N.R.; methodology, N.R.; writing—original draft preparation, N.R., J.P., J.H. and D.P.; writing—review and editing, N.R., J.P. and J.H.; visualization, J.P., D.P. and N.R. All authors have read and agreed to the published version of the manuscript.

Funding: This research was funded by the Ministry of Science, Technological Development and Innovation of the Republic of Serbia during the time of developing this review paper as an international collaborative project (Project number: 451-03-66/2024-03/200011) and the European Union's Horizon 2020 research and innovation program under the Marie Skłodowska-Curie grant agreement, MSCA-ITN-2018.

Data Availability Statement: Not applicable.

Acknowledgments: The authors express their sincere gratitude to Nenad Blagojević for his technical assistance in refining the manuscript.

Conflicts of Interest: The authors declare no conflicts of interest.

References

1. Yilmaz, B.; Müller, U. Catalytic applications of zeolites in chemical industry. *Top. Catal.* **2009**, *52*, 888–895. [CrossRef]
2. Garcia-Basabe, Y.; Rodriguez-Iznaga, I.; de Menorval, L.-C.; Llewellyn, P.; Maurin, G.; Lewis, D.W.; Binions, R.; Autie, M.; Ruiz-Salvador, A.R. Step-wise dealumination of natural clinoptilolite: Structural and physicochemical characterization. *Microporous Mesoporous Mater.* **2010**, *135*, 187–196. [CrossRef]
3. Farías, T.; Ruiz-Salvador, A.R.; Velazco, L.; de Ménorval, L.C.; Rivera, A. Preparation of natural zeolitic supports for potential biomedical applications. *Mater. Chem. Phys.* **2009**, *118*, 322–328. [CrossRef]
4. Godelitsas, A.; Armbruster, T. HEU-type zeolites modified by transition elements and lead. *Microporous Mesoporous Mater.* **2003**, *61*, 3–24. [CrossRef]
5. Dziedzicka, A.; Sulikowski, B.; Ruggiero-Mikolajczyk, M. Catalytic and physicochemical properties of modified natural clinoptilolite. *Catal. Today* **2016**, *259*, 50–58. [CrossRef]
6. Pabis-Mazgaj, E.; Gawenda, T.; Pichniarczyk, P.; Stempkowska, A. Mineral Composition and Structural Characterization of the Clinoptilolite Powders Obtained from Zeolite-Rich Tuffs. *Minerals* **2021**, *11*, 1030. [CrossRef]
7. Hrenovic, J.; Rozic, M.; Sekovanic, L.; Anic-Vucinic, A. Interaction of surfactant-modified zeolites and phosphate accumulating bacteria. *J. Hazard. Mater.* **2008**, *156*, 576–582. [CrossRef] [PubMed]
8. Rajic, N.; Stojakovic, D.; Jevtic, S.; Zabukovec, N.L.; Kovac, J.; Kaucic, V. Removal of aqueous manganese using the natural zeolitic tuff from the Vranjska Banja deposit in Serbia. *J. Hazard. Mater.* **2009**, *172*, 1450–1457. [CrossRef]
9. Rajic, N.; Stojakovic, D.; Daneu, N.; Recnik, A. The formation of oxide nanoparticles on the surface of natural clinoptilolite. *J. Phys. Chem. Solids* **2011**, *72*, 800–803. [CrossRef]
10. Stojakovic, D.; Hrenovic, J.; Mazaj, M.; Rajic, N. On the zinc sorption by the Serbian natural clinoptilolite and the disinfecting ability and phosphate affinity of the exhausted sorbent. *J. Hazard. Mater.* **2011**, *185*, 408–415. [CrossRef]
11. Stojakovic, D.; Milenkovic, J.; Daneu, N.; Rajic, N. A study of the removal of copper ions from aqueous solution using Clinoptilolite from Serbia. *Clays Clay Miner.* **2011**, *59*, 277–285. [CrossRef]
12. Rajić, N.; Logar, N.Z.; Rečnik, A.; El-Roz, M.; Thibault-Starzyk, F.; Sprenger, P.; Hannevold, L.; Andersen, A.; Stöcker, M. Hardwood lignin pyrolysis in the presence of nano-oxide particles embedded onto natural clinoptilolite. *Microporous Mesoporous Mater.* **2013**, *176*, 162–167. [CrossRef]
13. Pavlović, J.B.; Krogstad, T.; Rajić, N.Z. Applicability of zeolites in potassium and nitrate retention in different soil types. *J. Serbian Chem. Soc.* **2017**, *82*, 1303–1314. [CrossRef]
14. Šuligoj, A.; Pavlović, J.; Arčon, I.; Rajić, N.; Tušar, N.N. SnO_2-containing clinoptilolite as a composite photocatalyst for dyes removal from wastewater under solar light. *Catalysts* **2020**, *10*, 253. [CrossRef]
15. Jevtić, S.; Arčon, I.; Rečnik, A.; Babić, B.; Mazaj, M.; Pavlović, J.; Matijaševic, D.; Nikšić, M.; Rajić, N. The iron(III)-modified natural zeolitic tuff as an adsorbent and carrier for selenium oxyanions. *Microporous Mesoporous Mater.* **2014**, *197*, 92–100. [CrossRef]
16. Pavlović, J.; Šuligoj, A.; Opresnik, M.; Tušar, N.N.; Logar, N.Z.; Rajić, N. Studies of clinoptilolite-rich zeolitic tuffs from different regions and their activity in photodegradation of methylene blue. *Catalysts* **2022**, *12*, 224. [CrossRef]
17. Kalebić, B.; Pavlović, J.; Dikić, J.; Rečnik, A.; Gyergyek, S.; Škoro, N.; Rajić, N. Use of natural clinoptilolite in the preparation of an efficient adsorbent for ciprofloxacin removal from aqueous media. *Minerals* **2021**, *11*, 518. [CrossRef]
18. Pavlovic, J.; Popova, M.; Mihalyi, R.M.; Mazaj, M.; Mali, G.; Kovač, J.; Lazarova, H.; Rajic, N. Catalytic activity of SnO_2- and SO_4/SnO_2-containing clinoptilolite in the esterification of levulinic acid. *Microporous Mesoporous Mater.* **2019**, *279*, 10–18. [CrossRef]
19. Kaplanec, I.; Rečnik, A.; Mali, G.; Rajić, N. Study of the iron(III)-modified clinoptilolite in the adsorption of phosphate from aqueous medium: Mechanism and kinetics. *Desalin. Water Treat.* **2017**, *78*, 231–240. [CrossRef]
20. Tomić, S.; Rajić, N.; Hrenović, J.; Povrenović, D. Removal of Mg from spring water using natural clinoptilolite. *Clay Miner.* **2012**, *47*, 81–92. [CrossRef]
21. Tomic, S.; Knezevic, M.; Rajic, N.; Povrenovic, D. Removal of Magnesium in Spring Water Using the Natural Zeolite in a Continuous Flow System. *Hem. Ind.* **2014**, *68*, 475–482. [CrossRef]
22. Jovanovic, M.; Rajic, N.; Obradovic, B. Novel kinetic model of the removal of divalent heavy metal ions from aqueous solutions by natural clinoptilolite. *J. Hazard. Mater.* **2012**, *233–234*, 57–64. [CrossRef] [PubMed]

23. Stojakovic, D.; Milenkovic, J.; Stupar, S.; Velickovic, Z.; Rajic, N. Binary adsorption of nickel and zinc from aqueous solutions onto the Serbian natural clinoptilolite. *Desalin. Water Treat.* **2016**, *57*, 18748–18754. [CrossRef]
24. Pavlović, J.B.; Milenković, J.K.; Rajić, N.Z. Modification of natural clinoptilolite for nitrate removal from aqueous media. *J. Serbian Chem. Soc.* **2014**, *79*, 1309–1322. [CrossRef]
25. Kalebić, B.; Skoro, N.; Kovac, J.; Rajic, N. Regeneration of the ciprofloxacin-loaded clinoptilolite by non-thermal atmospheric plasma. *Appl. Surf. Sci.* **2022**, *593*, 153379. [CrossRef]
26. Rakić, V.; Rajić, N.; Daković, A.; Auroux, A. The adsorption of salicylic acid, acetylsalicylic acid and atenolol from aqueous solutions onto natural zeolites and clays: Clinoptilolite, bentonite and kaolin. *Microporous Mesoporous Mater.* **2013**, *166*, 185–194. [CrossRef]
27. Hrenovic, J.; Ivankovic, T.; Tibljas, D. The effect of mineral carrier composition on phosphate-accumulating bacteria immobilization. *J. Hazard. Mater.* **2009**, *166*, 1377–1382. [CrossRef]
28. Hrenović, J.; Tibljaš, D.; Dekić, S.; Rozman, T.I. Does bacterial surface hydrophobicity level influence their immobilization onto natural zeolite? In Proceedings of the 9th Croatian-Slovenian-Serbian Symposium on Zeolites, Split, Croatia, 23–25 September 2021.
29. Hrenović, J.; Tibljaš, D.; Dekić, S.; Venter, C. Mode of Acinetobacter baumannii immobilization onto natural zeolite in nutrient-poor and nutrient-rich water. In Proceedings of the 8th Serbian-Croatian-Slovenian Symposium on Zeolites, Belgrade, Serbia, 3–5 October 2019.
30. Hrenović, J.; Tibljaš, D.; Orhan, Y.; Büyükgüngör, H. Immobilisation of Acinetobacter calcoaceticus using natural carriers. *Water SA* **2005**, *31*, 261–266. [CrossRef]
31. Hrenović, J.; Büyükgüngör, H.; Orhan, Y. Use of natural zeolite to upgrade activated sludge process. *Food Technol. Biotechnol.* **2003**, *41*, 157–165.
32. Hrenovic, J.; Milenkovic, J.; Ivankovic, T.; Rajic, N. Antibacterial activity of heavy metal-loaded natural zeolite. *J. Hazard. Mater.* **2012**, *201–202*, 260–264. [CrossRef]
33. Hrenovic, J.; Milenkovic, J.; Daneu, N.; Kepcija, R.M.; Rajic, N. Antimicrobial activity of metal oxide nanoparticles supported onto natural clinoptilolite. *Chemosphere* **2012**, *88*, 1103–1107. [CrossRef]
34. Hrenovic, J.; Milenkovic, J.; Goic-Barisic, I.; Rajic, N. Antibacterial activity of modified natural clinoptilolite against clinical isolates of *Acinetobacter baumannii. Microporous Mesoporous Mater.* **2013**, *169*, 148–152. [CrossRef]
35. Milenkovic, J.; Hrenovic, J.; Matijasevic, D.; Niksic, M.; Rajic, N. Bactericidal activity of Cu-, Zn-, and Ag-containing zeolites toward *Escherichia coli* isolates. *Environ. Sci. Pollut. Res.* **2017**, *24*, 20273–20281. [CrossRef] [PubMed]
36. Hrenović, J.; Dekić, S.; Dikić, J.; Kazazić, S.; Durn, G.; Rajić, N. Metal-loaded zeolite remediation of soils contaminated with pandrug-resistant *Acinetobacter baumannii. Arh. Hig. Rada Toksikol.* **2020**, *71*, 146–151. [CrossRef] [PubMed]
37. Jevtić, S.; Grujić, S.; Hrenović, J.; Rajić, N. Surfactant-modified clinoptilolite as a salicylate carrier, salicylate kinetic release and its antibacterial activity. *Microporous Mesoporous Mater.* **2012**, *159*, 30–35. [CrossRef]
38. Milenkovic, J.; Hrenovic, J.; Goic-Barisic, I.; Tomic, M.; Djonlagic, J.; Rajic, N. Synergistic anti-biofouling effect of Ag-exchanged zeolite and D-Tyrosine on PVC composite against the clinical isolate of *Acinetobacter baumannii. Biofouling* **2014**, *30*, 965–973. [CrossRef] [PubMed]
39. Korkuna, O.; Leboda, R.; Skubiszewska-Zięba, J.; Vrublevs'ka, T.; Ryczkowski, J. Structural and physicochemical properties of natural zeolites: Clinoptilolite and mordenite. *Microporous Mesoporous Mater.* **2006**, *87*, 243–254. [CrossRef]
40. Šiljeg, M.; Cerjan Stefanović, Š.; Mazaj, M.; Novak Tušar, N.; Arčon, I.; Kovač, J.; Margeta, K.; Kaučić, V.; Zabukovec Logar, N. Structure investigation of As(III)- and As(V)-species bound to Fe-modified clinoptilolite tuffs. *Microporous Mesoporous Mater.* **2009**, *118*, 408–415. [CrossRef]
41. Dávila-Jiménez, M.M.; Elizalde-González, M.P.; Mattusch, J.; Morgenstern, P.; Pérez-Cruz, M.A.; Reyes-Ortega, Y.; Wennrich, R.; Yee-Madeira, H. In situ and ex situ study of the enhanced modification with iron of clinoptilolite-rich zeolitic tuff for arsenic sorption from aqueous solutions. *J. Colloid Interf. Sci.* **2008**, *322*, 527–536. [CrossRef]
42. Jiménez-Cedillo, M.J.; Olguína, M.T.; Fall, C. Adsorption kinetic of arsenates as water pollutant on iron, manganese and iron–manganese-modified clinoptilolite-rich tuffs. *J. Hazard. Mater.* **2009**, *163*, 939–945. [CrossRef]
43. Sanaei, L.; Tahmasebpoor, M. Physical appearance and arsenate removal efficiency of Fe(III)-modified clinoptilolite beads affected by alginate-wet-granulation process parameters. *Mater. Chem. Phys.* **2021**, *259*, 124009. [CrossRef]
44. Bilici Baskan, M.; Hadimlioglu, S. Removal of arsenate using graphene oxide-iron modified clinoptilolite-based composites: Adsorption kinetic and column study. *J. Anal. Sci. Technol.* **2021**, *12*, 22. [CrossRef]
45. Jeon, C.S.; Baek, K.; Park, J.K.; Oh, Y.K.; Lee, S.D. Adsorption characteristics of As(V) on iron-coated zeolite. *J. Hazard. Mater.* **2009**, *163*, 804–808. [CrossRef] [PubMed]
46. Mohseni-Bandpi, A.; Al-Musawi, T.J.; Ghahramani, E.; Zarrabi, M.; Mohebi, S.; Vahed, S.A. Improvement of zeolite adsorption capacity for cephalexin by coating with magnetic Fe_3O_4 nanoparticles. *J. Mol. Liq.* **2016**, *218*, 615–624. [CrossRef]
47. Mersin, G.; Açikel, Ü.; Levent, M. Efficient adsorption of Basic Blue 41 from textile wastewaters by natural and magnetically modified Manisa-Gördes clinoptilolite. *Chem. Eng. Process* **2021**, *169*, 108632. [CrossRef]
48. Noori, M.; Maryam Tahmasebpoor, M.; Foroutan, R. Enhanced adsorption capacity of low-cost magnetic clinoptilolite powders/beads for the effective removal of methylene blue: Adsorption and desorption studies. *Mater. Chem. Phys.* **2022**, *278*, 125655. [CrossRef]
49. Rouhani, M.; Ashrafi, S.D.; Taghavi, K.; Joubani, M.N.; Jaafari, J. Evaluation of tetracycline removal by adsorption method using magnetic iron oxide nanoparticles (Fe_3O_4) and clinoptilolite from aqueous solutions. *J. Mol. Liq.* **2022**, *356*, 119040. [CrossRef]
50. Chmielewská, E.; Tylus, W.; Drábik, M.; Majzlan, J.; Kravčak, J.; Williams, C.; Čaplovičová, M.; Čaplovič, L. Structure investigation of nano-FeO(OH) modified clinoptilolite tuff for antimony removal. *Microporous Mesoporous Mater.* **2017**, *248*, 222–233. [CrossRef]

51. Dalmaz, A.; Sivrikaya, S.Ö. Development of clinoptilolite zeolite-coated magnetic nanocomposite-based solid phase microextraction method for the determination of Rhodamine B in cosmetic products. *J. Chromatogr. A* **2022**, *1680*, 463433. [CrossRef]
52. Rodríguez-Iznaga, I.; Shelyapina, M.G.; Petranovskii, V. Ion exchange in natural clinoptilolite: Aspects related to its structure and applications. *Minerals* **2022**, *12*, 1628. [CrossRef]
53. Dimowa, L.T.; Petrov, O.E.; Djourelov, N.I.; Shivachev, B.L. Structural study of Zn-exchanged natural clinoptilolite using powder XRD and positron annihilation data. *Clay Miner.* **2015**, *50*, 41–54. [CrossRef]
54. Dakdareh, A.M.; Falamaki, C.; Ghasemian, N. Hydrothermally grown nano-manganese oxide on clinoptilolite for low-temperature propane-selective catalytic reduction of NO_x. *J. Nanopart. Res.* **2018**, *20*, 309. [CrossRef]
55. Matsuhashi, H.; Miyazaki, H.; Kawamura, Y.; Nakamura, H.; Arata, K. Preparation of a solid superacid of sulfated tin oxide with acidity higher than that of sulfated zirconia and its applications to aldol aondensation and benzoylation. *Chem. Mater.* **2001**, *13*, 3038–3042. [CrossRef]
56. Arata, K.; Matsuhashi, H.; Hino, M.; Nakamura, H. Synthesis of solid superacids and their activities for reactions of alkanes. *Catal. Today* **2003**, *81*, 17–30. [CrossRef]
57. Wang, S.; Peng, Y. Natural zeolites as effective adsorbents in water and wastewater treatment. *Chem. Eng. J.* **2010**, *156*, 11–24. [CrossRef]
58. Jahani, F.; Sadeghi, R.; Shakeri, M. Ultrasonic-assisted chemical modification of a natural clinoptilolite zeolite: Enhanced ammonium adsorption rate and resistance to disturbing ions. *J. Environ. Chem. Eng.* **2023**, *11*, 110354. [CrossRef]
59. Kragović, M.; Daković, A.; Sekulić, Ž.; Trgo, M.; Ugrina, M.; Perić, J.; Diego Gatta, G. Removal of lead from aqueous solutions by using the natural and Fe(III)-modified zeolite. *Appl. Surf. Sci.* **2012**, *258*, 3667–3673. [CrossRef]
60. Ugrina, M.; Čeru, T.; Nuić, I.; Trgo, M. Comparative study of mercury(II) removal from aqueous solutions onto natural and iron-modified clinoptilolite rich zeolite. *Processes* **2020**, *8*, 1523. [CrossRef]
61. Dimirkou, A. Uptake of Zn^{2+} ions by a fully iron-exchanged clinoptilolite. Case study of heavily contaminated drinking water samples. *Water Res.* **2007**, *41*, 2763–2773. [CrossRef]
62. Velarde, L.; Sadegh, M.N.; Escalera, E.; Antii, M.L.; Akhtar, F. Adsorption of heavy metals on natural zeolites: A review. *Chemosphere* **2023**, *328*, 138508. [CrossRef]
63. Vocciante, M.; De Folly D'Auris, A.; Finocchi, A.; Tagliabue, M.; Bellettato, M.; Ferrucci, A.; Reverberi, A.P.; Ferro, S. Adsorption of ammonium on clinoptilolite in presence of competing cations: Investigation on groundwater remediation. *J. Clean. Prod.* **2018**, *198*, 480–487. [CrossRef]
64. Wasieleweski, S.; Rott, E.; Minke, R.; Steinmetz, H. Application of natural clinoptilolite for ammonium removal from sludge water. *Molecules* **2021**, *16*, 114. [CrossRef] [PubMed]
65. Shibata, W.; Steff, K. Pb^{2+} exchange isotherms for zeolite Na-X at pH 5, 6, and 7. *Zeolites* **1997**, *19*, 87–89. [CrossRef]
66. Kim, Y.S.; Lee, Y.H.; An, B.; Choi, S.; Park, J.H.; Jurng, J.S.; Lee, S.H.; Choi, J.W. Simultaneous removal of phosphate and nitrate in u high-capacity anion-exchange resin. *Water Air Soil Pollut.* **2012**, *223*, 5959–5966. [CrossRef]
67. Akinnawo, S.O. Eutrophication: Causes, consequences, physical, chemical and biological techniques for mitigation strategies. *Environ. Chall.* **2023**, *12*, 100733. [CrossRef]
68. Devlin, M.; Brodie, J. Nutrients and eutrophication. In *Marine Pollution—Monitoring, Management and Mitigation*; Reichelt-Brushett, A., Ed.; Springer Textbooks in Earth Sciences, Geography and Environment; Springer: Cham, Switzerland, 2023. [CrossRef]
69. Guan, H.; Bestland, E.; Zhu, C.; Zhu, H.; Albertsdottir, D.; Hutson, J.; Simmons, C.T.; Ginic-Markovic, M.; Tao, X.; Ellis, A.V. Variation in performance of surfactant loading and resulting nitrate removal among four selected natural zeolites. *J. Hazard. Mater.* **2010**, *183*, 616–621. [CrossRef] [PubMed]
70. Schick, J.; Caullet, P.; Paillaud, J.L.; Patarin, J.; Mangold-Callarec, C. Batch-wise nitrate removal from water on a surfactant-modified zeolite. *Microporous Mesoporous Mater.* **2010**, *132*, 39–400. [CrossRef]
71. Schick, J.; Caullet, P.; Paillaud, J.L.; Patarin, J.; Mangold-Callarec, C. Nitrate sorption from water on a surfactant-modified zeolite. Fixed-bed column experiments. *Microporous Mesoporous Mater.* **2011**, *142*, 549–556. [CrossRef]
72. Zhan, Y.; Lin, J.; Zhu, Z. Removal of nitrate from aqueous solution using cetylpyridinium bromide (CPB) modified zeolite as adsorbent. *J. Hazard. Mater.* **2011**, *186*, 1972–1978. [CrossRef]
73. Reeve, P.; Fallowfield, H. The toxicity of cationic surfactant HDTMA-Br, desorbed from surfactant modified zeolite, towards faecal indicator and environmental microorganisms. *J. Hazard. Mater.* **2017**, *339*, 208–215. [CrossRef]
74. Du, G.; Li, Z.; Liao, L.; Hanson, R.; Leick, S.; Hoeppner, N.; Jiang, W.T. Cr(VI) retention and transport through Fe(III)-coated natural zeolite. *J. Hazard. Mater.* **2012**, *221–222*, 118–123. [CrossRef] [PubMed]
75. Mondal, M.; Biswas, B.; Garai, S.; Sarkar, S.; Banerjee, H.; Brahmachari, K.; Bandyopadhyay, P.K.; Maitra, S.; Brestic, M.; Skalicky, M.; et al. Zeolites enhance soil health, crop productivity and environmental safety. *Agronomy* **2021**, *11*, 448. [CrossRef]
76. Reháková, M.; Čuvanová, S.; Dzivák, M.; Rimár, J.; Gavaľová, Z. Agricultural and agrochemical uses of natural zeolite of the clinoptilolite type. *Curr. Opin. Solid State Mater. Sci.* **2004**, *8*, 397–404. [CrossRef]
77. Latifah, O.; Ahmed, O.H.; Muhamad, N.; Majid, A. Enhancing nitrogen availability from urea using clinoptilolite zeolite. *Geoderma* **2017**, *306*, 152–159. [CrossRef]
78. Souza, I.M.S.; Gurgel, G.C.S.; Medeiros, A.M.; Zonta, E.; Ruiz, J.A.C.; Paskocimas, C.A.; Motta, F.V.; Bomio, M.R.D. The use of clinoptilolite as carrier of nitrogened fertilizer with controlled release. *J. Environ. Chem. Eng.* **2018**, *6*, 4171–4177. [CrossRef]
79. Usman, A.R.A.; Kuzyakov, Y.; Lorenz, K.; Stahr, K. Remediation of a soil contaminated with heavy metals by immobilizing compounds. *J. Plant Nutr. Soil Sci.* **2006**, *169*, 205–212. [CrossRef]

80. Mahabadi, A.A.; Hajabbasi, M.A.; Khademi, H.; Kazemian, H. Soil cadmium stabilization using an Iranian natural zeolite. *Geoderma* **2007**, *137*, 388–393. [CrossRef]
81. Yi, N.; Wu, Y.; Fan, L.; Hu, S. Remediating Cd-contaminated soils using natural and chitosan-introduced zeolite, bentonite, and activated carbon. *Pol. J. Environ. Stud.* **2019**, *28*, 1461–1468. [CrossRef]
82. Colombani, N.; Mastrocicco, M.; Di Giuseppe, D.; Faccini, B.; Coltorti, M. Batch and column experiments on nutrient leaching in soils amended with Italian natural zeolitites. *Catena* **2015**, *127*, 64–71. [CrossRef]
83. Moraetis, D.; Papagiannidou, S.; Pratikakis, A.; Pentari, D.; Komnitsas, K. Effect of zeolite application on potassium release in sandy soils amended with municipal compost. *Desalin. Water Treat.* **2015**, *57*, 13273–13284. [CrossRef]
84. Eslami, M.; Khorassani, R.; Coltorti, M.; Malferrari, D.; Faccini, B.; Ferretti, G.; Di Giuseppe, D.; Fotovat, A.; Halajnia, A. Leaching behaviour of a sandy soil amended with natural and NH_4^+ and K^+ saturated clinoptilolite and chabazite. *Arch. Agron. Soil Sci.* **2018**, *64*, 1142–1151. [CrossRef]
85. Moreno-Tost, R.; Santamaría-González, J.; Rodríguez-Castellón, E.; Jiménez-López, A.; Autié, M.A.; González, E.; Carreras Glacial, M.; De las Pozas, C. Selective catalytic reduction of nitric oxide by ammonia over Cu-exchanged Cuban natural zeolites. *Appl. Catal. B Environ.* **2004**, *50*, 279–288. [CrossRef]
86. Ghasemian, N.; Falamaki, C. Zn^{2+}, Fe^{2+}, Cu^{2+}, Mn^{2+}, H^+ Ion-exchanged and raw clinoptilolite zeolite catalytic performance in the propane-SCR-NO_x Process: A comparative study. *Int. J. Chem. React. Eng.* **2018**, *16*, 20160192. [CrossRef]
87. Yordanova, I.; Hristov, S.; Kolev, H.; Todorova, S.; Naydenov, A. Cobalt-manganese ion-exchanged clinoptilolite supported catalysts for *n*-hexane oxidation. *Catal. Today* **2023**, *423*, 114267. [CrossRef]
88. Sobuś, N.; Król, M.; Piotrowski, M.; Michorczyk, B.; Czekaj, I.; Kornaus, K.; Trenczek-Zając, A.; Komarek, S. Conversion of dihydroxyacetone to carboxylic acids on pretreated clinoptilolite modified with iron, copper, and cobalt. *Catal. Commun.* **2022**, *171*, 106509. [CrossRef]
89. Fajdek-Bieda, A.; Wróblewska, A.; Miądlicki, P.; Tolpa, J.; Michalkiewicz, B. Clinoptilolite as a natural, active zeolite catalyst for the chemical transformations of geraniol. *React. Kinet. Mech. Catal.* **2021**, *133*, 997–1011. [CrossRef]
90. Özyağcı, B.; Şahin, V.; Karabakan, A. Esterification of 1-Octanol on clinoptilolite-supported TiO_2 catalysts. *Silicon* **2019**, *11*, 339–344. [CrossRef]
91. Popova, M.; Shestakova, P.; Lazarova, H.; Dimitrov, M.; Kovacheva, D.; Szegedi, A.; Mali, G.; Dasireddy, V.; Likozar, B.; Wilde, N.; et al. Efficient solid acid catalysts based on sulfated tin oxides for liquid phase esterification of levulinic acid with ethanol. *Appl. Catal. A Gen.* **2018**, *560*, 119–131. [CrossRef]
92. Popova, M.; Szegedi, A.; Lazarova, H.; Ristić, A.; Kalvachev, Y.; Atanasova, G.; Wilde, N.; Novak Tušar, N.; Gläser, R. Synthesis of biomass derived levulinate esters on novel sulfated Zr/KIL-2 composite catalysts. *Microporous Mesoporous Mater.* **2016**, *235*, 50–58. [CrossRef]
93. Bizani, E.; Fytianos, K.; Poulios, I.; Tsiridis, V. Photocatalytic decolorization and degradation of dye solutions and wastewaters in the presence of titanium dioxide. *J. Hazard. Mater.* **2006**, *136*, 85–94. [CrossRef]
94. Ajoudanian, N.; Nezamzadeh-Ejhieh, A. Enhanced photocatalytic activity of nickel oxide supported on clinoptilolite nanoparticles for the photodegradation of aqueous cephalexin. *Mater. Sci. Semicond. Process.* **2015**, *36*, 162–169. [CrossRef]
95. Nezamzadeh-Ejhieh, A.; Khorsandi, S. Photocatalytic degradation of 4-nitrophenol with ZnO supported nano-clinoptilolite zeolite. *J. Ind. Eng. Chem.* **2014**, *20*, 937–946. [CrossRef]
96. Nezamzadeh-Ejhieh, A.; Amiri, M. CuO supported clinoptilolite towards solar photocatalytic degradation of *p*-aminophenol. *Powder Technol.* **2013**, *235*, 279–288. [CrossRef]
97. Nezamzadeh-Ejhieh, A.; Moeinirad, S. Heterogeneous photocatalytic degradation of furfural using NiS-clinoptilolite zeolite. *Desalination* **2011**, *273*, 248–257. [CrossRef]
98. Dzinun, H.; Othman, M.H.D.; Ismail, A.F. Photocatalytic performance of TiO_2/Clinoptilolite: Comparison study in suspension and hybrid photocatalytic membrane reactor. *Chemosphere* **2019**, *228*, 241–248. [CrossRef] [PubMed]
99. Nezamzadeh-Ejhieh, A.; Zabihi-Mobarakeh, H. Heterogeneous photodecolorization of mixture of methylene blue and bromophenol blue using CuO-nano-clinoptilolite. *J. Ind. Eng. Chem.* **2014**, *20*, 1421–1431. [CrossRef]
100. Bahrami, M.; Nezamzadeh-Ejhieh, A. Effect of the supported ZnO on clinoptilolite nano-particles in the photodecolorization of semi-real sample bromothymol blue aqueous solution. *Mater. Sci. Semicond. Process.* **2015**, *30*, 275–284. [CrossRef]
101. Abdollahi, B.; Shakeri, A.; Aber, S.; Bonab, M.S. Simultaneous photodegradation of acid orange 7 and removal of Pb^{2+} from polluted water using reusable clinoptilolite–TiO_2 nanocomposite. *Res. Chem. Intermed.* **2018**, *44*, 1505–1521. [CrossRef]
102. Ullah, R.; Liu, C.; Panezai, H.; Gul, A.; Sun, J.; Wu, X. Controlled crystal phase and particle size of loaded-TiO_2 using clinoptilolite as support via hydrothermal method for degradation of crystal violet dye in aqueous solution. *Arab. J. Chem.* **2020**, *13*, 4092–4101. [CrossRef]
103. Sydorchuk, V.; Vasylechko, V.; Khyzhun, O.; Gryshchouk, G.; Khalameida, S.; Vasylechko, L. Effect of high-energy milling on the structure, some physicochemical and photocatalytic properties of clinoptilolite. *Appl. Catal. A Gen.* **2021**, *610*, 117930. [CrossRef]
104. Durham, D.R.; Marshall, L.C.; Miller, J.G.; Chmurny, A.B. Characterization of inorganic biocarriers that moderate system upsets during fixed-film biotreatment processes. *Appl. Environ. Microbiol.* **1994**, *60*, 3329–3335. [CrossRef]

materials

MDPI

Article

Characterization of Al-Containing Industrial Residues in the ESEE Region Supporting Circular Economy and the EU Green Deal

Emilija Fidanchevski [1], Katarina Šter [2], Maruša Mrak [2], Milica Rajacic [3], Bence David Koszo [4], Andrej Ipavec [5], Klemen Teran [6], Gorazd Žibret [6,*], Vojo Jovanov [1], Nikolina Stamatovska Aluloska [7], Mojca Loncnar [8], Lea Žibret [2] and Sabina Dolenec [2,9]

1 Faculty of Technology and Metallurgy, Ss. Cyril and Methodius University in Skopje, 1000 Skopje, North Macedonia; emilijaf@tmf.ukim.edu.mk (E.F.); vojo@tmf.ukim.edu.mk (V.J.)
2 Slovenian National Building and Civil Engineering Institute, 1000 Ljubljana, Slovenia; katarina.ster@zag.si (K.Š.); marusa.mrak@zag.si (M.M.); lea.zibret@zag.si (L.Ž.); sabina.dolenec@zag.si (S.D.)
3 Radiation and Environment Protection Department, Vinca Institute for Nuclear Science, National Institute of the Republic of Serbia, University of Belgrade, 11000 Belgrade, Serbia; milica100@vin.bg.ac.rs
4 Bay Zoltán Nonproft Ltd. for Applied Research, H-6726 Szeged, Hungary; bence.koszo@bayzoltan.hu
5 Alpacem Cement, d.d., Anhovo 1, 5210 Deskle, Slovenia; andrej.ipavec@alpacem.si
6 Geological Survey of Slovenia, 1000 Ljubljana, Slovenia; klemen.teran@geo-zs.si
7 Cementarnica "USJE"AD Titan Group, 1000 Skopje, North Macedonia; nikolinas@usje.mk
8 SIJ Acroni d.o.o., Cesta Borisa Kidriča 44, 4270 Jesenice, Slovenia; mojca.loncnar@acroni.si
9 University of Ljubljana, Faculty of Natural Sciences and Engineering, Department of Geology, Aškerčeva ulica 12, 1000 Ljubljana, Slovenia
* Correspondence: gorazd.zibret@geo-zs.si; Tel.: +386-41-370-740

Abstract: The increase in industrial waste generation presents a global problem that is a consequence of the needs of modern society. To achieve the goals of the EU Green Deal and to promote the concept of circular economy (CE), the valorization of industrial residues as secondary raw materials offers a pathway to economic, environmental, energetic, and social sustainability. In this respect, Al-containing industrial residues from alumina processing (red mud), thermal power plants (fly ash and bottom ash), and metallurgy (slag), as well as other industries, present a valuable mineral resource which can be considered as secondary raw materials (SRMs) with the potential to be used in construction, supporting the concept of circular economy. This paper focuses on the characterization of 19 secondary raw materials from the East South-East Europe (ESEE) region regarding their physical, chemical, mineralogical, and radiological characteristics. The goal is to provide a foundation for future innovations based on secondary raw materials, in alignment with the EU Green Deal and the principles of circular economy. The results showed that fly ash has the potential to be the best material among those analyzed to be used in the cement industry, mainly due to its favorable radiological and mineralogical properties. However, it is important to control the amount of free lime in the mixture, ensuring it remains below 10%. After evaluating secondary mineral raw materials for metal recovery, the results indicate that these materials are not viable sources for base metals or other technology-critical metals, such as REEs.

Keywords: Green Deal; Al-containing industrial residues; circular economy; fly ash; slag; red mud

1. Introduction

The EU Green Deal (EU GD) is a new growth strategy aimed at transforming the EU into a fair and prospective society. It focuses on creating a modern and competitive economy with the goal of completely reducing greenhouse gas emissions by 2050 in the whole European Union [1]. The mobilization of industries for a clean and circular economy is essential for achieving the new EU strategy. The circular economy action plan (ECAP) supports the transition from a linear model to a circular model, where waste becomes a valuable resource [2]. In addition, it recommends making use of mineral resources in more efficient way, i.e., to recycle and recover raw materials from any waste streams [3]. The concept of circular economy supports the utilization of secondary raw materials (SRMs), as these materials provide economic, energetic, and environmental benefits. Nonetheless, SRM still contains many valuable components and possesses characteristic physico-chemical properties for producing value-added products that, in recent years, have gained significant interest. In this regard, slag from the steel industry, red mud from alumina production, and fly ash and bottom ash from thermal power plants and the paper industry, among other wastes/by-products, present potential for valorizing as SRMs in cement plants [4] and in the production of bricks [5], glass–ceramics [6], adsorbents [7], zeolites [8], etc.

In 2021, the world iron slag production was estimated to be between 340 million and 410 million tons, and the steel slag production was estimated to be between 190 million and 280 million tons [9]. The most recent European statistics for 2020 indicate that 20.8 Mt of blast furnace slag (BFS) and 12.4 Mt of slag from steel production were produced in Europe [10]. The recycling of these slags provides a number of environmental benefits, including the preservation of natural resources, the recovery of valuable metals, and a reduction in the volume of solid waste. BFS is mainly used as a cement and concrete additive (approx. 85% of total BFS utilization); meanwhile, the utilization of steel slag is more diverse. It is estimated that 57% of the total steel slag is used in road construction, 25% for metallurgical use and internal storage, 5% as fertilizer, 4% for cement and concrete additives, and 2% for hydraulic engineering, while the rest is used for other purposes.

Red mud as a residue from the Bayer process (alumina production from bauxite ore) is estimated at 120–150 Mt of global annual production, and its accumulation is estimated at around 4 billion tones [11]. Red mud has been used for metal recovery, adsorption, soil amendment, in catalysis, and in oxidation reactions, as well as in the production of construction and building materials, but only in small portions [12].

According to the ECOBA (European Coal Combustion Product Association 2016), 15% and 9% of the total produced (40 million tons) coal combustion products (CCP) belong to fly ash and bottom ash, respectively. Fly ash has been widespread, used in cement and concrete production [13], bricks and blocks [14], glass–ceramics [15], etc. Bottom ash could be used as a supplementary material in cement or as aggregate in concrete [16]. Both fly ash and bottom ash can be used for innovative products such as aerogels, geopolymers, rare earth elements, zeolite, etc., supporting the green transition in construction [17].

Cement plants, as an energy-intensive industry, have invested serious efforts to achieve the goals defined in the EU GD and ECAP. Among many approaches towards decarbonization, utilizing secondary raw materials (SRMs) presents one of the ways to decrease the CO_2 footprint [18]. SRM can be incorporated in cement production as a raw meal of clinker (in the first stage of cement production) or as supplementary cementitious material (at the later stage of production, acting as a hydraulic or mineral additive) [19]. The incorporation of secondary raw materials in belite–sulfoaluminate cement (BCSA) was recently investigated. Namely, the recent investigations showed that bottom ash [20] and fly ash from thermal power plants [21], as well as steel slag [4], could be successfully embedded in BCSA. There is still potential to explore the use of characterized secondary raw materials for BCSA production and other applications. For example, studies have shown that alumina-rich wastes can be effectively utilized in geopolymers, with alkali-activated slag and fly ash geopolymers emerging as innovative, environmentally friendly alternatives to conventional OPC, especially as fire-resistant alkali-activated cementitious materials [22,23].

The potential utilization of SRM relies on its chemical and mineralogical composition, particularly the amount of amorphous phase contained within, its granulometry, and the content of natural radionuclides which defines these SRMs as NORMs (Natural Occurring Radioactive Materials). The Council Directive 2013/59/EUROATOM, 2013 [24], Article 75, defines the norms for SRM to be assessed from a radiological point of view as they will be used in standard building practice.

Most of the secondary raw materials presented in this study were collected and analyzed within the RIS-ALiCE project. Many of them are also available in the RIS-ALiCE registry [25]. This project aimed to collect data on Al-containing residues from the East South-East Europe (ESEE) region and to evaluate their potential use in the production of low-CO_2 mineral binders based on BCSA. This paper presents the results of the secondary raw materials characterized by their physical, chemical, mineralogical, and radiological aspects. The results present the basis for innovative solutions for utilizing secondary raw materials from the ESEE region, in line with the EU Green Deal and the circular economy model.

2. Materials and Methods

2.1. Materials

Approximately 50 kg of bulk composite sample was collected at each sampling site in the period between October 2019 and October 2020. Bulk samples were composed of a minimum of 10 subsamples of equal size sampled from randomly selected points of stockpiles or sedimentation ponds to achieve representatives of each sample. Distribution of the subsamples considered the shape and type of the residue deposit, possible gravitational segregation of the material, and size of the particles deposited.

Prior to further analyses, all collected samples (Figure 1) were air-dried at 20 °C to a constant mass, with humidity measurements taken both before and after drying. The only exception was the slag mineral residue (SL2) obtained from the processing of the mixture of EAF S slag and ladle slag. Due to its nature (wet sample), it was immediately dried after sampling at 105 °C to prevent phase transformation. For the reduction of the bulk sample (air dried), a coning and quartering protocol was used. The coning and quartering method was chosen to systematically reduce samples while preserving their representativeness. The process of mixing and quartering was repeated until the required size of the laboratory sample was obtained. Each homogenized sample was of such size that all the individual analyses could be carried out twice.

Figure 1. Samples of collected secondary raw materials.

Table 1 contains basic data on 19 samples, which represent red mud from alumina production (1 sample), different slags from the steel industry (5 samples), fly and bottom ashes from thermal power plants (8 samples), fly and bottom ash from paper mills (2 samples), and other industrial residues (3 samples). Additionally, in the frame of the project, two fly ashes and bottom ash from the thermal power plant REK Bitola, Republic of North Macedonia [26], five fly ash samples from the thermal power plant Nikola Tesla, Serbia [27], and red mud from Alumina Zvornik, Bosnia and Hercegovina [28] were characterized, as well as query and mine waste [29]. The collected samples are shown in Figure 1.

Table 1. Coding of the samples.

Sample ID	Date of Sampling	Origin	Type of Sample	Company/Location	Country
RM	October 2019	Alumina production	Red mud	Dobro Selo, Mostar	Bosnia and Hercegovina
SL1	October 2019	Steel industry	EAF C steel slag	SIJ Acroni, Jesenice	Slovenia
SL2	February 2020		Slag mineral residue (mixture of EAF S slag and ladle slag)	SIJ Acroni, Jesenice	Slovenia
SL3	October 2020		EAF C steel slag	SIJ Metal, Ravne	Slovenia
SL4	October 2020		Ladle slag	SIJ Metal, Ravne	Slovenia
SL5	October 2019		Blast furnace slag	Arcelor Mittal, Zenica	Bosnia and Hercegovina
FA1	November 2019	Thermal power plants	fly ash	Šoštanj power plant, Šoštanj	Slovenia
FA2	October 2019		Fly ash	Power plant Kakanj, Kakanj	Bosnia and Hercegovina
FA3	October 2019		Fly ash	Power plant Stanari, Stanari	Bosnia and Hercegovina
FA4	June 2020		Fly ash	TITAN Usje, Skopje	North Macedonia
FA5	September 2020		Fly ash	Anonymous	Hungary
FA6	September 2020		Fly ash		Hungary
BA1	November 2019		Bottom ash	Šoštanj power plant, Šoštanj	Slovenia
BA2	September 2020		Bottom ash	Anonymous	Hungary
PFA	September 2020	Paper mills	Fly ash	Vipap Videm, Krško	Slovenia
PBA	September 2020		Bottom ash	Vipap Videm, Krško	Slovenia
QS1	September 2020	Mining company for the production and processing of silica sands and the production of auxiliary casting material	By-product of quartz sand washing	Termit, Moravče	Slovenia
WJ1	September 2020	Steel industry	Waste water jet sand	SIJ Acroni, Jesenice	Slovenia
CW1	September 2020	Brick factory	Crushed brick rejects	Goriške opekarne, Renče	Slovenia

2.2. Methods

The characterization of collected samples includes determination of the physical characteristics (moisture content, granulometry, BET specific surface area, particle density, and bulk density), chemical composition (main oxides and LOI at 950 °C, trace/rare earth elements (REEs)/heavy elements), total organic content (TOC), mineralogical composition (including amorphous and crystalline non-quantifiable phases (ACns)), and radiological characteristics.

2.3. Physical Characterization

Moisture content was determined on the as-received samples. Samples (~50–100 g) were dried at 105 °C to a constant mass. The moisture content (mc) was determined according to the difference between the wet mass and mass after drying by Equation (1):

$$\% \text{ mcwb} = [(\text{mw} - \text{md})/\text{md}] \times 100 \quad (1)$$

where

mcwb is the moisture content of the sample;
mw is the mass of the wet sample;
md is the mass of the dry sample.

The particle density of samples (~20 g) was determined with the pycnometer method (helium pycnometer Quantachrome Ultrapyc 1200e, Anton Paar GmbH, Graz, AT) in accordance with the standard EN 1097-7, 2008 [30]. The bulk density of samples was measured according to standard JUSB.C8.023, 1982 [31]. The samples QS1 and CW1 were not analyzed due to the nature of the material (the samples were previously ground and therefore unsuitable for analysis). The specific surface area (SSA) was determined by the BET method according to the standard ISO 9277, 2010 [32] with Micromeritics ASAP-2020 (Norcross, GA, USA) using nitrogen adsorption at 77 K. The SSA was not determined for QS1 and CW1 due to the nature of the material (samples were previously ground and therefore unsuitable for analysis).

For all relevant samples, sieve analysis compliant with ISO 3310-2, 2013 [33] was performed using a HAVER EML digital plus device (test sieve shaker) and laboratory test sieves from ELE international (16 mm, 8 mm, 4 mm, 2 mm, 1 mm, 0.25 mm, 0.125 mm, 0.063 mm, and <0.063 mm). For the sieve analysis, we used 2.5 kg of individual representative samples. Some samples (FA1, FA2, FA3, QS1, and CW1) were not suitable for sieve analysis due to their nature (the samples in larger pieces were pre-ground, or all particles were smaller than 0.063 mm).

Particle size distribution (PSD), by laser granulometry, was determined for all samples. A Helos BR laser, by the manufacturer Sympatec (Clausthal-Zellerfeld, DE), was used for all samples except for SL3, SL4, PFA, PBA, QS1, WJ1, and CW1, for which a Microtrac SYNC Model 5001, Microtrac Retsch GmbH, Haan, DE was used. In total, ~100 g of dried samples was sieved through a 0.25 mm sieve prior to analysis. For the samples SL3, SL4, PFA, PBA, QS1, WJ1, and CW1, a 1 mm sieve was used. For PSD analysis, ~1 g of sample was loaded into the test cell and analyzed using a wet configuration in isopropanol. The sample QS1 had very fine particles, and therefore its PSD was determined using a wet configuration in demineralized water. SL5 was not suitable for laser granulometry due to its nature (the sample was in larger pieces that we had to pre-grind). Measurements were performed in parallel.

2.4. Chemical and Mineralogical Characterization

The main chemical oxides (SiO_2, Al_2O_3, Fe_2O_3, CaO, SO_3, MgO, Na_2O, and K_2O), Cl^-, and loss of ignition at 950 °C (LOI) of samples were determined by wet chemistry according to EN 196-2 [34], while P_2O_5 and TiO_2 were determined by X-ray fluorescence spectroscopy (XRF). For the XRF of most samples, WDXRF, Thermo Scientific ARL PERFORM'X, Thermo Fisher Scientific, Waltham, MA, USA; fused beads; and the UniQuant program were used, while for the samples SL3, SL4, PFA, PBA, QS1, WJ1, and CW1 S8, Tiger by BRUKER (Billerica, MA, USA), fused beads, and the "clinker" program were used. To prepare the fused beads, we used ~1 g of the previously ignited (950 °C) test sample. The samples were sieved and ground through a 90 μm sieve. The obtained sample was mixed with lithium tetraborate, which served as the flux, in a 1:10 ratio and afterwards, fused at 1100 °C to create beads.

Trace elements, rare earth elements (REEs), and heavy metals were determined by the ICP-ES/MS method (multi acid digestion, where the residue has been dissolved in

HCl). Based on the Hungarian standard MSZ 525-17, 2013 [35], 0.5 g sample, 1.16 g lithium metaborate, and 0.05 g ammonium nitrate were fused in a platinum crucible at 1000 °C for 60 min. A total of 44 ml of 5% hydrochloric acid and 30 ml of distilled water were used to dissolve the glass bed, and the solution was filled up to 100.0 mL with distilled water and analyzed with ICP OES (Perkin Elmer Avio200, Waltham, MA, USA). The plasma conditions were as follows: 1500 W; 12 L/min plasma gas; 0.2 L/min aux gas; and 0.65 L/min nebulizer gas. The sample flow was 1 ml/min. A MiraMist nebulizer and baffled cyclonic spray chamber were used. Lutetium was used as an internal standard at a 1 mg/L level. For samples where precipitation occurred at the dissolution step, more hydrochloric acid was added to the solution. In some cases, the fusion was repeated with a 0.3 g sample to avoid the formation of the precipitate.

For determination of the total mercury content (in solids and liquids), the solid sample of 100 mg without pre-treatment or pre-concentration was used. The system used was a LECO 254 Advanced Mercury Analyser, AMA—Atomic Absorption Spectrometer, available from Leco, UK.

The presence of organic matter, expressed by the total organic content (TOC), was analyzed in all samples (~200 mg) by an Analizator CW-800M "Multiphase" (Lahr, DE), ELTRA using the dry incineration method, and the detection of products with an IR detector.

The mineralogical composition of the samples was determined using an X-ray diffractometer (PANalytical Empyrean, Malvern Panalytical, Malvern, UK) equipped with CuKα radiation and a PIXcel 1D detector (Malvern panalytical, Malvern, UK,). Samples (~10 g) were ground in an agate mortar to a particle size below 0.063 mm. The ground powders were manually back-loaded into circular sample holders (27 mm in diameter), and data for each sample were collected from 4° to 70° (2θ) using a step size of 0.026° (2θ) and a scan time of 197s. Samples were measured at 45 kV and 40 mA and rotated during data collection with a revolution time of 2 s. The amount of crystalline phase and ACn were estimated by Rietveld refinement using the external standard method (alumina powder, Al_2O_3; NIST SRM 676a) and the PANalyticalX'Pert High Score Plus diffraction software, version 4.9 (Malvern Panalytical, Malvern, UK), using the structures for the phases from the ICDD PDF 4+2016 RDB powder diffraction files.

2.5. Radiological Characterization

The activity of radionuclides in the samples was determined by the gamma spectrometry method. Measurements were performed on HPGe detectors (Canberra, Sturbridge, MA, USA), with relative efficiencies of 20%, 18%, and 50% according to the IAEA TRS 295 method [36]. Efficiency calibration was performed using a certified radioactive standard (1035-SE-40845-17 [37]) and secondary reference materials produced from a radioactive solution (1035-SE-40844-17 [38]) from the Czech Metrology Institute, which contained ^{210}Pb, ^{241}Am, ^{57}Co, ^{60}Co, ^{137}Cs, ^{139}Ce, ^{85}Sr, ^{109}Cd, ^{88}I, and ^{51}Cr, and can be traced back to the International Bureau of Weights and Measures—BIPM (Bureau International des Poids et Mesures). The radioactive materials used were of a similar density and packing geometry as the samples.

After preparation, the samples (125 mL and 250 mL) were sealed with beeswax in the measurement geometry and left for at least one month before measurement, in order to establish a radioactive equilibrium between radon and its progeny. The measurements lasted 60,000 s, and the spectra were analyzed with the GENIE 2000 software package.

It is recommended (EC, Radiation protection 112) that the radiological hazard controls associated with exposure to ^{226}Ra, ^{232}Th, and ^{40}K be based on an effective dose of 1 mSv/year.

The annual effective dose rate (E) was calculated by Equation (2) [39], using a conversion coefficient of 0.7 Sv/Gy to convert the absorbed dose in the air into the effective dose in the human body. D is the absorbed dose in the air and $(p{\cdot}t)$ is the annual exposure time, where p is the percentage of years during which humans are exposed to radiation (occupancy factor) and t is 8.760 h (number of hours in the year).

$$E\ (mSv/y) = D\left(\frac{nGy}{h}\right)\cdot p\cdot t(h/y)\cdot 0.7\left(\frac{Sv}{Gy}\right)\cdot 10^{-6} \tag{2}$$

For estimating the outdoor effective dose rate (E_{out}, $E_{20\%}$), the outdoor occupancy factor pout is 0.2, while for estimating the indoor effective dose rate (E_{ind}, $E_{80\%}$), the calculation takes into account that people spend about 80% of time indoors (indoor occupancy factor p_{ind} is 0.8) [39].

The absorbed dose in the air, D (nGy/h), is estimated based on Equation (3) [39] where q_i is the specific dose rate for isotope "i" in (nGy/h)/(Bq/kg), and A_i is the activity concentration of isotope "i" in Bq/kg:

$$D = q_{226_{Ra}}\cdot A_{226_{Ra}} + q_{232_{Th}}\cdot A_{232_{Th}} + q_{40_K}\cdot A_{40_K} \tag{3}$$

The value of qi is usually calculated by simulating different cases. For the external terrestrial gamma radiation absorbed dose rate (D_{terr}) in the air at a height of 1 m above ground level, the q_i for ^{226}Ra, ^{232}Th, and ^{40}K are 0.462, 0.604, and 0.0417, respectively [39].

An Activity Concentration Index (ACI) is the most common screening method for assessing the dose caused by building materials. It is associated with gamma radiation exposure inside buildings that exceeds typical outdoor exposure, and is calculated using Equation (4) [40]. For superficial and other materials with restricted use, the dose criterion of 1 mSv is already satisfied at $ACI < 6$, but for materials used in bulk amounts, the value of ACI should be less than 1 [40].

$$ACI = \frac{A_{226_{Ra}}}{300} + \frac{A_{232_{Th}}}{200} + \frac{q_{40_K}}{3000} \tag{4}$$

The hazard indices are a screening method for the dose caused by the use of certain materials. The external hazard index (H_{ext}) reflects the external radiation hazard due to the emitted gamma radiation and it is calculated according to Equation (5) [41].

$$H_{ext} = \frac{A_{226_{Ra}}}{370} + \frac{A_{232_{Th}}}{259} + \frac{q_{40_K}}{4180} \tag{5}$$

The value of this index should be less than 1 in order to keep the radiation hazard insignificant [41], and the value of H_{ext} equal to 1 ensures $E_{terr,80\%}$ is less than 1 mSv.

In addition to external radiation, radon and its short-lived products are also hazardous to respiratory organs, and that is quantified by the internal hazard index, H_{int}, as in Equation (6) [41]. The value of H_{int} should be less than 1 for material used indoors. Due to H_{int} being stricter than H_{ext}, an H_{int} equal to 1 ensures an $E_{terr,80\%}$ less than 1 mSv.

$$H_{int} = \frac{A_{226_{Ra}}}{185} + \frac{A_{232_{Th}}}{259} + \frac{q_{40_K}}{4180} \tag{6}$$

The equivalent activity of radium (Ra_{eq}) [41] is equal to 370·Hext, so the criterion $Ra_{eq} < 370$ Bq/kg is equivalent to the criterion $H_{ext} < 1$. Also, the alpha index (I_α) < 1 requires that $A(^{226}\text{Ra}) < 200$ Bq/kg ($I_\alpha = A(^{226}\text{Ra})/200$) (Nordic 2000), which is a weaker criterion than $A(^{226}\text{Ra}) < 185$ Bq/kg, which requires $H_{int} < 1$. Due to the above, these parameters were not considered separately.

Although all the abovementioned parameters ensure the condition $E_{terr,80\%}$, for some cases they are too strictly defined (for example, if no one spends 80% of the hours in a year with the observed material). Therefore, these are only screening parameters, and for cases where their values are >1, it is necessary to estimate the dose for the situation in which the observed material is used.

To calculate the qi for estimating the absorbed dose of a building material, a standard model room [42] (a room of 20 m^2 and 3 m in height, with concrete of 20 cm thickness as the construction material) is the most frequently used, in a case where all structures including the floor, walls, and ceiling (D_{all}); the floor and walls (D_{fw}); and only the floor (D_f) were built with it, as well as a case where the material is used as the superficial material for all walls (D_{sup}) [42]. To estimate the annual effective dose rate for a standard room, p = 0.8. The value of q_i for approximated cases is used from Markkanen [42].

It should be noted that the dose reference value of 1 mSv/y refers to the excess gamma dose received outdoors, but the estimated absorbed dose in air based on Equation (3) (absorbed dose in air) is not an excess exposure to building materials, because concrete structures protect against gamma radiation from the undisturbed Earth's crust. Using the average value of the absorbed dose in the air for the background, the excess dose rate in the room is therefore (D D_{back}) nGy/h. Therefore, the annual excess effective dose to the occupant is as follows: E_{exc} = (D D_{back}) nGy/h 7000 h 0.7 Sv/Gy. An average outdoor background in Europe is 50 nGy/h [40]. To assess the potential health impact on the public due to exposure to the tested samples, the annual effective dose (E) of the total external absorbed gamma dose in air at a height of 1 m above ground level for outdoor and indoor cases and for four different standard rooms was calculated. Based on the results obtained for the observed materials, the activity concentration index (ACI), the external hazard index (H_{ext}), and the internal hazard index (H_{int}) were calculated.

3. Results and Discussion

3.1. The Results of Physical Characterization

The physical characteristics of the secondary raw materials are crucial parameters that determine their preparation processes before industrial use, including the need for drying, grinding, etc., which are heavily connected to energy consumption. The results of the measurements of the physical characteristics (moisture content, particle density, bulk density, and specific surface area) of the collected secondary raw materials are presented in Table 2.

The moisture content varies between 0 and 20.8 wt.%, and it is heavily influenced by the type of material. Generally, the red mud, slags, and fly ashes and bottom ashes from paper industry and thermal power plants were found to contain relatively small amounts of moisture (between 0 and 2.83 wt.%, except SL2). The samples WJ1 and CW1 have slightly higher moisture content, up to 8 wt.%. In the case of sample SL2, the increased moisture content was mainly because a wet process was used to extract metal prior to disposal. The by-product of quartz sand production (sand washing residues, QS1), contained the highest measured moisture (20.8 wt.%) in this study, which is probably heavily influenced by the treatment and storage processes at the producer of this secondary raw material.

The particle density (solid phase only, pores are excluded) of the sampled materials varies between 1.91 and 4.14 g/cm^3, while the bulk density (including pores) varies between 478 and 2206 kg/m^3. The highest particle density was determined for waste water jet sand (WJ1), while the lowest was for fly ash from the thermal power plant (FA3). The density of the supplementary cementitious materials (SCMs) is important information as it allows a more accurate calculation of the proportions of concrete mixtures by volume instead of by mass [43]. Comparing the average particle density of the collected secondary raw materials with those of soils, which is 2.65 g/cm^3 [44], it can be concluded that the density of the materials from this study is in a similar size range to that of soils. Particle density comparison between the sampled materials and rocks revealed that the particle densities of slags and fly and bottom ashes from the paper industry are comparable to those of igneous rocks, the particle density of bottom ash from the thermal power plant is comparable to that of limestone, and the particle densities of red mud and fly ash from the thermal power plant are comparable to those of shales or sandstones [45]. Particle density plays a crucial role in optimizing concrete mixtures, particularly in terms of strength development and hydration, as it influences the packing density and performance of supplementary cementitious materials (SCMs) like fly ash, slag, and silica fume. It helps determine the volume and distribution of particles within the mix, enabling the fine-tuning of the particle size distribution (PSD) to improve its workability and hydration efficiency. High-density particles can effectively fill voids, enhancing concrete strength by minimizing excess pore spaces, while denser materials like slag and certain fly ashes reduce water demand, contributing to long-term durability [46–48].

Table 2. Moisture content, particle density, bulk density, and BET specific surface area (SSA) of investigated samples. n.a.—not analyzed.

Type of Sample	Sample ID	Moisture Content [%]	Particle Density [g/cm^3]	Bulk Density [kg/m^3]	BET SSA [m^2/g]
Red mud (RM)	RM	0.21	3.58	1100	13.05
Slag (S)	SL1	0.3	2.66	1810	1.56
	SL2	17.5	3.25	1281	3.51
	SL3	2.83	3.39	1835	2.98
	SL4	0.00	3.04	1138	0.61
	SL5	0.08	2.62	1032	0.58
Fly ash—paper industry (FAPI)	PFA	0.17	3.05	728	51.98
Fly ash—thermal power plant (FATP)	FA1	0.01	2.34	1035	2.41
	FA2	0.05	2.47	1100	0.97
	FA3	1.89	2.40	478	6.11
	FA4	0.03	1.91	612	7.26
	FA5	0.00	2.62	1278	1.80
	FA6	0.00	2.58	620	14.33
Bottom ash—paper industry (BAPI)	PBA	0.15	2.76	529	5.23
Bottom ash—thermal power plant (BATP)	BA1	0.12	2.52	698	10.37
	BA2	0.003	2.64	1261	0.28
Other (O)	QS1	20.8	2.66	n.a.	n.a.
	WJ1	5.65	4.14	2206	0.76
	CW1	8.14	2.76	n.a.	n.a.

The bulk density of soils varies between 1500 and 1700 kg/m^3 [44], so it can be concluded that sampled materials (except WJI) are less dense than average soils for most of the materials. Only some slags and water jet sand (WJ1) have a higher bulk density than soils. The higher density of water jet sand (WJ1) in comparison to soil is due to the presence of garnets. The EAF slag, formed largely by a content of oxides of iron, has high density and low porosity [49]. The bulk density of supplementary cementitious materials (SCMs) is critical for optimizing concrete mix design as it influences packing density, reduces voids, and improves both the strength and durability of the material. High-density SCMs enhance workability and lower water demand, creating a cohesive mix while also supporting sustainability by enabling greater clinker replacement and reducing CO_2 emissions during production. Proper understanding and control of SCM bulk density ensure precise mix proportioning, minimizing unreacted materials and maximizing performance [50–52].

The specific surface areas (SSAs) of the collected samples vary much more than the other measured physical properties. The largest SSA was measured for fly ash from the paper industry (PFA) and exceeds the average SSA values of red mud 4-fold, the average SSA of fly ash from the thermal power plant 7-fold, the average SSA of bottom ash 10-fold, the average SSA of slag 28-fold, and the average waste water jet sand (WJ1) SSA 68-fold. Significant variations of SSA were also observed within individual types of secondary raw material samples. It looks like specific conditions during the process of generation of secondary raw materials play a crucial role in the value of the SSA parameter, which cannot be generalized but shall be measured individually. The specific surface area (SSA) of cement plays a critical role in determining the water demand during cement hydration. As the SSA increases (such as with finer cement particles), the surface area available for reaction with water also increases, which, in turn, raises the water demand for a given mass of cement. This is because more water is required to wet the finer particles and facilitate the hydration process. Studies show that the water demand for cement rises significantly with finer grinding, which increases SSA. This effect is closely linked to the particle size distribution (PSD), as finer particles tend to compact more densely, enhancing hydration but requiring more water [53]. Additionally, the inclusion of supplementary cementitious materials (SCMs), such as fly ash or slag, often changes the particle size and SSA of the binder system, affecting both the packing density and water demand. For example, increased use of SCMs typically leads to a higher SSA, which requires more water for the mixture to achieve the desired fluidity [54]. In practical terms, controlling the SSA through particle size optimization can help manage water demand while balancing the fluidity and strength of cement-based mixtures [53]. Finer particles, which have a larger surface area, generally exhibit higher reactivity. This is due to the increased surface contact between the cement particles and water, enhancing the hydration process. Studies on various types of cement have shown that the SSA correlates with compressive strength, with finer cement (higher SSA) often leading to stronger concrete due to better hydration and chemical activity [55].

The granulometric composition of samples determined by sieve analysis is presented in Figure 2.

RM contained particles in size between 1 and <0.063 mm, while most of them were less than 0.25 and 0.125 mm. In the slag, the smallest particles (<0.125 mm) were in SL2 (mineral product) and SL4 (ladle slag), where 70.2 wt.% were less than 0.063 mm. The largest particles were in SL3 (EAF C slag), where most of them passed the 8, 4, 2, and 1 mm sieves. SL1 (EAF C slag) and SL5 (blast furnace slag) have the most particles between 1 and 0.5 mm. In PFA (fly ash from the paper industry), 79.0 wt.% of the particles were less than 0.063 mm. In FA6 and FA5 (fly ashes from thermal power plant), the particles were a bit larger—most of them were less than 0.25 and 0.063 mm. In PBA (bottom ash from the paper industry), the most particles passed sieves between 1 and 0.063 mm. In BA1 (bottom ash from the thermal power plant), the most particles passed sieves between 2 and 0.25 mm, while in BA2, 68.1 wt.% were less than 0.5 mm. In sample WJ1 (waste water jet sand), most of the particles passed sieves of 0.125 and 0.063 mm.

The results of the PSD (Figure 3) showed that most of the samples of secondary raw materials have a unimodal distribution, while the samples of RM, FA1, FA2, and FA6 have a bimodal distribution. According to the characteristic values D_{10}, D_{50}, and D_{90} (Figure 4), the largest mean diameter, D_{50}, occurs in the sample of WJ1 (138.3 μm), while the smallest is for the red mud. In general, the samples contain a higher number of larger particles (30–200 μm) as the curves are moved to the right side. However, the red mud, RM, contains a high number of smaller particles (0.7–10 μm). The reactivity of cement is often enhanced by optimizing its PSD. Finer particles with higher surface areas react more readily with water, increasing the rate of hydration. This is particularly evident when using finely ground supplementary cementitious materials (SCMs) such as slag, silica fume, or fly ash, which can improve reactivity by filling voids between larger particles and increasing the contact area for hydration reactions [54,56]. Optimizing the packing density through careful control of the PSD helps reduce voids within the cement matrix, which not only improves the material's strength but also minimizes its porosity. Studies show that well-distributed particle sizes contribute to a denser, more uniform packing of the cement particles. This reduces the amount of unreacted water and enhances the chemical reactivity of the cement, leading to stronger and more durable concrete [56].

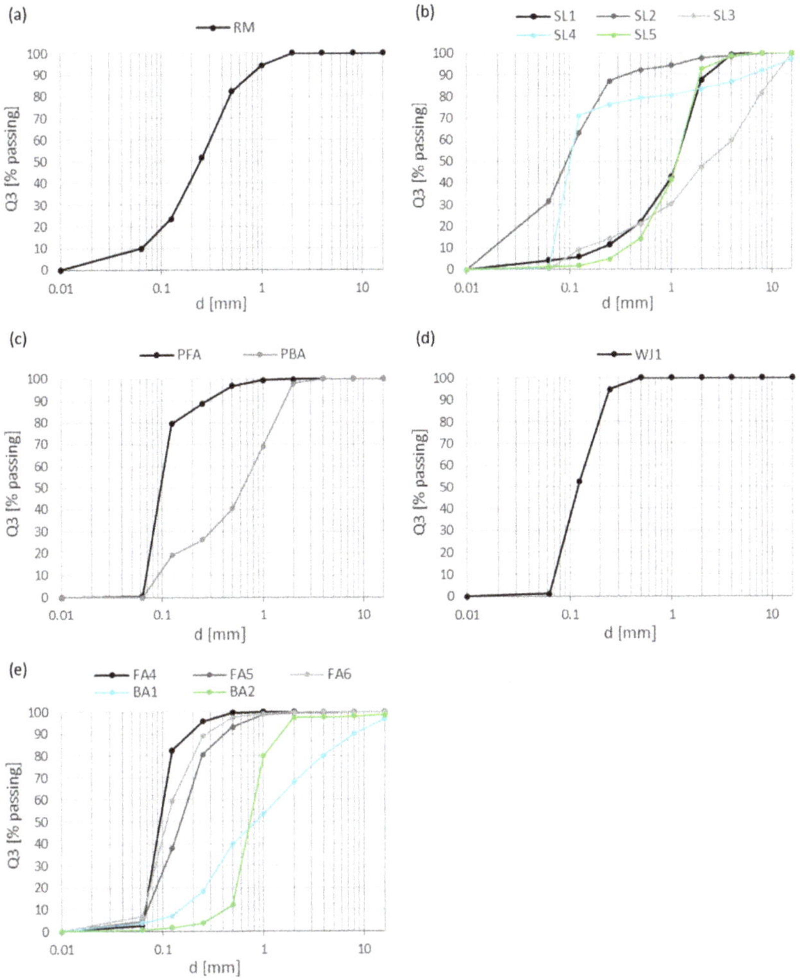

Figure 2. Sieve analysis for Al-containing industrial residues: (**a**) red mud, (**b**) steel slags, (**c**) ashes from the paper industry, (**d**) waste water jet sand, and (**e**) ashes from thermal power plants.

3.2. Chemical Characterization

Table 3 presents the results of the chemical composition of the collected secondary raw materials. Loss of ignition is considered to be low for the majority of samples, except for the red mud (RM), fly ash, and bottom ash (PFA and PBA) from the paper industry and fly ash (FA6) from the thermal power plant. The most important parameter for the valorization of these materials for the production of Al-rich types of cement is the content of Al_2O_3. The most abundant materials with Al_2O_3 are the fly ashes (FA1, FA2, FA3, and FA4) from thermal power plants, with Al_2O_3 contents around 23 wt.%, with only two exceptions where Al_2O_3 levels were low, around 5 wt.% (FA5 and FA6). A similar situation also occurs with the bottom ash (BA1) sample from the thermal power plant, which has a relatively high Al_2O_3 content, while the BA2 sample contains it in much lower levels. It can be speculated that the increased Al_2O_3 levels in the fly ashes are due to the clay admixtures in coal or waste paper. Red mud has a 16.8 wt.% of Al_2O_3 in this study. Waste water jet sand (WJI) and waterjet abrasive from the steel industry, and crushed brick rejects (CW1) from the brick factory have between 16 and 21 wt.% of Al_2O_3. Other materials have lower

Al_2O_3 levels, around 10 wt.%, which means they were depleted even compared with the average Al_2O_3 content in the Earth's upper crust.

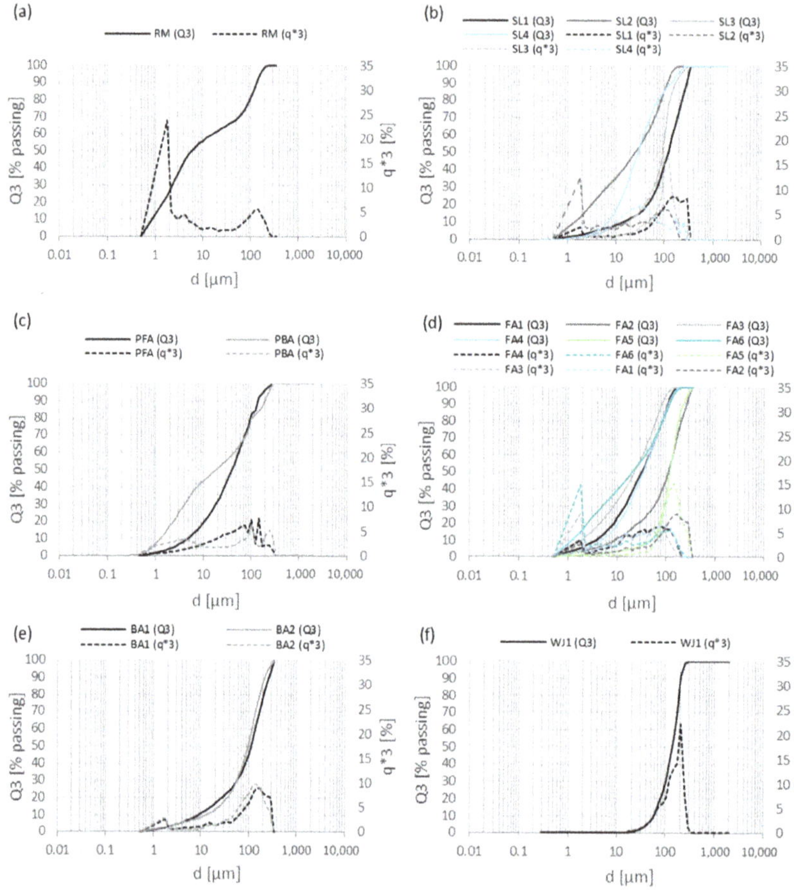

Figure 3. PSDs of (**a**) red mud from alumina production, (**b**) slags from still production, (**c**) fly and bottom ashes from the paper industry, (**d**) fly ashes from thermal power plants, (**e**) bottom ashes from thermal power plants, and (**f**) waste water jet sand.

In comparison to Ordinary Portland cement (OPC), the most used binder in the world, for the production of the low-CO_2 non-Portland binders (belite–CSA (BCSA) binders), an approx. three times higher Al_2O_3 content is needed. In particular, the use of alumina-rich wastes can lower the manufacturing costs of BCSA cement mainly depending on the use of the "expensive" bauxite [57].

Considering the content of SiO_2 in the samples, data show that all samples are depleted in this component compared to the average Earth's upper crust. However, all kinds of ashes collected in this study have higher SiO_2 levels compared to the red mud and slag samples (except SL5). The Fe_2O_3 levels vary a lot in samples from this study—from 46.68 wt.% to less than percent. The highest Fe_2O_3 content is in the red mud sample RM, followed by WJ1 and SL1, while the other samples contain less than a 10 wt.% of Fe_2O_3. However, a significant portion of ash samples were still enriched with Fe_2O_3 compared to the Earth's upper crust.

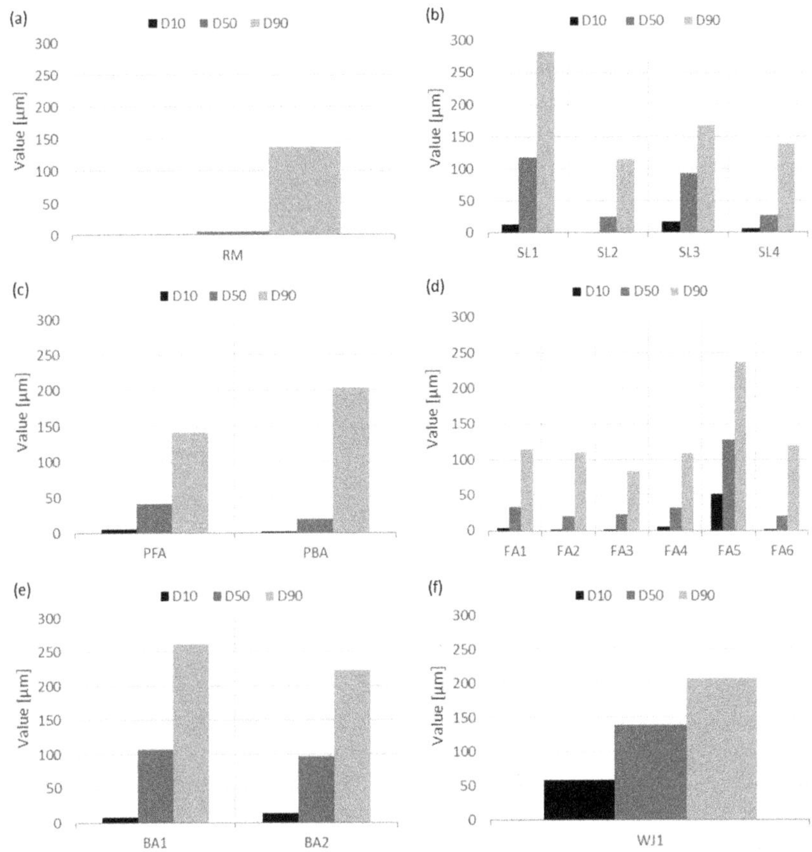

Figure 4. PSDs (D10, D50, and D90) of (**a**) red mud from alumina production, (**b**) slags from still production, (**c**) fly and bottom ashes from the paper industry, (**d**) fly ashes from thermal power plants, (**e**) bottom ashes from thermal power plants, and (**f**) waste water jet sand.

The majority of samples have elevated CaO levels compared to the Earth's crust. Considering the CaO levels in samples from this study, all slags, ashes from the paper industry, and fly ashes from thermal power plants (FA6 and FA2) can be regarded as CaO-rich materials, with a CaO content between 20 and 47 wt.%, while red mud and the rest of the ashes from the coal power plant can be regarded as materials with a lower CaO content. The studied ashes FA4, FA5, and BA2 are classified as siliceous fly ash, with less than 10% calcium oxide (CaO), whereas the calcareous ash samples (PFA, FA1, FA2, FA3, FA6, PBA, and BA1) contain more than 10% CaO [58]. According to EN 450-1 [59], the sum of SiO_2, Al_2O_3, and Fe_2O_3 must not be less than 70%, a requirement met by all the siliceous fly ashes (FA1, FA3, FA4, and FA5) and the calcareous fly ash FA2. The sulfate contents of all the studied ashes were below the specified limit (<3%), as were the total alkali contents (Na_2Oeq < 5%), phosphate contents (<5%), and MgO contents (<4%). The category C LOI (<5%) criterion was met by all the fly ashes except for the calcareous fly ash FA6, which also exceeded the broader limit of category C LOI < 9%. Similar conclusions can also be made for the MgO content in the waste materials from this study, with the exception that the MgO levels in many samples are lower than those from the Earth's upper crust. According to the EN 197-1 standard [58], the content of MgO in cement shall not exceed more than 5 wt.%. All the investigated slags are characterized by a higher MgO content (in the range from 5.29 wt.% for SL5 to 11.32 wt.% for SL2), as well as WJ1. Magnesium oxide (MgO) in cementitious materials serves as an expansive agent, helping to control shrinkage, reduce cracking, and enhance durability by improving the microstructure and reducing porosity [60,61]. However, excessive MgO can cause overexpansion, cracking, and long-term reductions in compressive strength, with these effects influenced by the curing conditions and dosage levels [61,62]. The P_2O_5, K_2O, SO_3, and Na_2O levels are generally very low, but this is not so significant for the cement industry. Alkalis (Na_2O and K_2O) play a significant role in the alkali–silica reaction (ASR), with the maximum alkali content in concrete often limited to $\leq$3.0 kg/m^3 of Na_2Oeq (calculated as $Na_2Oeq = Na_2O + 0.658\ K_2O$) in cement and the aggregates for exposure conditions susceptible to ASR. Some national annexes or complementary standards suggest limiting cement alkalis to $\leq$0.60% Na_2Oeq by mass when reactive aggregates are present [63]. Also, EN 450-1 [59] limits the alkali content of Class F fly ash used in concrete to 1.5% Na_2O_3eq to control the risk of ASR, although its effectiveness also depends on factors such as ash reactivity and the type of aggregates used. In this respect, only the RM shows a higher content in comparison to the other SRMs. The highest TiO_2 levels were measured in the red mud, while the other samples have TiO_2 levels within the range of those in the Earth's upper crust. Titanium dioxide (TiO_2) enhances cementitious materials by offering photocatalytic properties, which enable self-cleaning, air-purifying, and antimicrobial features, making it valuable in eco-friendly construction. It also influences hydration by accelerating gel formation at early stages and improves microstructure, though excessive amounts may slow down the hydration process by reducing the availability of water to cement particles, potentially affecting the final setting time and overall durability. Additionally, a high TiO_2 content may increase the risk of material degradation in environments with high UV exposure [64].

Table 3. Chemical composition of samples.

Type of Sample	Sample ID	LOI 950 °C	Al_2O_3	SiO_2	Fe_2O_3	CaO	MgO	P_2O_5	K_2O	SO_3	Na_2O	TiO_2
RM	RM	14.05	16.21	3.28	46.68	12.91	0.45	0.10	0.32	0.06	2.56	3.46
S	SL1	0.48	8.88	9.82	30.96	32.68	10.04	0.34	0.01	0.24	0.06	0.36
	SL2	5.12	9.70	16.84	10.33	39.12	11.32	0.12	0.03	0.35	0.13	0.68
	SL3	5.77	11.38	15.91	20.82	21.61	10.15	0.13	0.15	0.23	0.26	0.21
	SL4	0.00	19.51	17.16	1.03	52.82	7.89	0.01	0.01	0.16	0.09	0.11
	SL5	0.00	8.68	40.75	1.12	39.75	5.29	0.03	0.74	0.22	0.26	0.26
FAPI	PFA	13.86	9.55	22.29	8.86	35.86	5.3	0.29	0.84	1.99	0.46	0.96
FATP	FA1	0.57	23.23	44.60	9.86	13.38	2.68	0.45	1.8	1.46	0.92	0.83
	FA2	0.26	21.34	43.76	7.49	20.01	2.3	0.32	1.42	1.69	0.36	0.65
	FA3	3.33	23.74	48.49	7.42	11.52	3.12	0.06	3.2	1.39	0.11	1.21
	FA4	1.48	23.07	53.30	8.23	4.89	2.14	0.20	2.69	0.39	1.11	0.89
	FA5	1.89	5.78	72.30	2.20	8.91	1.56	0.61	3.76	0.38	0.82	0.56
	FA6	12.32	5.69	38.34	1.64	26.37	3.65	1.95	7.93	2.11	0.64	0.35
BAPI	PBA	15.41	8.42	13.89	0.49	58.89	1.94	0.28	0.31	0.33	0.30	0.29
BATP	BA1	2.93	22.14	41.44	10.38	16.01	2.86	0.49	1.40	0.63	0.70	0.86
	BA2	0.59	1.84	79.74	0.55	9.61	1.39	0.74	3.44	0.14	0.14	0.09
O	QS1	2.56	6.30	86.61	1.12	1.51	0.36	0.02	0.88	0.09	0.09	0.24
	WJ1	0.00	21.03	35.12	35.09	2.58	5.84	0.04	0.01	0.05	0.04	1.99
	CW1	1.49	15.96	63.29	7.52	5.17	1.79	0.11	2.09	0.23	0.87	0.84

3.3. *Trace Elements, REEs/Heavy Elements of Samples*

The results of minor, trace, and rare earth elements (REEs) for the investigated samples are presented in Table 4.

Table 4. Results of minor, trace, and rare earth elements in the samples of red mud (RM), slag (S), fly ash from the paper industry (FAPI), fly ash from thermal power plants (FATP), bottom ash from the paper industry (BAPI), bottom ash from thermal power plants (BATP), and other samples (O). The last column B (background) indicates the averages in Earth's upper crust [65].

Sample ID		RM	S					FAPI	FATP						BAPI	BATP		O			B
		RM	SL1	SL2	SL3	SL4	SL5	PFA	FA1	FA2	FA3	FA4	FA5	FA6	PBA	BA1	BA2	QS1	WJ1	CW1	
Minor and trace elements (mg/kg) *	Pb	184	<50	82	92.7	11.2	<20	196.5	55	79	<20	30.9	<30	22.7	109.1	<20	<30	13.8	9.2	33.8	17
	As	157	<25	<20 [b]	3.3	<0.2	<10	17	54	108	<10	67.5	<30	<20	2.4	<10	<30	4.1	2.3	13.4	4.8
	Zn	258	88	460	686.4	52.9	33	1733.6	157	215	50	103	112	338	174.6	83	130	37.1	241.8	109.5	67
	Co	60.3	4	22	16	9.3	<1	48.5	16	25	48	22.1	<10 [e]	4	4.1	16	<10	3.8	42.3	27.7	17.3
	Cd	14.7	<5	<2	1.5	<0.02	<2	1.8	<2	<2	<2	<2	<3	3.3	0.5	<2	<3	0.1	0.5	0.2	0.09
	Cr	1360	16,860	20,755	>10,000	914	83	163	89	311	531	97.2	53	49	33	325	215	51	1082	147	92
	Ni	435	324	1050	259.5	42.1	19	97.7	40	303	534	60.3	19.3	31	13	93	76.5	15.3	253.5	113.4	47
	Sb	<50	<50	<40 [b]	3.1	2.7	<10	6.2	<10	<10	<10	<10	<15	<10	2.7	<10	<15	0.6	0.4	0.8	0.4
	Mn	3020	31,053	15,206	>10,000	802	24,049	1155	1717	416	990	721	805	1632	195	2634	500	81	7005	1683	774.5
	Cu	101	307	253	232.5	95.9	54	335.9	71	123	161	81.6	64	112	482	77	54.6	9.1	80.6	61	28
	Ag #	<25	<25	<10	- [c]	96	<10	680	<10	<10	<10	<10	<15	<10	820	<10	<15	46	<20	81	53
	Ba	104	1001	351	188	291	1411	161	513	469	420	674	267	417	360	510	223	195	11	312	628
	Sr	180	316	253	292	460	429	624	572	350	256	285	230	596	820	650	247	23	10	138	320
	Ga	60.9	<50 [a]	<15	6.8	0.5	<15	11	45	53	54	34.7	<15	<10	10.5	39	<15	5.5	10.7	19.5	17.5
	Nb	77.9	84	346	204.6	5	<20	7.5	<20	<20	<20	18.8	<15	<10	5.6	<20	<15	3.4	38	11.3	12
	Ta	<50	<50	<20	2.6	0.2	<20	0.6	<40	<20	<20	<20	<30	<20	0.5	<20	<30	0.2	2.6	0.7	0.9
	U	<75	<125	<20	3.3	4.9	<20 [d]	2.3	<50	<50	<50	<30	<45	<30	2.9	<50	<45	2	1.5	1.1	2.7
	V	812	525	404	3265	312	11	65	186	197	254	132	30.8	30.9	15	201	12.6	26	227	148	97
	Zr	779	249	536	618.6	412.7	100	97	152	158	310	159	248	120	109.9	147	43.1	16.1	39.2	34.9	193
	Hg	0.098	0.003	0.024	0.002	0.002	0.005	0.258	0.02	0.116	0.824	0.047	0.000	0.000	0.000	0,002	0.015	0.014	0.000	0.001	0.05
REEs ** (mg/kg)	Ce	499	<30 [a]	84	20.7	18.7	61	47	94	60	91	112	31	29	59.8	70	7.7	27.1	36.9	56.3	63
	Dy	27.1	<10	<5	1.1	1.2	15	3.1	<5	<5	6.5	6.44	<7.5	<5	1.7	6.2	<7.5	1.5	32.3	2.8	3.9
	Er	<25	<20	<5	0.8	0.8	<10	1.4	<10	<10	<10	<10	<7.5	<5	0.8	<10	<7.5	0.6	27.3	1.2	2.3
	Eu	8.31	<5	<2	0.2	0.2	<3	0.8	<3	<3	3.3	<3	<5	<3	0.7	<3	<5	0.5	0.2	0.8	1
	Gd	41.5	<25	<10	1.2	1.5	17	3.1	<10	<10	<10	<10	<15	<10	2.5	<10	<15	2	13.2	4.1	4
	Ho	<12.5	<12.5	<5	0.2	0.3	<5	0.5	<5	<5	<5	<5	<7.5	<5	0.3	<5	<7.5	0.2	7.8	0.5	0.83
	La	179	15	143	13.5	14.3	28	23.2	37	33	46	55.3	14	13.9	28.8	35	<5	14.3	16.1	23.5	31
	Nd	166	16	<10	8	7.6	32	19	33	34	42	44.5	12.6	12.8	22	36	<10	12.4	17.2	24.3	27
	Pr	<25	<25	<10	2.1	2.3	<10	5.2	<10	<10	<10	<10	<15	<10	5.9	<10	<15	3.1	4.3	5.9	7.1
	Sm	42.6	<25	<10	1.6	1.3	12	4.1	10	11	13	10.4	<7.5	<5	4.4	12	<7.5	2.9	5.2	4.6	4.7
	Sc	108	<2.5	1.9	2.9	2.6	16	6.7	21	20	34	17.1	5.6	4.1	2.5	21	<2	2.9	66.3	13.8	14
	Tb	<12.55	<25	<10	0.1	<0.1	<10	0.4	<10	<10	<10	<5	<7.5	<5	0.3	<10	<7.5	0.1	3.6	0.4	0.7
	Tm	<12.5	<12.5	<5	0.1	0.1	<5	0.2	<5	<5	<5	<5	<7.5	<5	0.1	<5	<7.5	<0.1	4.1	0.2	0.3
	Yb	16.7	<5	<0.5	0.9	0.8	6.2	1.4	<5	<5	<5	4.16	<2	1.1	0.8	<5	<2	0.7	26.7	1.3	1.96
	Y	136	2.9	5.4	8	8.9	69	16.3	32	32	48	36.4	15.3	12.2	8.9	34	4.5	6.2	182.1	12.2	21

Table 4. *Cont.*

Sample ID		RM	S					FAPI	FATP						BAPI	BATP		O			B
		RM	SL1	SL2	SL3	SL4	SL5	PFA	FA1	FA2	FA3	FA4	FA5	FA6	PBA	BA1	BA2	QS1	WJ1	CW1	
Other elements (mg/kg)	Li	-	-	-	15.9	8.7	-	22.4	-	-		-	-	-	11.4	-	-	15.8	6	66.1	24
	Be	9.11	<5	2.3	<1	<1	6.4	2	2.7	2.3	2.4	5.43	<3	<2	<1	2.70	<3	<1	<1	2	2.1

Notes: # Ag expressed in μg/kg; * selenium and thallium failed in the recovery test, so these values are not given; ** lutetium was used as an internal standard, so it was not measured in samples. Sample specific comments: [a] SL1: The cerium emission lines were strongly disturbed by the matrix, so higher LOQ values had to be used. Similarly, the high chromium content disturbed the Ga 294.364 nm line, so a higher LOQ value had to be used. In the case of tantalum at the first parallel measurement, the more sensitive emission line (Ta 240.063 nm) was saturated by Fe 239.924 nm. Later on, this effect was eliminated. [b] SL2: The usual and more sensitive As 188.979 line was disturbed by the matrix, so that a higher LOQ value had to be used. Both of the utilized antimony lines (Sb 206.836 and Sb 217.582 nm) were disturbed by the matrix, so higher LOQ values had to be used. These LOQ values are lower compared to other samples. This is the result of the method development between the different sets of samples. [c] SL3: The analytical result of Ag could not be provided due to unusually high levels of interference from other elements. [d] SL5: The uranium LOQ value is lower compared to the other samples. This is the result of the method development between the different sets of samples. [e] FA5: The cobalt spectrum was strongly disturbed, so a higher LOQ had to be used.

3.4. Comparison of Different Industrial Residues Based on Minor and Trace Elements

Steel industry slags (Ss) stand alone among all the analyzed industrial residues since they contain high average levels of Cr and Mn (14,600 and 16,300 mg/kg, respectively). Cr and Mn are alloying elements that enhance the quality of steel. Minor amounts of alloying elements (Cr, Mn, etc.) can be incorporated in the slag, which originates from recycled scrap or from secondary metallurgy. In comparison to the Earth's upper crust are steel slags enriched with Cr and Mn [65] (Figure 5). Both these elements are incorporated in EAF C slags and mixtures that contain EAF S slag (Table 4—samples SL1, SL2, and SL3). The detected levels are comparable to the other EAF slags [66,67]. High concentrations of Mn were also detected in blast furnace slag (Table 4—sample SL5). Chemical speciation and mineralogical phases that contain these elements play crucial roles in the evaluation of the reactivity and availability of elements. Mineralogical analyses of the collected samples showed that chromite and Mg–chromite are the main carriers of Cr, while Mn is probably incorporated in a solid solution of (Fe, Mg, and Mn)O. Additionally, in the case of the magnesium chromite partial replacement of Cr^{3+} by Al^{3+}, Mg^{2+} by Mn^{2+} can also occur [67]. Besides Cr and Mn, slags also show considerable enrichments with Nb, Ni, V, Cu, Sb, and Zn.

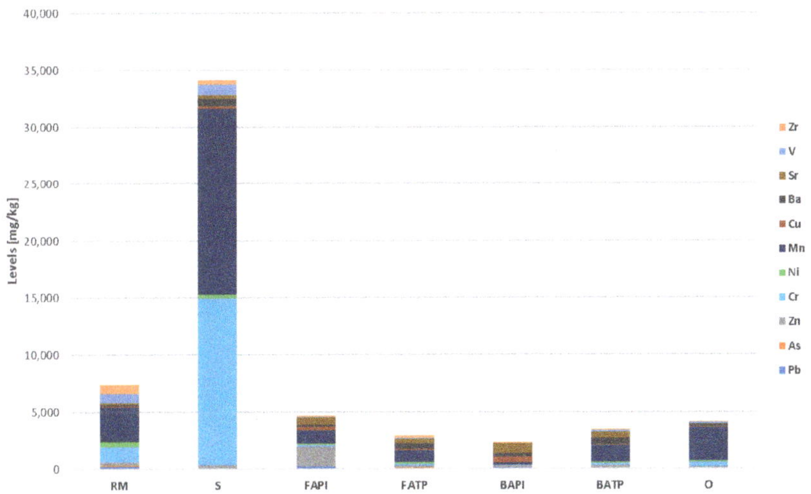

Figure 5. Average levels of selected minor and trace elements in the analyzed industrial wastes. Units in mg/kg.

Red mud (RM), as the material with the second highest enrichment with minor and trace elements, shows much lower elemental levels in comparison to the steel industry slags. The highest levels among the analyzed elements were detected for Mn, followed by Cr, Ni, V, and Zr (Figure 5). In comparison to the averages in Earth's upper crust [65], RM contains 163-fold higher levels of Cd, 33-fold higher levels of As, 15-fold higher levels of Cr, and 10- to 8-fold higher levels of Pb, Ni, and V (Table 4). Similar trace element levels were also detected by Radusinović and Papadopoulos [68] for red mud in Podgorica, Montenegro. One of the issues of red mud application in the cement industry is the conversion of total Cr into water soluble Cr^{6+} [69].

In fly ashes from paper mills (FAPIs), Zn is the predominant and the most enriched trace element (Figure 5). Also considerably enriched in comparison to the averages in Earth's upper crust [65] are Cd, Sb, Ag, Cu, Pb and Hg, showing 20- to 5-fold higher levels. In bottom ashes from paper mills (BAPIs), Sr, Cu, and Ba prevail, but only Cu is significantly enriched in comparison to the averages in Earth's upper crust. Other enriched elements are Ag, Cd, and Pb (Table 5). The fly ash from paper mills generally shows higher enrichments of the listed elements than the bottom ash from paper mills. The source of the listed elements is the high-temperature combustion of hog fuel in boilers. This fuel can also be mixed with grass plants, sugar and oil crops, agricultural residues, residues from the food and paper industries, municipal green wastes, sewage and de-inking sludge, and organic wastes and residues [70], which may contribute to the enrichment of the above listed elements. In our case, the sample PFA is combustion residue of burning the mixture of de-inking fiber paper sludge, waste wood, bark, coal, and sewage sludge, while the sample PBA is the combustion residue of de-inking and sewage sludge and natural gas [71]. Higher enrichments of PFA with trace elements might be connected with the application of coal in the burning mixture.

Table 5. Results of TOC measurements.

Type of Sample	Sample ID	TOC (%)
RM	RM	0.16
S	SL1	0.05
	SL2	0.28
	SL3	0.22
	SL4	0.02
	SL5	0.04
FAPI	PFA	10.07
FATP	FA1	0.16
	FA2	0.35
	FA3	0.30
	FA4	3.61
	FA5	0.29
	FA6	1.3
BAPI	PBA	0.45
BATP	BA1	1.32
	BA2	0.05
O	QS1	0.11
	WJ1	0.02
	CW1	0.03

Fly ashes from thermal power plants (FATPs) and bottom ashes from thermal power plants (BATPs) contained the largest amounts of Mn, Ba, and Sr (Figure 5), but they do not show any significant enrichment in comparison to the Earth's upper crust. The most enriched elements in FATP are As, Cd, Cu, and Ni, while BATP shows slight enrichment with Zn. The samples BA2 and FA3 also showed high enrichments of Hg (42-fold and 16-fold, respectively), while the other samples contain Hg levels comparable to the natural background. Slight enrichments were also detected for Be, Co, Ga, Pb, Sr, Th, V, and Zn. Those enrichments are characteristic of the coal fired thermal plants [72] and reflect the composition of the coal which is used as fuel [73].

Among the other samples (O), only WJ1—waste water jet sand from a waterjet cutting machine—shows significant enrichments of Cr, Mn, Ni, and some REEs (Table 4), which are consequences of the almandine (garnet group) sand used in the process and steel particles mixed in the material. The residuals from quartz sand washing (QS1) and brick rejects (CW1) do not contain any enrichments in comparison to the Earth's upper crust.

Minor elements like chromium (Cr), manganese (Mn), copper (Cu), and zinc (Zn) affect clinkerization by influencing the burning process and phase formation. Cr typically raises the temperature at which the liquid phase forms, potentially increasing the clinkerization temperature. In contrast, Mn and Zn lower the melting point of the raw mix, promoting early liquid phase formation and reducing the clinkerization temperature. However, excessive Zn may cause issues like kiln coating or ring formation [74,75]. Cu and Zn influence clinker phase formation by altering the crystallization of key minerals such as

C_3S and C_3A. A high Cu content promotes the decomposition of C_3S into C_2S and free lime, while Zn reduces C_3A content by forming alternative compounds. These changes impact the performance of the cement, underscoring the need for controlled levels of these minor elements in the raw materials [76,77]. Minor elements such as chromium (Cr), manganese (Mn), and zinc (Zn) influence cement hydration by altering the phase formation and hydration kinetics. High concentrations of these elements can interact with calcium silicates (C_3S and C_2S) and calcium hydroxide, affecting the rate of heat generation, setting time, strength development, and durability. Mn accelerates clinker hydrolysis, improving early hydration compressive strength but reducing strength after 80 days due to combined water formation, while Cr slows hydration, decreasing both the hydration degree and compressive strength after 28 and 80 days. Cu and Zn significantly retard early hydration and strength development within the first day, but exert an accelerating effect at later ages [78–80].

3.5. REE Levels in Industrial Residues

The highest REE levels (including Sc) among the analyzed industrial residues were detected for RM, reaching 1268 mg/kg (Figure 6). Light rare earth elements (LREEs; La, Ce, Pr, Nd, Pm, Sm, and Eu) present almost a 70% share of the REEs in RM. The detected levels are somehow lower in comparison to those measured in the RM in Podgorica, Montenegro, where RM contains between 1535 and 1646 mg/kg REEs [68]. The composition of red mud also depends on the composition of the processed bauxite. Karstic bauxites in the Balkan region contain between 200 and 3500 mg/kg REEs, but bauxite with a REE content below 1000 mg/kg predominates [68,81].

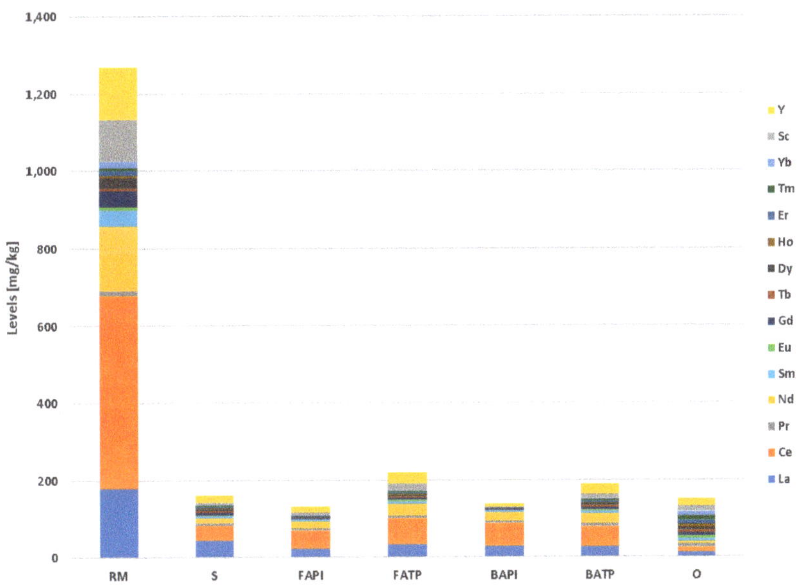

Figure 6. Average levels of REEs in the analyzed industrial wastes. Units in mg/kg.

The REEs in RM are present in ferrotitanite, phosphate, and carbonate minerals or are adsorbed on goethite and cancrinite—the latter two are most probably the main carriers of REEs in RM [82], which also corresponds well with the mineralogical composition of our RM sample. REEs can be extracted from RM using multistage extraction with different combinations of acid leaching, acid roasting, and high temperature smelting [8]. The main objective of the process is to remove or isolate major elements such as Fe, Al, and Ca from the REEs, which can then be selectively leached. The efficiency of such multistage extractions is between 40% and 80% [11,83,84]. However, this process was tested only at a laboratory scale, and pilots in real environments have not been utilized yet due to its economic feasibility, and the environmental impact of the process is still questionable. However, RM is an appropriate target for REE extraction when methods for industrial extraction will be available.

The average levels of REEs in the other industrial residues are 161 mg/kg for slag (S), 132 mg/kg for paper industry fly ash (FAPI), 220 mg/kg for thermal power plant fly ash (FATP), 140 mg/kg for paper industry bottom ash (BAPI), 188 mg/kg for thermal power plant bottom ash (BATP), and 150 mg/kg for other (O) materials. The content of REEs in these materials does not represent any economic potential nor environmental issue since they are directly comparable to their averages in Earth's upper crust, which is 183 mg/kg [65].

3.6. Presence of Organic Matter (TOC)

Another important parameter that defines the chemical composition of raw material is the presence of organic matter, expressed by the total organic content (TOC) analyzed (Table 5). It affects mainly emissions, as in unstable operating conditions the presence of TOC in clinker raw mixture can contribute to the CO_2 emissions. Research on the specific impact of TOC on clinker has shown that elevated levels can hinder proper sintering or lead to incomplete reactions, impacting the cement's strength and durability. Furthermore, excess organic carbon might contribute to carbon emissions during the production process, which is a significant concern for the cement industry's environmental footprint [85]. A balance must be struck, as excessive organic matter (above a certain threshold) can lead to lower strength gain and even longer-term deterioration. Some studies suggest that organic matter content above approx. 5% in raw materials can begin to degrade cement performance by disrupting clinker chemistry, increasing energy consumption during combustion, and introducing impurities that weaken the final product while also exacerbating environmental emissions [86]. However, the specific threshold can vary depending on the type of cement and other compositional factors in the raw materials.

The results showed (Table 5) that the highest values of TOC in secondary raw materials (red mud, slag, fly and bottom ash, etc.) are between 0.02 and 3.6%. Some samples contained a higher amount of TOC, especially fly ash from the paper industry (PFA) which contains 10.07%, which is prescribed as the remains of cellulose.

3.7. Mineralogical Composition

The mineralogical composition of the investigated samples is presented in Tables 6–10. Figure 7 presents the XRD patterns of the selected Al-containing industrial residues.

As can be seen from Table 6, the sample of red mud (RM) contained hematite, cancrinite, and goethite as major mineral phases, but also ilmenite, katoite, rutile, illite, and quartz in minor quantities.

Table 6. Mineralogical composition of red mud.

Sample ID	Phases, wt.%											
	A	C	H	Ca	Ka	Il	Go	Bo	I	Q	R	Sum
RM	14.2	27.1	27.0	16.4	2.6	2.1	13.1	5.5	1.5	1.1	0.2	100

A—amorphous; C—calcite; H—hematite; Ca—cancrinite; Ka—katoite; Il—ilmenite; Go—goethite; Bo—boehmite; I—illite; Q—quartz; R—rutile.

Table 7. Mineralogical composition of slags.

Sample ID	Phases, wt.%																
	A	β-B	γ-B	F	C	W	M	P	Ma	Ch	M-C	D	Q	H	Ga	Me	SUM
SL1	4.7	38.7	0.4	25.9	/	14.1	11.1	2.2	1.9	1.0	/	/	/	/	/	/	100.0
SL2	35.8	24.0	/	9.6	11.2	0.4	/	/	10.4	0.6	/	6.2	0.8	/	/	/	100.0
SL3	*	29.9	3.5	5.8	11.2	/	/	12.8	/	/	14.0	11.2	10.2	2.3	1.1	/	100.0
SL4	*	53.0	11.7	/	/	/	0.5	0.3	25.5	/	/	2.4	/	0.3	4.0	2.3	100.0
SL5	100	/	/	/	/	/	/	/	/	/	/	/	/	0		/	100.0

A—amorphous; β-B—β-belite; γ-B—γ-belite; F—ferrite; C—calcite; W—wuestite; M—magnetite; P—periclase; Ma—mayenite; Ch—chromite; M-C—Mg-chromite; D—dolomite; Q—quartz; H—hematite; Ga—galenite; Me—merwinite; *—amorphous phase was not considered.

Table 8. Mineralogical composition of collected fly ash samples.

Sample ID	Phases, wt.%																		
	A	Ge	F	Pl	C	Q	O	Ma	L	Mu	H	K	An	Mf	Et	I/M	Do	Po	P	SUM
PFA	43.8	6.0	14.8	3.9	1.9	9.4	0.4	0.8	8.3	/	/	/	/	/	/	/	/	/	/	100.0
FA1	79.5	0.6	/	7.9	/	4.1	/	/	0.2	6.7	0.6	0.4	5.5	/	/	/	/	/	/	100.0
FA2	81.9	/	/	/	/	5.0	/	/	2.6	/	4.5	/	/	0.5	/	/	/	/	/	100.0
FA3	80.4	/	/	2.8	/	10.4	/	/	/	5.3	0.2	/	/	/	0.9	/	/	/	/	100.0
FA4	64.9	/	/	18.0	/	9.0	/	/	/	/	0.6	/	/	/	/	7.3	0.2	/	/	100.0
FA5	35.4	/	/	11.1	0.8	48.1	/	/	0.3	/	/	4.3	/	/	/	/	/	/	/	100.0
FA6	51.7	/	/	5.4	18.0	19.0	/	/	1.2	/	/	1.0	/	/	/	/	1.1	1.4	1.2	100.0

A—amorphous; Ge—gehlenite; F—ferrite; Pl—plagioclase; C—calcite; Q—quartz; O—orthoclase; Ma—mayenite; L—lime; Mu—mullite; H—hematite; K—K-feldspar; An—anhydrite; Mf—magnesioferrite; Et—ettringite; I/M—illite/muscovite; Do—dolomite; Po—portlandite; P—periclase.

Table 9. Mineralogical composition of collected bottom ash samples.

Sample ID	Phases, wt.%													
	A	C	Po	Q	L	T	B	My	An	Ge	Pl	H	K	B	SUM
PBA	26.5	23.0	9.8	0.3	12.0	0.7	20.1	1.6	0.4	5.6	/	/	/	/	100.0
BA1	65.1	/	/	2.8	/	/	/	1.7	/	1.4	27.9	0.6	0.5	/	100.0
BA2	42.4	/	/	56.4	0.5	/	/	/	/	/	/	/	/	0.7	100.0

A—amorphous; C—calcite; Po—portlandite; Q—quartz; L—lime; T—talc; B—belite; My—mayenite; An—anhydrite; Ge—gehlenite; Pl—plagioclase; H—hematite; K—K-feldspar; B—belite.

Table 10. Mineralogical composition of other collected industrial residues.

Sample ID	Phases, wt.%								
	A	Q	I/M	Kao	Al	Ilm	Pl	H	C	SUM
QS1	24.4	72.2	1.9	1.5	/	/	/	/	/	100.0
WJ1	23.2	1.0	/	/	85.0	0.8	/	/	/	100.0
CW1	51.5	31.3	7.4				8.1	1.1	0.6	100.0

A—amorphous; Q—quartz; I/M—illite/muscovite; Kao—kaolinite; Al—almandine; Ilm—ilmenite; Pl—plagioclase; H—hematite; C—calcite.

In all steel slag samples (SL1, SL2, SL3, and SL4), the main phase was belite, as shown in Table 7. Other phases that occurred in higher amounts in most samples were ferrite, calcite, mayenite, and dolomite. Namely, the other phases present in SL1 were ferrite, wuestite and magnetite. In SL3, besides belite, the other phases present were Mg-chromite, periclase, calcite, dolomite, quartz, and ferrite. In SL2 (mixture of EAF C and ladle slag), in addition to belite, the other phases were calcite, mayenite, ferrite, and dolomite. The ladle slag SL4 contained over 50 wt.% of belite, and the other phases were mayenite, γ-belite, and gehlenite. In some steel slag samples, traces (<3 wt.%) of dolomite, merwinite, galenite, hematite, chromite, periclase, mayenite, γ-belite, gehlenite, wuestite, quartz, and magnetite were also present. The blast furnace slag sample SL5 consisted only of amorphous content (100.0 wt.%).

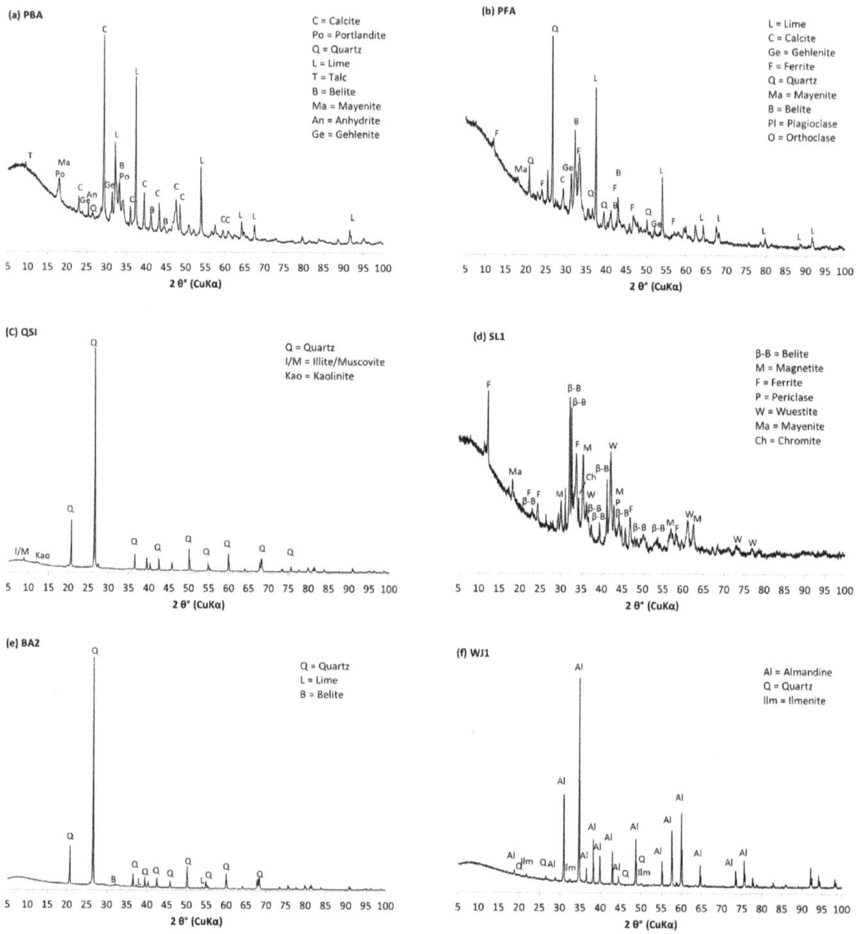

Figure 7. XRD patterns for selected samples. (**a**) Bottom ash from the paper industry (PBA), (**b**) coal fly ash (PFA), (**c**) by-product of washing quartz sand (QSI), (**d**) EAF C steel slag (SL1), (**e**) coal bottom ash (BA2), and (**f**) waste water jet sand (WJ1).

As regarding fly ashes, shown in Table 8, in fly ash from the paper industry the most abundant phases were ferrite and belite, followed by quartz, lime, and gehlenite. The other phases were albite, calcite, mayenite, and orthoclase.

In samples of fly ash from the thermal power plant, the predominant phase in most materials was plagioclase (FA1 and FA4) or quartz (FA3, FA5, and FA6). The other phases determined were, e.g., hematite, mullite, illite/muscovite, and in small amounts, gehlenite, lime, anhydrite, feldspar, calcite, dolomite, portlandite, and periclase.

In bottom ash from the paper industry (PBA), as shown in Table 9, the main phases were calcite and belite, followed by lime, portlandite, and gehlenite. In small amounts, mayenite, talc, anhydrite, and quartz were present.

Quartz was the main phase in the bottom ash from the thermal power plant, BA2, while in BA1 the main phase was plagioclase. The dominant phases in PBA were calcite, followed by lime, portlandite, and gehlenite. Other minor phases present in some samples of bottom ash in a quantity of less than 3 wt.% were belite, lime, K-feldspar, hematite, anhydrite, talc, quartz, mayenite, and gehlenite.

Other collected industrial residues have different compositions, as shown in Table 10. In a sample of by-product of washing quartz sand, QS1, the main phase was quartz, followed by small amounts of illite/muscovite and kaolinite. In the sample WJ1, the main phase was almandine, with small amounts of quartz and ilmenite. The sample CW1 contained quartz, plagioclase, and illite/muscovite, and small amounts of hematite and calcite.

Gamma belite, a form of calcium silicate, has a lower reactivity than alite, which can slow hydration and hinder early strength development when used as an additive in cementitious materials. It is more effective as a raw material in clinker production, where its presence can be better controlled to avoid negative impacts on hydration and strength development [21,87]. Periclase, composed of magnesium oxide, can disrupt the clinkerization process and slow hydration rates when used as an additive in cement, negatively affecting early strength and durability. Its instability under certain conditions can also lead to expansion issues, compromising the long-term integrity of cement-based materials [88]. Therefore, while periclase might be useful in specific contexts, its presence in cement needs to be carefully controlled to avoid these potential problems.

Free lime in ashes, such as fly ash, can contribute to additional cementitious reactions, improving strength and durability. However, too much can reduce strength and cause issues like shrinkage and increased water absorption. For optimal performance, free lime content in ash should generally be kept below 10%, as higher levels can lead to expansion and durability problems [89,90].

3.8. The Results of Radiological Characterization

Table 11 shows the results of the radiological characterization of the secondary raw materials. Except for ^{137}Cs, which was detected in a few samples, all detected radionuclides have a natural origin. The measured concentrations of ^{137}Cs are low and are a consequence of the Chernobyl accident in 1986. The measured values of ^{226}Ra, ^{232}Th, and 40K are characteristic for the appropriate type of samples and correspond to the values measured in other tests [36,91–93].

Based on the results shown in Table 11, the external hazard index (H_{ext}), internal hazard index H_{int}, and the activity concentration index (*ACI*) were calculated, and their values are presented in Table 12.

It must be noted that the permitted limits of radioactive elements depend on the purpose of the material and are limited based on the estimated doses. To assess the potential health impact on the public due to exposure to the tested samples for which the *ACI* is greater than 1, the annual effective dose (*E*) of the total external absorbed gamma dose in air at a height of 1 m above ground level for outdoor and indoor cases (p = 0.2, $E_{terr,20\%}$ and p = 0.8, $E_{terr,80\%}$) and for four different cases of standard rooms (all walls, floor and ceiling, only floor and walls, and only floor made of them, as well as only superficial materials) was calculated and is shown in Table 13.

It can be concluded from a radiological point that with the exception of RM (red mud), FA1 (fly ash from the thermal power plant), and BA1 (bottom ash from the thermal power plant), all the other tested samples met both of the strictest criteria: *ACI* < 1 and *Hint* < 1. Accordingly, the criteria of R_{aeq} < 370, $H_{ex}t$ < 1, and I_α < 1 are also met for them, as well as the effective dose for a standard model room being less than 1 mSv/y in the case that all walls, the floor, and the ceiling are built with them. From the point of view of radioactivity, they can be freely used as building materials. For samples with *ACI* > 1, the effective dose was calculated for the external exposure of gamma rays in air at the height of 1 m above ground level for outdoor and indoor cases ($E_{terr,20\%}$, and $E_{terr,80\%}$) and four different cases of standard rooms (E_{all}, E_{fw}, E_f, and E_{sup}). Considering that all the calculated doses $E_{terr,20\%}$ are less than 1, all the tested materials are safe from the aspect of external terrestrial exposure, with the limitation of the duration of exposure being less than 20% of the hours per year. Also, all the tested materials can be used to make only the floor, and as a surface material (E_f and E_{sup} are less than 1).

Table 11. Radiological characteristics of the investigated samples.

Type of Sample	Sample ID	^{210}Pb	^{226}Ra	^{232}Th	^{40}K	^{137}Cs	^{238}U	^{235}U
RM	RM	107 ± 11	176 ± 7	397 ± 25	32 ± 6	<0.1	164 ± 15	10.3 ± 1.5
S	SL1	<2	19.2 ± 1.3	3.3 ± 0.5	<1	<0.04	14.1 ± 1.7	1.00 ± 0.08
	SL2	5 ± 2	22.6 ± 1.1	6.6 ± 0.8	11.7 ± 1.7	0.6 ± 0.1	23 ± 4	1.7 ± 0.3
	SL3	6.8 ± 2.1	29 ± 2	14.7 ± 12	36.7 ± 3.2	<0.03	28.6 ± 3.0	1.8 ± 0.2
	SL4	<0.7	34.9 ± 1.5	8.2 ± 0.6	<0.7	<0.02	44.5 ± 5.6	2.5 ± 0.3
	SL5	<13	124 ± 7	24 ± 4	160 ± 20	<0.6	88 ± 13	7.0 ± 0.7
FAPI	PFA	42.4 ± 4.8	42 ± 2	32 ± 3	242 ± 17	9.9 ± 0.7	39.2 ± 5.3	1.9 ± 0.2
FATP	FA1	362 ± 24	406 ± 16	57.6 ± 4.6	562 ± 36	<0.08	364 ± 45	24 ± 3
	FA2	55.6 ± 6	24 ± 2	19 ± 3	170 ± 20	<0.2	33 ± 5	2.2 ± 0.3
	FA3	17 ± 4	25 ± 2	23 ± 3	90 ± 10	<0.2	20 ± 4	1.5 ± 0.2
	FA4	93 ± 8	99 ± 4	78 ± 5	670 ± 40	<0.09	114 ± 10	5.4 ± 0.4
	FA5	96 ± 7	30.0 ± 1.6	22.5 ± 2.2	1260 ± 77	8.9 ± 0.7	31.5 ± 5.4	2.0 ± 0.2
	FA6	271 ± 18	42 ± 2	23 ± 3	1810 ± 110	16.3 ± 1.2	33.6 ± 6.7	1.5 ± 0.2
BAPI	BAS2	23.5 ± 3.7	33 ± 2	42 ± 3	90 ± 8	1.2 ± 0.3	34.7 ± 4.6	1.8 ± 0.2
BATP	BA1	736 ± 87	420 ± 24	56 ± 4	469 ± 31	<0.1	408 ± 27	23 ± 1
	BA2	13 ± 3	20 ± 1	10.9 ± 1.4	1030 ± 64	1.2 ± 0.2	6.8 ± 2.7	0.30 ± 0.05
O	QS1	45.5 ± 5.2	24 ± 1	13 ± 1	281 ± 18	<0.04	38.9 ± 4.7	1.8 ± 0.2
	WJ1	14.2 ± 4.3	29.6 ± 2.0	103 ± 7	8.2 ± 1.8	<0.03	26.2 ± 3.9	1.6 ± 0.2
	CW1	45.5 ± 6.9	35 ± 1.6	48 ± 3	653 ± 41	<0.04	44 ± 6	2.4 ± 0.2

Table 12. Screening indices for an assessment of the radiological load by the investigated samples.

Type of Sample	Sample ID	H_{ext}	H_{int}	ACI
RM	RM	2.02	2.49	2.58
S	SL1	0.06	0.12	0.08
	SL2	0.09	0.15	0.11
	SL3	0.14	0.22	0.18
	SL4	0.13	0.22	0.16
	SL5	0.47	0.80	0.59
FAPI	PFA	0.29	0.41	0.38
FATP	FA1	1.45	2.55	1.83
	FA2	0.18	0.24	0.23
	FA3	0.18	0.25	0.23
	FA4	0.73	1.00	0.94
	FA5	0.47	0.55	0.63
	FA6	0.64	0.75	0.86
BAPI	PBA	0.27	0.36	0.35
BATP	BA1	1.46	2.60	1.84
	BA2	0.34	0.40	0.46
O	QS1	0.18	0.25	0.24
	WJ1	0.48	0.56	0.62
	CW1	0.44	0.53	0.57

Table 13. Estimation of the annual effective dose for different uses of the investigated samples.

Type of Sample	Sample ID	$E_{terr,80\%}$ (mSv)	$E_{terr,20\%}$ (mSv)	E_{all} (mSv)	E_{fw} (mSv)	E_f (mSv)	E_{sup} (mSv)
RM FATP BATP	RM FA1 BA1	>1	<1	>1	>1	<1	<1
Other	/	<1, due to ACI < 1					

If the hypothetical situation occurs where the annual effective dose for all situations is greater than one, whether the material can be disposed of in the environment is examined (the criteria are defined by competent institutions in the specific country). If it turns out that the material is not safe for disposal, it is reported to a competent institution (usually the radiological inspection), which passes measures on further action.

4. Conclusions

These characterized (from their physical, chemical, mineralogical, and radiological aspects) 19 samples of industrial Al-containing residues from the ESEE region present a base for their advantages as alternative raw materials that can influence improvement of the properties/add value on the final products as well as of the cost-effectiveness and environmental stability of cements. This paper's results tackle two aspects of secondary raw materials, the environmental (heavy metals) and radiological aspects, which are mostly separately considered in practice.

The characterized Al-containing residues from the ESEE region showed the potential to be used in construction and beyond. Up to now, globally, fly ash has been mostly used in cement and concrete production, but also the other characterized Al-containing residues could be successfully utilized either in the construction and building industries or in wider applications for traditional building products, like bricks, glass–ceramics, tiles, etc., or for products supporting the green transition like geopolymers, aerogels, zeolites, REEs, etc. The proportions of each Al-containing residue should be calculated based on the needs of the final material/product to achieve the necessary goal. In this respect, pre-treatment of such Al-containing residues could have a beneficial effect.

Future investigations need to be performed for creative applications of the characterized Al-containing residues, as they seem to be particularly promising in relation to environmental benefits, such as lowering CO_2 emissions, controlling greenhouse emissions, and reducing environmental contamination.

Author Contributions: Conceptualization: S.D. and E.F.; methodology: S.D.; project administration: S.D. and K.Š.; formal analysis and investigation: K.Š., M.M., M.R., B.D.K., A.I., K.T., G.Ž., V.J., N.S.A., M.L., L.Ž. and S.D.; writing—original draft preparation: S.D., E.F., K.Š., M.M., M.R., B.D.K., A.I., K.T., G.Ž., V.J., M.L. and L.Ž., writing—review and editing: S.D., E.F. and G.Ž.; funding acquisition: S.D. and L.Ž.; resources: K.Š., M.M., M.R., B.D.K., A.I., K.T., G.Ž., V.J., N.S.A., M.L., L.Ž. and V.J.; supervision: S.D. and E.F. All authors have read and agreed to the published version of the manuscript.

Funding: The study has received funding from the European Institute of Innovation and Technology (EIT), a body of the European Union, under Horizon 2020, the EU Framework Programme for Research and Innovation (RIS-ALiCE, project no. 18258). It was also partly supported by the Slovenian Research and Innovation Agency (ARIS) via the core research funding programs P2–0273 "Building structures and materials" and P1-0025 "Mineral resources".

Institutional Review Board Statement: Not applicable.

Informed Consent Statement: Not applicable.

Data Availability Statement: The original contributions presented in the study are included in the article, further inquiries can be directed to the corresponding author.

Acknowledgments: The Metrology Institute of the Republic of Slovenia is acknowledged for the use of XRF.

Conflicts of Interest: Author Bence David Koszo was employed by the Bay Zoltán Nonproft Ltd. for Applied Research. Author Andrej Ipavec was employed by the Alpacem Cement, d.d. Author Nikolina Stamatovska Aluloska was employed by the Cementarnica "USJE"AD Titan Group. Author Mojca Loncnar was employed by the SIJ Acroni d.o.o. The remaining authors declare that the research was conducted in the absence of any commercial or financial relationships that could be construed as a potential conflict of interest.

References

1. European Commission. Communication No. 640 The European Green Deal. 2019. Available online: https://eur-lex.europa.eu/legal-content/EN/TXT/?uri=COM:2019:640:FIN (accessed on 15 January 2023).
2. European Commission. Communication No. 98 A New Circular Economy Action Plan for a Cleaner and More Competitive Europe. 2020. Available online: https://eur-lex.europa.eu/legal-content/EN/TXT/?uri=COM:2020:98:FIN (accessed on 15 January 2023).
3. European Commission. Communication No. 29 Monitoring Framework for the Circular Economy. 2018. Available online: https://eur-lex.europa.eu/legal-content/EN/TXT/?uri=COM:2018:29:FIN (accessed on 15 January 2023).
4. Žibret, L.; Šter, K.; Borštnar, M.; Loncnar, M.; Dolenec, S. The Incorporation of Steel Slag into Belite-Sulfoaluminate Cement Clinkers. *Appl. Sci.* **2021**, *11*, 1840. [CrossRef]
5. Gencel, O.; Munir, M.J.; Kazmi, S.M.S.; Sutcu, M.; Erdogmus, E.; Velasco, P.M.; Quesada, D.E. Recycling industrial slags in production of fired clay bricks for sustainable manufacturing. *Ceram. Int.* **2021**, *47*, 30425–30438. [CrossRef]
6. Liu, X.; Li, B.; Wu, Y. The pretreatment of non-ferrous metallurgical waste slag and its research progress in the preparation of glass-ceramics. *J. Clean. Prod.* **2023**, *404*, 136930. [CrossRef]
7. Yan, H.; Zhang, G.; Wang, Y.; Liu, J.; Li, G.; Zhao, Y.; Xu, Y.; Lv, Y. A green synthesis strategy for low-cost multi-porous solid CO_2 adsorbent using blast furnace slag. *Fuel* **2022**, *329*, 125380. [CrossRef]
8. Lin, S.; Jiang, X.; Zhao, Y.; Yan, J. Zeolite greenly synthesized from fly ash and its resource utilization: A review. *Sci. Total. Environ.* **2022**, *851*, 158182. [CrossRef]
9. U.S. Geological Survey, Mineral Commodity Summaries. Iron and Steel Slag. 2022. Available online: https://pubs.usgs.gov/periodicals/mcs2022/mcs2022-iron-steel-slag.png (accessed on 15 January 2023).
10. The European Steel Association. European Steel in Figures. 2021. Available online: https://www.eurofer.eu/assets/publications/brochures-booklets-and-factsheets/european-steel-in-figures-2023/FINAL_EUROFER_Steel-in-Figures_2023.png (accessed on 15 January 2023).
11. Rivera, R.M.; Ulenaers, B.; Ounoughene, G.; Binnemans, K.; Van, G.T. Extraction of rare earths from bauxite residue (red mud) by dry digestion followed by water leaching. *Miner. Eng.* **2018**, *119*, 82–92. [CrossRef]
12. Mahinroosta, M.; Karimi, Z.; Allahverdi, A. Recycling of Red Mud for Value-Added Applications: A Comprehensive Review. In *Encyclopedia of Renewable and Sustainable Materials*; Elsevier Ltd.: Amsterdam, The Netherlands, 2020. [CrossRef]
13. Shahrokhi, A.; Tettey-Larbi, L.; Akuo-Ko, E.O.; Tóth-Bodrogi, E.; Kovács, T. The New Conception of Radiological Sustainability, Possibilities by Reutilization of Residues Products and Building Materials. *Sustainability* **2023**, *15*, 10647. [CrossRef]
14. Mo, K.H.; Ling, T.C. Utilization of Coal Fly Ash and Bottom Ash in Brick and Block Products, Low Carbon Stabilization and Solidification of Hazardous Wastes 2022. Available online: https://doi.org/10.1016/b978-0-12-824004-5.00026-8 (accessed on 15 January 2023).
15. Angjusheva, B.; Jovanov, V.; Fidanchevski, E. Conversion of coal fly ash into glass-ceramics by controlled thermal treatment. *Maced. J. Chem. Chem. Eng.* **2021**, *40*, 307–319. [CrossRef]
16. Tamanna, K.; Raman, S.N.; Jamil, M.; Hamid, R. Coal bottom ash as supplementary material for sustainable construction: A comprehensive review. *Constr. Build. Mater.* **2023**, *389*, 131679. [CrossRef]
17. Gollakota, A.R.K.; Volli, V.; Shu, S.M. Progressive utilization prospects of coal fly ash: A review. *Sci. Total Env.* **2019**, *672*, 951–989. [CrossRef]
18. Scrivener KL, Vanderley M, John VM, Ellis M, Gartner EM, Eco-efficient cements: Potential economically viable solutions for a low-CO_2 cement-based materials industry. *Cem. Concr. Res.* **2018**, *114*, 2–26. [CrossRef]
19. Gartner, E.; Sui, T. Alternative cement clinkers. *Cem Concr Res.* **2018**, *114*, 27–39. [CrossRef]
20. Žibret, L.; Ipavec, A.; Dolenec, S. Microstructural characteristics of belite-sulfoaluminate cement clinkers with bottom ash. *Constr. Build. Mater.* **2022**, *321*, 126289. [CrossRef]
21. Kramar, S.; Žibret, L.; Fidanchevska, E.; Jovanov, V.; Angjusheva, B.; Ducman, V. Use of fly ash and phosphogypsum for the synthesis of belite-sulfoaluminate clinker. *Mater. Constr.* **2019**, *69*, 176. [CrossRef]
22. Pan, Z.; Tao, Z.; Cao, Y.-F.; Wuhrer, R. Measurement and prediction of thermal properties of alkali-activated fly ash/slag binders at elevated temperatures. *Mater. Struct.* **2018**, *51*, 108. [CrossRef]
23. Cao, Y.-F.; Tao, Z.; Pan, Z.; Wuhrer, R. Effect of calcium aluminate cement on geopolymer concrete cured at ambient temperature. *Constr. Build. Mater.* **2018**, *191*, 242–252. [CrossRef]

24. Directive 2013/59/EURATOM 5-December 2013, Official European Union 17/01/2014. 2013. Available online: http://data.europa.eu/eli/dir/2013/59/oj (accessed on 15 January 2023).
25. Žibret, G.; Teran, K.; Žibret, L.; Šter, K.; Dolenec, S. Building of the Al-containing Secondary Raw Materials Registry for the Production of Low CO_2 Mineral Binders in South-Eastern European Region. *Sustainability* **2021**, *13*, 1535. [CrossRef]
26. Fidanchevski, E.; Angjusheva, B.; Jovanov, V.; Murtanovski, P.; Vladiceska, L.; Aluloska, N.S.; Nikolic, J.K.; Ipavec, A.; Šter, K.; Mrak, M.; et al. Technical and radiological characterisation of fly ash and bottom ash from thermal power plant. *J. Radioanal Nucl Chem.* **2021**, *330*, 685–694. [CrossRef]
27. Mirković, M.; Kljajević, L.; Nenadović, S.; Dolenec, S.; Šter, K.; Žibret, L.; Rajačić, M. Fly ash as a raw material for low-carbon cement clinkers and its radiological properties. *J. Radioanal. Nucl. Chem.* **2021**, *328*, 1391–1398. [CrossRef]
28. Kljajević, L.M.; Mirković, M.M.; Dolenec, S.; Ster, K.; Hadzalic, M.; Vukanac, I.S.; Nenadović, M.T. Radiological and physico-chemical characterization of red mud as an Al-containing precursor in inorganic binders for the building industry. *Nucl. Technol. Radiat.* **2021**, *36*, 182–191. [CrossRef]
29. Fidanchevski, E.; Šter, K.; Mrak, M.; Kljajević, L.; Žibret, G.; Teran, K.; Poletanovic, B.; Fidanchevska, M.; Dolenec, S.; Merta, I. The Valorisation of Selected Quarry and Mine Waste for Sustainable Cement Production within the Concept of Circular Economy. *Sustainability* **2022**, *14*, 6833. [CrossRef]
30. *EN1097-7:2008*; Tests for mechanical and physical properties of aggregates.Determination of the particle density of filler. Pycnometer method. Association Française de Normalisation: Paris, France, 2008.
31. *JUS B.C8.023:1982*; Methods of Testing the Physical Properties of Cement. Savezna Komisija za Standardizaciju: Beograd, Socialitstička Federativna Republika Jugoslavija, 1982.
32. *ISO 9277:2010*; Determination of the specific surface area of solids by gas adsorption—BET method. ISO: Geneva, Switzerland, 2010.
33. *ISO 3310-2:2013*; Test sieves—Technical Requirements and Testing—Part 2: Test Sieves of Perforated Metal Plate. ISO: Geneva, Switzerland, 2013.
34. *EN196-2:2013*; Method of Testing Cement—Part 2: Chemical Analysis of Cement. British Standards Institution: London, UK, 2013.
35. *MSZ525-17:2013*; Chemical Analysis of Cement. Part 17: Determination of Total Chrome Content as Cr_2O_3. Magyar Szabványügyi Testület: Budapest, Hungary, 2013.
36. *Measurement of Radionuclides in Food and the Environment*; Technical Reports Series No. 295, a Gudiebook; IAEA: Vienna, Austria, 1989.
37. *1035-SE-40845-17*; Radioactive Material Homogeneously Dispersed in Silicone Resin, Type CBSS 2. Czech Metrology Institute: Prague, Czech Republic, 2017.
38. *1035-SE-40844-17*; Radioactive Solution in 5 mL Glass Ampoule, Type ER X. Czech Metrology Institute: Prague, Czech Republic, 2017.
39. UNSCEAR. Sources and Effects of Ionizing Radiation. United Nations Scientific Committee on the Effects of Atomic Radiation. UNSCEAR 2000 Report (Vol. I) to the General Assembly, with Scientific Annexes (2000). Available online: https://www.unscear.org/docs/publications/2000/UNSCEAR_2000_Report_Vol.I.png (accessed on 15 January 2023).
40. European Commission. *Radiological Protection Principles Concerning the Natural Radioactivity of Building Materials*; Publications Office: Brussels, Belgium, 1999. Available online: https://op.europa.eu/en/publication-detail/-/publication/988f3243-5259-43a5-b621-fbff662deeb0/language-en (accessed on 17 April 2024).
41. Beretka, J.; Mathew, P.J. Natural radioactivity of Australian building materials, industrial wastes and by products. *Health Phys.* **1985**, *48*, 87. [CrossRef] [PubMed]
42. Markkanen, M. Radiation Dose Assessments for Materials with Elevated Natural Radioactivity, Report STUK-B-STO 32. Radiation and Nuclear Safety Authority-STUK. Finnish Centre for Radiation and Nuclear Safety, 1995. Volume 28 No. 6, ISSN 1235-6719. Available online: https://inis.iaea.org/collection/NCLCollectionStore/_Public/28/021/28021209.png (accessed on 17 April 2024).
43. Helsel, M.A.; Ferraris, C.F.; Bentz, D. Comparative study of methods to measure the density of Cementious powders. *J. Test Eval.* **2016**, *44*, 2147–2154. [CrossRef] [PubMed]
44. Mukhopadhyay, S.; Masto, R.E.; Tripathi, R.C.; Srivastava, N.K. Application of Soil Quality Indicators for the Phytorestoration of Mine Spoil Dumps. In *Phytomanagement of Polluted Sites*; Pandey, V.C., Bauddh, K., Eds.; Elsevier: Amsterdam, The Netherlands, 2019; pp. 361–388. [CrossRef]
45. Sharma, P.V. *Environmental and Engineering Geophysics*; Cambridge University Press: Cambridge, UK, 1997; p. 475. [CrossRef]
46. Navarrete, I.; Kurama, Y.; Escalona, N.; Lopez, M. Impact of physical and physicochemical properties of supplementary cementitious materials on structural build-up of cement-based pastes. *Cem. Concr. Res.* **2020**, *130*, 105994. [CrossRef]
47. Arvaniti, E.C.; Juenger, M.C.G.; A Bernal, S.A.; Duchesne, J.; Courard, L.; Leroy, S.; Provis, J.L.; Klemm, A.; De Belie, N. Determination of particle size, surface area, and shape of supplementary cementitious materials by different techniques. *Mater. Struct.* **2015**, *48*, 3687–3701. [CrossRef]
48. Arvaniti, E.C.; Juenger, M.C.G.; Bernal, S.A.; Duchesne, J.; Courard, L.; Leroy, S.; Provis, J.L.; Klemm, A.; De Belie, N. Physical characterization methods for supplementary cementitious materials. *Mater. Struct.* **2015**, *48*, 3675–3686. [CrossRef]
49. Colangelo, F.; Cioffi, R.; Farina, I. *Handbook of Sustainable Concrete and Industrial Waste Management*; Elsevier Ltd.: Amsterdam, Netherlands, 2022.
50. The Euclid Chemical Company, Supplementary Cementitious Materials. Available online: https://www.nahb.org/-/media/NAHB/nahb-community/docs/councils/bsc/supplementary-cementitious-materials.png (accessed on 26 November 2024).

51. Kline, J. Optimising SCM Use in Cement and Concrete, Global Cement Magazine. 2020. Available online: https://globalcement.com/magazine/articles/1190-optimising-scm-use-in-cement-and-concrete (accessed on 26 November 2024).
52. Hanson, K. SCMs in Concrete, NPCA Blog: Production. 2017. Available online: https://precast.org/blog/scms-in-concrete/ (accessed on 26 November 2024).
53. Chen, B. Effect of particle size distribution on properties of cement. In Proceedings of the 4th International Symposium on Traffic Transportation and Civil Architecture (ISTTCA), Suzhou, China, 12–14 November 2021; pp. 710–713. [CrossRef]
54. Mehdipour, I.; Khayat, K.H. Effect of Particle-Size Distribution and Specific Surface Area of Different Binder Systems on Packing Density and Flow Characteristics of Cement Paste. *Cem. Concr. Compos.* **2017**, *78*, 120–131. [CrossRef]
55. Degefa, A.B.; Jeon, G.; Choi, S.; Bak, J.; Park, S.; Yoon, H.; Park, S. Data-Driven Insights into Controlling the Reactivity of Supplementary Cementitious Materials in Hydrated Cement. *Int. J. Concr. Struct. Mater.* **2024**, *18*, 39. [CrossRef]
56. Velichko, E.G.; Vatin, N.I. Reactive Powder Concrete Microstructure and Particle Packing. *Materials* **2022**, *15*, 2220. [CrossRef] [PubMed]
57. Telesca, A.; Matschei, T.; Marroccoli, M. Study of Eco-Friendly Belite-Calcium Sulfoaluminate Cements Obtained from Special Wastes. *Appl. Sci.* **2020**, *10*, 8650. [CrossRef]
58. *EN197–1:2011*; Cement—Part 1: Composition, Specifications and Conformity Criteria for Common Cements. British Standards Institution: London, UK, 2011.
59. *EN450-1, 2012*; Fly Ash for Concrete—Part 1: Definition, Specifications and Conformity Criteria. Comite Europeen de Normalisation: Brussels, Belgium, 2012.
60. Sequeira, L.; Cantero, B.; Bravo, M.; de Brito, J.; Medina, C. The Influence of Recycled Cement, Fly Ash, and Magnesium Oxide on the Mechanical Performance of Sustainable Cementitious Materials. *Materials* **2023**, *16*, 2760. [CrossRef] [PubMed]
61. Ye, Y.; Liu, Y.; Shi, T.; Hu, Z.; Zhong, L.; Wang, H.; Chen, Y. Effect of Nano-Magnesium Oxide on the Expansion Performance and Hydration Process of Cement-Based Materials. *Materials* **2021**, *14*, 3766. [CrossRef] [PubMed]
62. Mahmood, A.; Kaish, A.B.M.A.; Ab Gulam, N.F.B.; Raman, S.N.; Jamil, M.; Hamid, R. Effects of MgO-Based Expansive Agent on the Characteristics of Expansive Concrete. *Eng. Proc.* **2021**, *11*, 14. [CrossRef]
63. *EN206:2013+A2:2021*; Concrete—Specification, Performance, Production and Conformity. British Standards Institution: London, UK, 2021.
64. Diamantopoulos, G.; Katsiotis, M.; Fardis, M.; Karatasios, I.; Alhassan, S.; Karagianni, M.; Papavassiliou, G.; Hassan, J. The Role of Titanium Dioxide on the Hydration of Portland Cement: A Combined NMR and Ultrasonic Study. *Molecules* **2020**, *25*, 5364. [CrossRef] [PubMed]
65. Rudnick, R.L.; Gao, S. The Composition of the Continental Crust. In *Treatise on Geochemistry*; Holland, H.D., Turekian, K.K., Eds.; Elsevier: Amsterdam, The Netherlands, 2003; Volume 3, pp. 1–64. [CrossRef]
66. Horckmans, L.; Möckel, R.; Nielsen, P.; Kukurugya, F.; Vanhoof, C.; Morillon, A.; Algermissen, D. Multi-Analytical Characterization of Slags to Determine the Chromium Concentration for a Possible Re-Extraction. *Minerals* **2019**, *9*, 646. [CrossRef]
67. Loncnar, M.; Mladenovic, A.; Zupancic, M.; Bukovec, P. Comparison of the mineralogy and microstructure of EAF stainless steel slags with referense to the cooling treatment. *J Min Met. Sect. B Met.* **2017**, *53*, 19–29. [CrossRef]
68. Radusinović, S.; Papadopoulos, A. The Potential for REE and Associated Critical Metals in Karstic Bauxites and Bauxite Residue of Montenegro. *Minerals* **2021**, *11*, 975. [CrossRef]
69. Liu, X.; Zhang, N. Utilization of red mud in cement production: A review. *Waste Manag. Res. J. Sustain. Circ. Econ.* **2011**, *29*, 1053–1063. [CrossRef]
70. Cherian, C.; Siddiqua, S. Pulp and Paper Mill Fly Ash: A Review. *Sustainability* **2019**, *11*, 4394. [CrossRef]
71. Pavšič, P.; Mladenovič, A.; Mauko, A.; Kramar, S.; Dolenec, M.; Vončina, E.; Vrtač, K.P.; Bukovec, P. Sewage sludge/biomass ash based products for sustainable construction. *J. Clean. Prod.* **2014**, *67*, 117–124. [CrossRef]
72. Altıkulaç, A.; Kurnaz, A.; Turhan, S.; Kutucu, M. Natural Radionuclides in Bottled Mineral Waters Consumed in Turkey and Their Contribution to Radiation Dose. *ACS Omega* **2022**, *7*, 34428–34435. [CrossRef] [PubMed]
73. Markic, M. Process innovation: A precondition for business excellence. *Int. J. Innov. Learn.* **2006**, *3*, 455–467. [CrossRef]
74. Stephan, D.; Mallmann, R.; Knöfel, D.; Härdtl, R. High intakes of Cr, Ni, and Zn in clinker. Part 1: Influence on burning process and formation of phases. *Cem. Concr. Res.* **1999**, *29*, 1949–1957. [CrossRef]
75. Bhatty, I.J. Effect of Minor Elements on Clinker and Cement. Available online: https://www.cementequipment.org/cement-chemistry/effect-minor-elements-clinker-cement-javed-bhatty/ (accessed on 26 November 2024).
76. Stephan, D.; Mallmann, R.; Knöfel, D.; Härdtl, R. High intakes of Cr, Ni, and Zn in clinker. Part 2: Influence on the hydration properties. *Cem. Concr. Res.* **1999**, *29*, 1959–1967. [CrossRef]
77. Gineys, N.; Aouad, G.; Sorrentino, F.; Damidot, D. Incorporation of Trace Elements in Portland Cement Clinker: Threshold limits for Cu, Ni, Sn or Zn. *Cem. Concr. Res.* **2011**, *41*, 1177–1184. [CrossRef]
78. Wang, J.; Chen, F.; Yu, R.; Fan, D.; Zhang, T. Effect of heavy metal (Mn, Pb and Cr) on the properties and hydration in low water/binder cement-based composites (LW/B-CC). *Constr. Build. Mater.* **2023**, *386*, 131567. [CrossRef]
79. Liu, M.; Zhao, Y.; Yu, Z.; Cao, Z. Binding of Cu(II) and Zn(II) in Portland cement immobilization systems: Effect of C-A-S-H composition. *Cem. Concr. Compos.* **2022**, *131*, 104602. [CrossRef]
80. Han, W.; Han, F.; Zhang, K. Influence of Copper and Zinc Tailing Powder on the Hydration of Composite Cementitious Materials. *Materials* **2022**, *15*, 5612. [CrossRef]

81. Tomašić, N.; Čobić, A.; Beđeković, M.; Miko, S.; Ilijanić, N.; Gizdavec, N.; Matošević, M. Rare Earth Elements Enrichment in the Upper Eocene Tošići-Dujići Bauxite Deposit, Croatia, and Relation to REE Mineralogy, Parent Material and Weathering Pattern. *Minerals* **2021**, *11*, 1260. [CrossRef]
82. Tóth, T.M.; Schubert, F.; Raucsik, B.; Fintor, K. Mineralogical and Geochemical Constraints of the REE Accumulation in the Almásfüzitő Red Mud Depository in Northwest Hungary. *Appl. Sci.* **2019**, *9*, 3654. [CrossRef]
83. Li, W.; Li, Z.; Wang, N.; Gu, H. Selective extraction of rare earth elements from red mud using oxalic and sulfuric acids. *J. Environ. Chem. Eng.* **2022**, *10*, 108650. [CrossRef]
84. Borra, C.R.; Mermans, J.; Blanpain, B.; Pontikes, Y.; Binnemans, K.; Van Gerven, T. Selective recovery of rare earths from bauxite residue by combination of sulfation, roasting and leaching. *Miner. Eng.* **2016**, *92*, 151–159. [CrossRef]
85. Bouchenafa, O.; Hamzaoui, R.; Florence, C.; Mansoutre, S. Cement and Clinker Production by Indirect Mechanosynthesis Process. *Constr. Mater.* **2022**, *2*, 200–216. [CrossRef]
86. Cement Environmental Directive, 2020, Holcim. Available online: https://www.holcim.com/sites/holcim/files/documents/lafargeholcim_cement_environmental_directive_revison_2020.png (accessed on 26 November 2024).
87. Iliushchenko, V.; Kalina, L.; Sedlačík, M.; Cába, V.; Másilko, J.; Novotný, R. Effect of Alkali Salts on the Hydration Process of Belite Clinker. *Materials* **2022**, *15*, 3424. [CrossRef]
88. Yan, M.; Deng, M.; Wang, C.; Chen, Z. Effect of Particle Size of Periclase on the Periclase Hydration and Expansion of Low-Heat Portland Cement Pastes. *Adv. Mater. Sci. Eng.* **2018**, *2018*, 1307185. [CrossRef]
89. American Coal Ash Association. Fly Ash Facts for Highway Engineers. 2003. Available online: https://www.fhwa.dot.gov/pavement/recycling/fach03.cfm (accessed on 26 November 2024).
90. Saint-Astier. Mineralogy of Binders & the Effects of Free Lime Content and Cement Addition in Lime Mortars. Available online: https://www.stastier.co.uk/mineralogy-of-binders-and-the-effects-of-free-lime-content-and-cement-addition-in-mortars/ (accessed on 26 November 2024).
91. Nuccetelli, C.; Pontikes, Y.; Leonardi, F.; Trevisi, R. New perspectives and issues arising from the introduction of (NORM) residues in building materials: Acritical assessment on the radiological behaviour. *Constr. Build. Mater.* **2015**, *82*, 323–331. [CrossRef]
92. European Commission. *Enhanced Radioactivity of Building Materials*; European Commission: Luxembourg, 1999. Available online: https://op.europa.eu/en/publication-detail/-/publication/7de4275f-d1a2-11e6-ad7c-01aa75ed71a1 (accessed on 17 April 2024).
93. Todorović, N.; Bikit, I.; Krmar, M.; Mrdja, D.; Hansman, J.; Nikolov, J.; Todorovic, S.; Forkapic, S.; Jovanovic, N.; Bikit, K.; et al. Assessment of radiological significance of building materials and residues. *Rom. J. Phys.* **2017**, *62*, 817.

materials

MDPI

Review

Recovery of Phosphorus and Metals from the Ash of Sewage Sludge, Municipal Solid Waste, or Wood Biomass: A Review and Proposals for Further Use

Sara Tominc [1,*], Vilma Ducman [1], Wolfgang Wisniewski [1], Tero Luukkonen [2], Gunvor M. Kirkelund [3] and Lisbeth M. Ottosen [3]

[1] Laboratory for Cements, Mortars and Ceramics, The Department of Materials, Slovenian National Building and Civil Engineering Institute (ZAG), Dimičeva ulica 12, 1000 Ljubljana, Slovenia; vilma.ducman@zag.si (V.D.); wolfgang.wisniewski@zag.si (W.W.)

[2] Faculty of Technology, Fibre and Particle Engineering, University of Oulu, P.O. Box 8000, 90570 Oulu, Finland; tero.luukkonen@oulu.fi

[3] Department of Environmental and Resource Engineering, Technical University of Denmark (DTU), Brovej, 2800 Lyngby, Denmark; gunki@dtu.dk (G.M.K.); limo@dtu.dk (L.M.O.)

* Correspondence: sara.tominc@zag.si; Tel.: +386-70-480-223

Abstract: This review provides an overview of methods to extract valuable resources from the ash fractions of sewage sludge, municipal solid waste, and wood biomass combustion. The resources addressed here include critical raw materials, such as phosphorus, base and precious metals, and rare earth elements for which it is increasingly important to tap into secondary sources in addition to the mining of primary raw materials. The extraction technologies prioritized in this review are based on recycled acids or excess renewable energy to achieve an optimum environmental profile for the extracted resources and provide benefits in the form of local industrial symbioses. The extraction methods cover all scarce and valuable chemical elements contained in the ashes above certain concentration limits. Another important part of this review is defining potential applications for the mineral residues remaining after extraction. Therefore, the aim of this review is to combine the knowledge of resource extraction technology from ashes with possible applications of mineral residues in construction and related sectors to fully close material cycle loops.

Keywords: critical raw materials; extraction; sewage sludge ash; municipal solid waste incineration ash; wood biomass ash

1. Introduction

The European Commission (EC) has recognized the necessity to recover critical raw materials (CRMs), not only from primary but also from secondary sources as much as possible. In 2015, the document "Closing the loop—An EU action plan for the Circular Economy" expressed the EC's aim to support the Raw Materials Information System, which would provide data on secondary raw materials (SRMs). SRMs can be sources used instead of primary raw materials. SRMs can also be a source for CRMs, which are of high economic importance due to their expected scarcity in the EU market or a high risk of supply disruption. The EU thus wishes to promote recycling these materials, especially the recovery of CRMs, as part of the move towards the circular economy [1]. This review focuses on various incineration ashes as SRMs and recovering CRMs and other resources from them.

One CRM is phosphorus (P), which is an irreplaceable resource and an essential nutrient for the growth of organisms [2,3]. Most P is currently extracted from phosphate rocks, which are the primary and non-renewable sources occurring in a limited number of deposits worldwide [4]. Most of the world's phosphate rock deposits are located in Morocco, while the rest are found in China, the United States, South Africa, and Jordan [5].

These global P resources will be depleted within a few decades, which is why the EC has added P and phosphate rock to the list of CRMs shown in Figure 1 [6]. Solutions for a P recovery from various secondary resources such as ashes are hence required. Optimized use of P in agriculture and soil stabilization to prevent erosion is additionally encouraged. Rare earth elements (REEs) are another limited resource, and the EC placed them in the CRMs list due to the increasing demand for them in current industrial production (e.g., permanent magnets required for electric motors or wind energy). They are sorted into the groups of light and heavy REEs (LREEs and HREEs, respectively) [6,7].

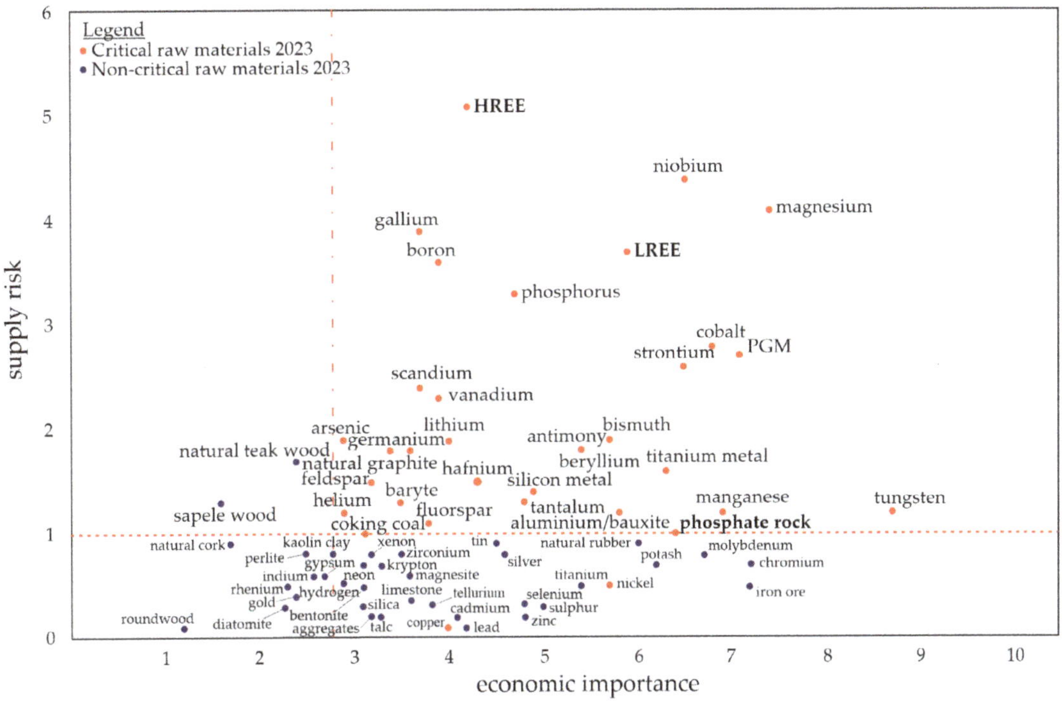

Figure 1. Raw materials (CRMs list from 2023) in relation to their economic importance and supply risk, redrawn based on an EU report [6]. The "critical area" is defined by a supply risk ≥ 1 and an economic importance ≥ 2.8.

Even apart from CRMs, there is a fundamental dependence on metals in different industries. Non-ferrous base metals, such as Cu, Zn, and Pb, are some of the most used metals worldwide, only exceeded by Fe and Al. All these elements are necessary for our current society and infrastructure. Projections predict an increase in the demand for Cu by 140%, Zn by 81%, and Pb by 46% compared to the 2010 demand until 2050 [8]. By the end of the 21st century, the projected demand is even higher for Cu and Zn (330% for Cu and 130% for Zn) [8]. At the same time, the expected depletion years for these elements (the year when the cumulative primary production exceeds the reserves) can be reached before 2025–2038 without the peak year of primary production being reached [8]. Consequently, used ore grades become lower, which typically leads to an increased energy demand for extraction, e.g., for fossil fuels, resulting in increasing greenhouse gas emissions for constant production [9,10]. This calls for developing new economically viable mining methods and extraction strategies applicable to SRMs. The use of elements in products usually dilutes their concentration [11], whereas incinerating end-of-life products causes the accumulation of elements in incineration residues, which can hence be an important source for improving the circularity of elements otherwise deposited in landfills.

Table 1 compares the concentrations of P and selected heavy metals reported in various ashes with typical concentrations in mined ores. Sewage sludge ash (SSA) is a rich source of P, while municipal solid waste incineration fly or bottom ashes (MSWI-FA, MWSI-BA) are rich in metals, especially Zn, Pb, and Cu. Wood biomass fly or bottom ashes (WB-FA, WB-BA), on the other hand, do not have a very high P content but contain essential micronutrients for plants as well as Na [12]. The composition of the biomass varies depending on the combusted biomass type [12–15]. REEs are generally not enriched in these ashes in comparison to the concentrations found in the Earth's crust [7]; combined REE concentrations of 88–124 mg/kg were found in MSWI-BA and averaged 54 mg/kg in MSWI-FA sampled in Italy [16]. Biomass ashes contain even lower combined REE concentrations of, e.g., ~27 mg/kg [17]. Recovering REEs from such ashes may still become economical as they can be co-extracted with the other resources in Table 1, and their demand is expected to reach a ~4.4% annual growth rate globally by 2026 [18] and rise by 2600% over the next 25 years [19]. Despite extensive, mostly lab-scale research efforts on REE recycling, less than 1% were recycled in 2011 due to inefficient collection, technological problems, and a lack of incentives [20]. Achieving a circular economy and a truly sustainable society must include the recovery of REEs from SRMs with low concentrations [7].

Table 1. Minimum and maximum contents of P and selected heavy metals in treated SSA, MSWI-FA, MSWI-BA, and wood biomass fly and bottom ashes (WB-FA, WB-BA) compared to typical ore concentrations.

Element	SSA [2,4,21–25]	MSWI-FA [15,26,27]	MSWI-BA [26]	WB-FA [12,28,29]	WB-BA [28]	Typical Ore Concentration [30,31]
P (g/kg) *	35–99	4–5	-	10–23	8–17	110–160
Zn (mg/kg)	895–2823	9000–70,000	610–7800	446–2274	74–234	50,000–150,000
Pb (mg/kg)	70–460	5300–26,000	100–13,700	11–177	5–80	300,000–400,000
Cu (mg/kg)	423–839	600–3200	190–8200	89–161	65–111	5000–20,000
Cr (mg/kg)	78–460	140–1100	23–3200	18–101	25–70	310,000
Cd (mg/kg)	4–126	50–450	0.3–70	7–16	0.1–0.5	1000–10,000

* Calculated based on the P_2O_5 content.

Legislation can be a powerful tool to promote recycling. Following the legislative developments in Switzerland and Germany, Austria has also adopted legislation for a mandatory P recovery from SSA. In 2022, the Swiss Federal Council published an Ordinance on the Avoidance and the Disposal of Waste (SR 814.600), where P must be recovered and recycled from SSA (Section 3, Art. 15, in force since 1st Jan. of 2019) and metals recovered from MSWI filter ash (Section 3, Art. 32) [32]. There are also requirements and limit values for using waste as a raw material in this Ordinance (Annex 4, in force since 1 January 2022) [32]. Denmark has the Statutory Order No. 1672/2016 on the use of residual products, where MSWI-BA is on the list to replace primary raw materials [26]. However, the ash must meet quantitative criteria regarding the content and leachability of certain inorganic substances [26]. The Statutory Order No. 732/2019 on the application of biomass ash in agriculture additionally determines the extent to which biomass ash can be requested as a replacement for commonly used fertilizers or soil improvers (i.e., which types of biomass ash are allowed, the limit values for the content of heavy metals, the maximum amount of allowed ash and its reactivity) [26]. The Finnish legislation on fertilizers has also been modernized, as a new Fertilizer Act 711/2022 came into force in July 2022 [33].

2. Recovery of Resources from Selected Ashes

The extraction of specific components from ashes is a wide field as the ash properties vary depending on the fuel type and composition, ash fraction, and power plant processes, and various components are of interest for extraction. The ashes selected for this review are rich in siliceous phases, macronutrients such as P, K, Ca, Mg, or S necessary for plant growth [34], and contain trace amounts of Fe, Zn, and/or Cu. The content of potentially toxic elements is also very important, but some of these are also necessary micronutrients for plant growth. The main elements of interest in this review are P and the potentially toxic elements Zn, Pb, Cd, Cr, and Cu. They are usually extracted using various organic/inorganic acids, chelating agents, or basic solvents, as well as electrochemical methods such as electrodialytic separation (EDS).

2.1. Overview of Extraction Technologies

Various techniques, such as wet extraction, thermochemical, and electrochemical methods, have been developed to extract or recover metals from different ashes [35,36]. This overview also covers processes at the technology readiness level (TRL) 7 (see Section 2.2) using wet extraction methods. Wet chemical extraction is the most widely used method for extracting P from various ashes due to its high recovery rate, low cost, and procedure simplicity. Choosing the right extractant is very important. Common examples are inorganic acids such as sulfuric acid (H_2SO_4) [2,4,22–25,37–39], nitric acid (HNO_3) [4,23,25,37,39], and hydrochloric acid (HCl) [4,15,40,41]; organic acids and chelating agents such as citric acid (CA) [25,37–39], oxalic acid (OA) [4,22,25,37–39], lactic acid (LA) [37], ethylenediaminetetraacetic acid (EDTA) [25,38,39], and ethylenediamine tetra (methylene phosphonic acid) (EDTMP) [25,38,39]; and bases such as NaOH [4]. Novel chelating agents for ashes could also be bisphosphonates [42].

The most commonly used inorganic acid and the cheapest extractant on today's market is H_2SO_4 [23]. Its main advantages are easy transportation due to its low volatility, the possibility of concentrating up to 98%, and ensuring less co-dissolution of heavy metals, especially Pb [2,4]. Other inorganic acids, such as HNO_3 [4,23,25,37,39], HCl [4,15,40,41], and H_3PO_4 [5], have also been used. HCl may facilitate the occurrence of unwanted complexation reactions [2], and H_3PO_4 is comparably expensive [5]. Organic acids are usually chosen in research for their reduction properties (especially OA, which is the strongest naturally occurring organic acid) and for their environmentally friendly production (CA and LA) [22,25,37–39]. OA is the most efficient P extractant among the organic acids as it combines a high P extraction efficiency with a relatively low co-extraction of heavy metals; however, H_2SO_4 has an economic advantage over OA due to the lower costs for optimal P extraction [4,25].

Chelating agents have a marginal effect on the morphology and particle size distribution of ashes and are less effective for P recovery compared to inorganic and organic acid extractants due to their high affinity to metal ions, resulting in partial dissolution of P [34,39]. EDTA performed better than EDTMP for trace elements such as Zn, Pb, or Cd, so EDTA could be used ahead of the P extraction to remove significant quantities of metals without leaching P [38]. Fang et al. also published a study where a combination of EDTA and H_2SO_4 was used but was ineffective for P extraction [43]. The use of Cyanex, a highly stable P-based chelating extractant, is effective for the extraction of heavy metals (Zn, Pb, and Cd) from leachates after wet extraction [27,44]. However, the selective separation of REEs requires very specific kinds of chelating agents, where bisphosphonates could be a promising option [42]. They contain a carbon center with two phosphonate groups ($PO(OH)_2$) and two other substituents (e.g., a hydroxyl group and a carbon chain with a primary amine group) [45]. Bisphosphonates have been immobilized on nanoporous silicon and then applied to selectively recover Sc from a highly complex ore sample leaching solution [42]; a similar approach could be applied to ashes. The desorption of Sc from bisphosphonates can be conducted by using acids such as H_3PO_4 and H_2SO_4 [42]. Bisphosphonates can also be used for a relatively selective recovery of other REEs, but information remains scarce [46].

Alkali-metal bases such as NaOH dissolve almost no heavy metals, mainly due to the high pH of around 13 at the end of the extraction procedure. They are also ineffective for P extraction because Ca-phosphates show poor solubility in alkaline environments (especially when the molar P/Ca ratio is lower than one), while Al- and Fe-phosphates are highly soluble in such media [4,24].

Optimal process conditions require combining the highest P extraction efficiency, the lowest possible co-extraction of heavy metals, and the lowest possible operating costs. Additionally, variables such as the extractant type and its concentration, the contact time (optimally 2 h [2,22,24,38,39]), the liquid/solid ratio (optimally 20:1 [2,23,24,38]), the extraction temperature and ash composition significantly affect the P extraction efficiency [2,22,24]. Longer extraction times (e.g., one week compared to 2 h) result in a lower P extraction efficiency and more heavy metal leaching [23]. It is also necessary to consider the variability of certain elemental contents, e.g., the P content in SSA can be partially attributed to differences in wastewater treatment systems and incineration conditions [47]. The sampling period also appears to influence its content, as, e.g., a lower P content has been measured in the summer months while higher contents were measured in February and March [48]. This can be explained by the different food habits and leisure activities of people in different seasons [48]. The recovery of P by wet chemical extraction can be effectively applied to different types of SSA, as they contain higher amounts of P [2,4,22–25,37–39]. Ashes from wood biomass and MSWI are not as rich in P as SSA, but they are rich in Zn, which is also important to recover [40,44,49–51].

An overview of the extractants used to extract P, Zn, Pb, Cu, Cr, and/or Cd from the selected ashes is presented in Table 2, along with the respective achieved extraction rates. In the case of P, the highest extraction rate of 100% was achieved using H_2SO_4 [23], HNO_3 [23,37], and OA [22], while the most effective extractant for Zn was HCl, reaching 86% [41]. The highest reported extraction rate in the analyzed literature for Cr was 58% [39], 74% for Cu [38], 62% for Pb [40], and 97% for Cd [40]. Although chemical extraction achieves high efficiencies, it requires further purification and the treatment of insoluble acid residues [22,34,38,52,53]. It also often requires undesirably large amounts of acids, encouraging researchers to develop alternative methods [35]. One alternative for achieving high extraction rates of Zn (around 90%), Pb, and Cd from ash is the thermochemical method, but concerns regarding its operating costs, high energy input, and equipment lifetimes have been voiced [52–54]. KCl, $MgCl_2$, and $CaCl_2$ are often added in the thermochemical process, where high concentrations of chlorine compounds can be extremely corrosive [53,55,56]. Increasing the treatment temperature led to higher Pb and Zn removal rates [55], and above 1400 °C, the thermal removal of heavy metals also enabled the separation of Fe, increasing the bioavailability of P in the ash [53]. The most promising one-step extraction method is EDS, whose main advantage over other techniques is the ability to separate P from the remaining waste and remove heavy metals from ash in one process [3,57–60]. Here, electromigration separates the component by transporting P towards the anode and heavy metals mainly towards the cathode, a very important aspect considering mixed component wastes [57]. After the treatment, P is recovered from the anolyte by filtration to separate the liquid from the remaining solids, and the heavy metals are solubilized in the catholyte [59]. However, the EDS process is time consuming [57,58,60,61], and the operating costs are relatively high [52,57,62].

Table 2. Summary of the extractants used to extract P, Zn, Pb, Cu, Cr, and/or Cd from selected ashes (marked with the x) with corresponding extraction rates (in %). The analyzed literature does not contain such extraction rates for WBA.

Extractant	Conc. (mol/L)	P	Zn	Pb	Cu	Cr	Cd	SSA	MSWI	WBA	Reference
H_2SO_4	0.05	>95						x			[22]
	0.1	88.3						x			[24]
	0.19	100						x			[23]
	0.19	~88						x			[23]
	0.2	92						x			[2]
	0.2	94						x			[38]
	0.25	93	~36	~1	~38	~5	~28	x			[4]
	0.4	96.4						x			[37]
	0.5	>70	~42	38.4	~40	57.7	50	x			[39]
	0.5	74	~42		~40			x			[25]
HNO_3	0.3	89	~32	~24	~36	~5	~27	x			[4]
	0.4	100						x			[37]
	0.5	>70	~36	40	~38	~52	~6	x			[39]
	0.5	~71	~36					x			[25]
	1.5	~80	16	56				x			[23]
	1.5	100	71	47				x			[23]
HCl	0.3	98.8	~32	~30	40	~5	~28	x			[4]
	0.5	>95						x			[41]
	1.0		75	1			71		x		[40]
	1.0		~58	~1			40		x		[15]
HCl + H_2O_2	1.0 + 9.8		68	62			97		x		[40]
Citric acid ($C_6H_8O_7$)	0.2	~80						x			[38]
	0.4	59.3						x			[37]
	0.5	>70	~23	13.3	~16	~25	~7	x			[39]
	0.5	72	~23		~16			x			[25]
Oxalic acid ($C_2H_2O_4$)	0.05	100						x			[22]
	0.2	>95						x			[38]
	0.55	95.4	37	~1	37	~8	~13	x			[4]
	0.4	100						x			[37]
	0.5	>70	56.9	4	65.8	~53	~13	x			[39]
	0.5	74	56.9		65.8			x			[25]
Lactic acid ($C_3H_6O_3$)	0.4	28.4						x			[37]
EDTA ($C_{10}H_{16}N_2O_8$)	0.02	~20						x			[38]
	0.05	<30	~14	37	~5	~42	~6	x			[39]
	0.05	~24	~14		~5			x			[25]
	0.05	~40						x			[38]
EDTMP ($C_6H_{20}N_2O_{12}P_4$)	0.05	<30	~10	~22	~9	~26	~6	x			[39]
	0.05	~13	10		~9			x			[25]
	0.05	~25						x			[38]
NaOH	0.5	40	~3	~3	~2	~2	~4	x			[4]

2.2. *Methods Applied on an Industrial Scale*

Prototype plants are already implemented or under construction (i.e., at TRL7) for the wet extraction of P from SSA (EasyMining, Uppsala, Sweden), salts and Zn from MSWI ashes (Stena Recycling, Gothenburg, Denmark), Zn from MSWI ashes (RENOVA, Göteborg, Sweden) and the full-scale commercially available process FLUWA/FLUREC operating in Switzerland for the recovery of metals. Additional plants recovering P from SSA using H_3PO_4 are operating in Germany, Switzerland, and Japan. These wet extraction processes at the high TRL level are based on using waste acid from nearby industries or from the wet scrubber of the incineration plant itself. A wet extraction method using HCl and lime for recovering commercial P, Fe, and Al products called Ash2Phos was developed by EasyMining, Sweden. The process has recovery rates of 90–95% for P, 60–80% for Al, and 10–20% for Fe from SSA [63]. Simultaneously, the heavy metal content in connection to P is reduced by at least 96%, making it a very pure and clean fertilizer product [63]. SSA is dissolved in HCl at 40 °C, and the P, Fe, and Al are separated as pure $Ca_3(PO_4)_2$, $FeCl_3$, and $NaAlO_2$ [63]. The separation process is based on chemical precipitation steps in a unique combination, and the solution is later neutralized to remove heavy metals. CaO is used during the precipitation steps and for neutralization. The produced phosphorus–calcium-rich product ($Ca_5(PO_4)_3OH$) contains a minimum of 16.5% P and 35% Ca and can be used as raw material for feeds or fertilizer applications [63]. The Fe and Al products can be reused in wastewater treatment plants [63]. After the treatment, non-dissolved SSA is filtered, washed, neutralized, and called "silicate sand" (48.3% SiO_2, 22.9% Fe_2O_3, 7.2% Al_2O_3), which is potentially useable as a partial cement replacement in mortars after milling [64]. A full-scale plant able to annually treat 30,000 tons of SSA is under construction in Sweden, and plants in Germany are also under development [63].

The recovery of metals from MSWI-FA is achieved by the FLUWA/FLUREC processes developed in Switzerland by AIK Technik AG, as well as by the HaloSep process in Denmark developed by Stena Recycling. In 2018, >60% of the MSWI-FA in Switzerland was treated with the FLUWA process [39], which is based on wet extraction by adding acidic (HCl) and neutral (NaCl) waste scrub water to MSWI-FA where 60–80% Zn, 80–95% Cd, 50–85% Pb, 50–85% Cu can be extracted [40,65]. The metal-enriched filtrate obtained after FLUWA needs to be further processed to recover the metals, either by leading the filtrate to a wastewater treatment plant to precipitate a metal hydroxide sludge that can be recovered in smelting plants or by the FLUREC process, which allows a high-purity Zn recovery. Cd, Pb, and Cu are separated by reductive separation (cementation) using Zn powder as a reducing agent [65]. This cement, with a high Pb content of 50–70%, can be sent directly to a lead smelter where metals are recovered in the Pb production process [65]. Zn is removed from the remaining liquid by solvent extraction, followed by electrowinning to recover high-grade Zn (>99.99% Zn), which can be sold [66]. The remaining FA particles (filter cake) are currently landfilled.

The HaloSep process is another wet extraction method using a HCl scrubber liquid and MSWI-FA, which produces brine and a neutralized and washed FA. The resulting residues from the process are a stabilized FA, a metal fraction, and a brine solution. The metals are precipitated from the brine into a filter cake containing up to 38–40% Zn, which can be recovered at smelters. The remaining brine contains salt products ($CaCl_2$, NaCl, KCl) useful for industrial applications. The treated FA complies with the European leaching limits for acceptance in landfills [65] but can also be used in construction [67]. A full-scale HaloSep plant is operating at the incineration plant Vestforbrænding in Copenhagen, Denmark, and plants in other countries are under exploration [67].

The RENOVA process also uses a HCl scrubber liquid for Zn leaching but differs from HaloSep by using an acidic pH in the process and re-incineration of the leached ash to destroy dioxins. Up to 70% Zn was leached in pilot scale studies, and NaOH-precipitated filter cake contained 80% $Zn(OH)_2$ [68]. Re-incineration studies showed that more than 90% of the leached ash was converted into bottom ash [68]. There are plans to build a demonstration plant in Sweden [69].

Additional plants recovering P from SSA using H_3PO_4 are, e.g., a sewage sludge incineration plant in Werdohl, Germany, which uses the Remondis TetraPhas process [52]. It consists of leaching P from SSA by H_3PO_4 and purifying the P-concentrated acid leachate, allowing an 80% P extraction. The product, called RePacid, mainly contains H_3PO_4 and can be directly used by the industry [52]. Another solvent-extraction process called Phos4life was designed in the canton of Zürich, Switzerland, where the main product is technical H_3PO_4 (74%). Here, P is extracted from SSA by H_2SO_4, and more than 95% of it can be recovered from SSA in the form of H_3PO_4 [52]. Another well-known P production company is Nippon Phosphoric Acid Co., Ltd. (NPA) in Achi, Japan, where the H_3PO_4 is also obtained through a wet extraction process followed by filtration and purification. Gypsum ($CaSO_4{\cdot}2H_2O$), with possible applications in cement, plasterboard, or soil improvement, is a by-product of this process [52].

2.3. Economic Assessment of Alternative Extraction Methods

In addition to achieving an optimal environmental profile for the extracted resources, the economic costs of recovering secondary raw materials are also of great importance, as both environmental and economic costs influence their acceptance in society [35]. Although high percentages of P extraction can be achieved by acid wet extraction of P, the acid demand is very high [70]. An example of acid demand is given by Donatello et al., where an optimized procedure showed that 368 kg of 98% H_2SO_4 was required for 80–100% P extraction from 1 ton of SSA [71]. Handling and transporting such large quantities of acid is not easy in all urban environments [70]. Moreover, the simultaneous extraction of heavy metals and P [71] requires a second separation step to support the European Green Deal, which promotes the recycling of limited resources while emphasizing the goal of zero pollution [70]. For this purpose, the development of alternative technologies, such as thermochemical methods, is supported. Products recovered by thermochemical methods are still ash, and the separated heavy metals represent only a small part of the ash, which is suitable for transport [72]. However, the energy consumption and high capital costs of the alternative methods still need to be optimized compared to the traditional wet chemical method [3,34,36]. In 2016, Egle et al. reported that the cost of recovering P from SSA using the wet chemical method is about 5–6 EUR/kg of P, whereas using the thermochemical method, the cost is about 2–3 EUR/kg of P [73]. However, the thermochemical method is closely related to the energy price, and with the current increasing energy prices, the cost of the thermochemical method has certainly increased [72]. Therefore, alternative extraction methods, such as microwave-assisted acid extraction, can reduce the total cost by up to 76% for MSWI-FA and up to 52% for MSWI-FA compared to traditional metal extraction by heating [35]. This cost analysis considered the cost of acid/chemicals, energy consumption, miscellaneous costs, and other laboratory costs for processing a given amount of ash [35]. In addition, alternative technologies, such as electrochemical methods, are still in the development and optimization phase, so there are not many analyses of their economic performance.

3. Potential Uses of Solid Residues of Ashes

Waste incineration is steadily increasing in Europe, but there are environmental concerns about the solid residues that require pre-treatment and are usually landfilled [74]. There are possibilities to use as-received or pre-treated ashes in agriculture, soil stabilization, and the building sector as the following:

- Supplementary cementitious materials (SCMs);
- Precursors for alkali-activated materials (AAMs);
- Artificial fillers or fine aggregates;
- Additives in clay-based materials;
- Precursors for carbonated products.

The use of waste ash in construction materials has attracted many researchers in recent years [74]. However, a good understanding of their chemical, physical, and microstructural characteristics is necessary for their full-scale use [75]. The question of how to keep the ash characteristics constant when heterogeneous materials such as sewage sludge, MSW, or varying types of biomasses are incinerated is especially important for their large-scale utilization.

AAMs/geopolymers are synthetic materials obtained by alkaline activation of Si- and Al-rich materials and specific industrial wastes [76]. AAMs are alternative cementitious or ceramic-like binders used as alternative construction materials and for the solidification/stabilization of various waste streams [77]. As cement production is among those human activities most generating CO_2 emissions, sustainable development of the building industry requires three approaches: using renewable energy, using recycled products, and replacing cement [78]. Every 600 kg of cement causes about 400 kg of CO_2 to be released into the atmosphere [79–83]. Therefore, the potential applications of different ashes as SCMs have been studied for decades [84] but recently with a focus on bio-based ashes [80].

Another very promising research topic in the cement and concrete industry is carbonation utilizing CO_2 sequestration: a low-tech approach to the carbon capture and storage process that mainly involves the reaction of CO_2 with Ca-containing materials to form Ca carbonates [85–87]. Potential sources from waste streams are ashes containing a certain amount of Ca and Mg compounds, especially WBA [88–90].

3.1. Potential Use of SSA

Sewage sludge is the most common and continuously generated by-product of wastewater treatment, containing the 2nd highest amount of P after bone meal [3,5,47]. It has a great potential for P recovery after an appropriate thermal treatment [2,4,22–25,37,52,53]. Sewage sludge has been directly used as agricultural fertilizer for decades, but its limitations are increasing all over the world due to the high contents of heavy metals, organic pollutants, and micro/nanoplastics. Hence, its incineration is considered to be the best way for disposal [53]. The incineration of sewage sludge at about 850 °C is widely used in the EU and is currently the most efficient method, reducing the volume by 90% (the mass by 70%) and removing organic pollutants and pathogens [47,52,59]. The resulting SSA contains 4–12 wt% of P, usually in the form of $AlPO_4$ and $Ca_3(PO_4)_2$, which are poorly bioavailable [26]. SSA also contains Fe and potentially toxic trace elements such as Zn, Pb, Ni, Cr, and Cd and is mostly landfilled [5,23,48]. Pre-treatment is required to prevent the loss of this potential P source and aims to increase the bioavailability of P and remove heavy metals, which often exceed the legal limits for fertilizer production, see Table 3 [24,48,74,91]. With, for example, innovative EDS, 80–90% of the P can be recovered while also achieving a low content of heavy metals [58–60]. A high concentration of CaO and SiO_2 in the SSA after P extraction is the main reason for using SSA as a building material component [5].

Table 3. Legal limits for trace elements in EU fertilizing products (in mg/kg), adapted from the EU regulation [86].

Element	Organic Fertilizer	Organo-Mineral Fertilizer	Inorganic Fertilizer
As	40	40	40
Cd	1.5	3	3
Cr	2	2	2
Cu	300	600	600
Hg	1	1	1
Ni	50	50	100
Pb	120	120	120
Zn	800	1500	1500

SSA is a material comparable to lightweight sand and is less dense than Portland cement [47]. It consists of porous particles with irregular shapes, which is not ideal for its classification as a potential cementitious material [47]. It should be noted that the extraction of P with H_2SO_4 produces $CaSO_4$, which negatively affects the cement properties [4]. Using OA as an extractant produces Ca oxalate, which does not have this negative effect [4]. SSA typically contains an elevated amount of about 14% Al_2O_3 compared to the ca. 5% in Portland cement, indicating a natural suitability for use in aerated concrete [47]. The high Al_2O_3 content in SSA may also benefit the chloride attack resistance in concrete applications due to the chloride binding capacity of amorphous Al_2O_3 [47]. SSA can be used as a possible cement replacement material, but it requires a pre-treatment due to the undesirable effects of the contained heavy metals and P recovery. It has the potential to replace cement in mortars [23,92] or partially replace clay in bricks [59]. Due to the small grain size, SSA is also suitable as a filler or fine aggregate component in mortar and concrete, where the effects on strength performance have been shown to be manageable for SSA contents up to 15 wt% [47,49,93]. Research on the reuse of SSA as an aluminosilicate precursor material for alkali activation/geopolymerization has also recently begun [77,94–96].

3.2. Potential Use of MSWI Ashes

MSW contains wood but also paper, plastic, glass, and textile scrap material, which cannot be degraded naturally. In the last few decades, the total mass of MSW has increased drastically due to rapid urbanization and an increased world population. This has encouraged many countries to properly dispose of this waste [74,97]. The so-called "green economy" has begun, encouraging waste reduction, reuse of materials through recycling or recovery, and supporting sustainability [97,98]. Several different treatments of MSW have been developed, as shown in Figure 2. The gases produced by the natural decomposition of MSW in landfills represent 18% of the energy production from biogas in the EU [97,99]. Nowadays, one effective and popular method is the incineration of MSW due to the volume reduction in MSW by 90% recovery of heat/energy. Two main residues are produced by incineration: around 80 wt% MSWI-BA and around 20 wt% MSWI-FA [27,51,97,100–102].

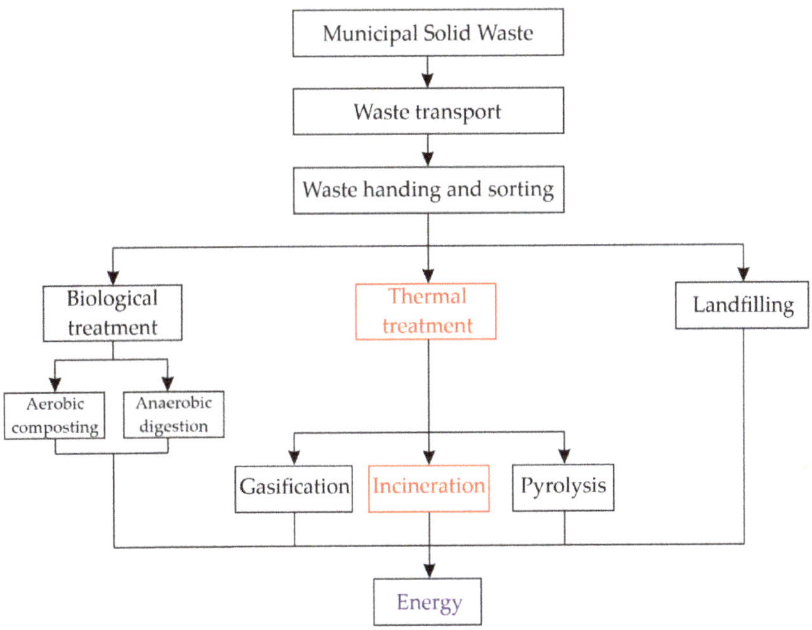

Figure 2. Different treatments for MSW management; redrawn based on [97].

MSWI-BA is classified as non-hazardous waste and mainly consists of amorphous SiO_2, Al_2O_3, and CaO; its exact composition varies from incineration plant to incineration plant and even from batch to batch within a single incineration plant [97,100]. MSWI-BA is commonly utilized in road construction [103] and can be an alternative lightweight aggregate [97] or an alternative material for cement production [104,105]. One of the advantages of using MSWI ash as cement raw material is the reduction in CO_2 emissions [104]. MSWI-FA is classified as a hazardous waste as it contains soluble salts, dioxins, and a significant amount of heavy metals such as Zn, Pb, and Cd [40]. Due to the presence of potentially leachable contaminants harmful to the environment and human health, landfill sites are becoming fewer, and the possibility of utilizing MSWI-FA has attracted many researchers [27,35,50,51,61,101,102,106]. Sekito et al. reported a 2-fold higher content of

Zn and Pb in MSWI-FA compared to MSWI-BA, while the content of Cd was even 13-fold higher [107]. Therefore, MSWI-FA must be pre-treated before further use, and much research has focused on how to extract and recover various metals from it. In MSWI-FA, the pH has a significant effect on the removal efficiency of heavy metals [50]. Many metals have a high solubility at low pH levels, so using strong acids as the extractant is necessary. As MSWI-FA is alkaline, alternative methods are desired to avoid the consumption of large acid volumes. A new microwave-assisted acid extraction method has recently been developed [35]. Significant advantages of this method are lower costs, shorter processing times, and better efficiency of metal extraction compared to conventional heating [35,108–110].

Electrodialytic (ED) treatment is another innovative method also used for contaminated SSA [59]. It reduces the content of heavy metals and salts and increases the reactivity of Si and the Si/Al ratio [61,111]. Such a pre-treatment method can make MSWI-FA into a potential precursor in geopolymers based on AAMs that can naturally trap heavy metals inside its matrix. As MSWI-FA is alkaline (having a pH of around 11), ED treatment results in an acidic pH, similar to the common natural precursor in geopolymers, i.e., metakaolin. The combination of MSWI-FA pre-treatment and initiating up to 20 wt% of MSWI-FA in geopolymers achieves the lowest metal leaching and a high compressive strength, making it a potential construction material [61,111]. Thus, the use of geopolymerization for hazardous waste not only contributes to the best technological practices and legal provisions but is also ecologically efficient [112]. Studies have shown that the use of raw MSWI-FA is also possible but achieves lower compressive strengths than that of AAMs prepared with slag or pulverized FA [101]. The reason for the low compressive strength could be its variable mineral composition because MSWI-FA contains less SiO_2 and Al_2O_3 than slag or pulverized FA and has a large specific surface area due to a high proportion of small particles [101].

MSWI-FA has also been studied as a potential replacement for cementitious materials, but adding it to cement-based products means that technical and environmental requirements such as sufficient strengths, disabilities, and leaching limits of heavy metals must be met [113]. The main problem with using MSWI-FA as a cement substitute is the presence of leachable toxic heavy metals and a high salt content. It is beneficial to use water washing and mechanochemical [113] or ED pre-treatments [114] to improve its performance before it is used in mortar, concrete, or bricks. The mechanochemical processes can stabilize the heavy metals and activate the MSWI-FA, allowing it to partially replace Portland cement in building materials [113], while an ED treatment can remove heavy metals and soluble salts from the MSWI-FA suspension which is thus decontaminated [96,115]. As an ultrafine material, MSWI-FA is also a potential substitute for clay in bricks, which should stabilize the heavy metals, reduce raw material imports, and, at the same time, conserve primary clay resources [115]. Studies have shown that fired bricks with an addition of 2.5–5 wt% treated MSWI-FA may be feasible [96,116,117].

3.3. Potential Use of WBA

WBA results from wood biomass combustion generating WB-BA, collected from the bottom of a combustion chamber, and WB-FA, which is subdivided into fine fly ash (particle size <1 μm, collected from electrostatic or bag house filters) and coarse fly ash (particle size >1 μm, collected from the cyclone or boilers) [83,118,119]. CaO and SiO_2 are generally the major chemical components in WBA, while other compounds such as Al_2O_3, Fe_2O_3, K_2O, Na_2O, MgO, P_2O_5, or SO_3 occur in lower amounts [15,75,120]. Minor contents of As, Cd, Cr, Cu, Pb, and Ni have also been detected [15,75]. Significant differences in the content of volatile heavy metals occur amongst the ash types and are the main concern when using WBA. A higher concentration of heavy metals was measured in WB-FA compared to WB-BA, as heavy metals are more concentrated in smaller particle size fractions (<75 μm) [12,14,118,119]. The particles of the fine fly ash fraction are also lighter and smaller, making them easy to inhale and a health risk, e.g., Cd accumulates in the kidneys and affects bone density [121].

The chemical composition of WB-FA also differs from coal FA, as WB-FA usually contains more alkali elements and less Al [118]. Nutrients, such as P and Mg, are primarily found in the WB-BA and coarse WB-FA. WB-FA shows significantly lower Cd concentrations compared to MSWI-FA [12,15,28,29]. The chemical and physical properties of WBA depend on the combustion technology, the heat treatment temperature, the tree species, and the geographic location; however, other factors, such as soil conditions, climate characteristics, and storage methods, also influence its properties [12–15,82,83,118,121,122].

Since wood biomass is considered to be a CO_2-neutral renewable energy source, it is environmentally desirable to use WBA in the construction industry [119,123]. This would not only reduce rising disposal costs, 70% of WBA still ends up in landfills, but also preserve natural resources and reduce greenhouse gas emissions [119,124]. WBA has the potential to be used in various construction areas: as a partial replacement of aggregates or mineral admixtures in concrete [118,124], as a partial replacement of raw materials for clinker [118,119], as a filler/partial sand replacement material in cement-based materials [13,28], in brick production [13], road construction [13] and others. Proper storage and transport conditions are important for WBA use in cementitious composites, as carbonation and hydration can occur suddenly during these procedures in wet circumstances and thus strongly determine the quantity of CaO and other carbonate elements [13]. It is also very important to find the optimal cement/ash ratio so that the strength of the cement composites remains sufficient [13,125]. Replacing up to 45% of cement with WB-FA has been described as suitable for construction purposes; however, WB-FA has more potential as a filler material than as a cement replacement material in construction [13,82]. Most studies report that the optimal content of WB-FA and WB-BA to replace part of the cement in mortars is 10 wt% [13,79,80,126]. It has been reported that adding WBA generally lowered the mechanical properties of bricks [127,128], but many clay mixtures still fulfilled the required strength parameters. Another emerging use of WBA is as a feedstock material for CO_2 utilization via mineral carbonation [129,130]. Even though the CO_2 uptake capacity is limited, it can be improved as, e.g., mechanochemical activation has doubled its capacity [131]. Biomass ash is also a low-cost medium that does not need to be transported to a plant and can be used for a low-tech sequestration approach [86].

The suitability of coal combustion FA for the production of AAMs has been proven many times due to its suitable chemical composition and the large volume available [132,133]. It is influenced by the coal properties, the combustion technique, and the waste handling. WB-FA can similarly be used as a complete or partial replacement material when preparing geopolymer mortars, which could reduce the cost of geopolymer source materials and avoid the cost of WBA landfilling [76,132]. Replacing coal combustion FA by up to 20% WBA in the binder mass resulted in better strength and porosity properties than the control mixture after 3 and 7 days of aging. After 28 days, the geopolymer containing 10% WBA was the only one to show a higher compressive strength and a lower total porosity than the control mixture [132].

Using WBA as a forest fertilizer also has potential; however, future research should focus on the effect of trace element solubility on the natural leaching processes in forest soil [12,134]. Another important factor preventing the use of WBA on certain soil typologies is its alkaline pH (usually higher than 12) [135]. Accordingly, Pasquali et al. proposed a technology to stabilize heavy metals in WBA and lower their pH based on the use of other by-products (coal FA, rice husk ash, and MSWI-FA) [135]. MSWI-FA has a similar pH as WBA and is a source of leachable heavy metals, while its P concentration is low. Ca-rich coal FA was used in the stabilization procedure, while rice husk ash was chosen as a heavy metal stabilizer due to its amorphous silica content. Wolffers et al. recently reported the recovery of heavy metals from WB-FA based on acid leaching, a process also applied to MSWI-FA [15]. In Switzerland, the disposal of WB-FA in landfills will be prohibited in 2023 due to the elevated concentrations of very toxic Cr(VI) and other heavy metals [15]. The FLUWA process represents a promising method for managing WB-FA [15].

4. Conclusions

This literature review focused on the technologies available to recover valuable elements, such as P and the selected heavy metals Zn, Pb, Cd, Cr, and Cu, from MSWI, SSA, and WBA. The wet chemical extraction method is used for P extraction, with H_2SO_4 as the most common extractant. Other methods are needed to extract resources from ashes in locations where waste acid is not easily available. Electrochemical technologies are beneficial alternatives because their major input is an electrical direct current, which can be directly gained from renewable sources, and the extraction can be performed in periods with excess grid energy. Nevertheless, future studies should deal with the duration of the electrodialytic process and reduce energy consumption.

An overview of the potential use of incineration ash as secondary raw material was provided, confirming their potential in various fields, with or without a pre-treatment. Pre-treated SSA can be used as a substitute material for cement, as a filler or fine aggregate component in mortar and concrete, can partially replace clay in bricks, or can be reused as a precursor material for geopolymerization. Non-hazardous MSWI-BA is commonly used in the construction industry as an alternative lightweight aggregate, as an alternative material for cement production, or in road construction. The hazardous MSWI-FA must be pre-treated before further use in geopolymers or as a substitute material for cementitious material or clay in bricks. For WBA, it is environmentally desirable to use it in the construction industry or for CO_2 utilization through mineral carbonation.

Author Contributions: Conceptualization, S.T., V.D., W.W., G.M.K. and L.M.O.; data curation, S.T.; investigation, S.T. and W.W.; writing—original draft preparation, S.T.; writing—review and editing, V.D., W.W., G.M.K., L.M.O. and T.L.; supervision, V.D., G.M.K. and L.M.O. All authors have read and agreed to the published version of the manuscript.

Funding: This research was EU-funded by the ASHCYCLE project, Grant Number 101058162.

Institutional Review Board Statement: Not applicable.

Informed Consent Statement: Not applicable.

Data Availability Statement: Not applicable.

Conflicts of Interest: The authors declare no conflict of interest.

References

1. European Commission. Closing the Loop—An EU Action Plan for the Circular Economy, Communication from the Commission to the European Parliament, the Council, the European Economic and Social Committee and the Committee of the Regions, COM(2015) 614 Final. 2015. Available online: https://eur-lex.europa.eu/legal-content/EN/TXT/?uri=CELEX:52015DC0614 (accessed on 6 September 2022).
2. Boniardi, G.; Turolla, A.; Fiameni, L.; Gelmi, E.; Malpei, F.; Bontempi, E.; Canziani, R. Assessment of a simple and replicable procedure for selective phosphorus recovery from sewage sludge ashes by wet chemical extraction and precipitation. *Chemosphere* **2021**, *285*, 131476. [CrossRef] [PubMed]
3. Kwapinski, W.; Kolinovic, I.; Leahy, J.J. Sewage sludge thermal treatment technologies with a focus on phosphorus recovery: A review. *Waste Biomass Valorization* **2021**, *12*, 5837–5852. [CrossRef]
4. Luyckx, L.; Geerts, S.; Van Caneghem, J. Closing the phosphorus cycle: Multi-criteria techno-economic optimization of phosphorus extraction from wastewater treatment sludge ash. *Sci. Total Environ.* **2020**, *713*, 135543. [CrossRef] [PubMed]
5. Cieślik, B.; Konieczka, P. A review of phosphorus recovery methods at various steps of wastewater treatment and sewage sludge management. The concept of "no solid waste generation" and analytical methods. *J. Clean. Prod.* **2017**, *142*, 1728–1740. [CrossRef]
6. European Commission. Study on the Critical Raw Materials for the EU 2023-Final Report. 2023. Available online: https://op.europa.eu/en/publication-detail/-/publication/57318397-fdd4-11ed-a05c-01aa75ed71a1/language-en (accessed on 27 March 2023).
7. Lima, A.T.; Kirkelund, G.M.; Ntuli, F.; Ottosen, L.M. Screening dilute sources of rare earth elements for their circular recovery. *J. Geochem. Explor.* **2022**, *238*, 107000. [CrossRef]
8. Watari, T.; Nansai, K.; Nakajima, K. Major metals demand, supply, and environmental impacts to 2100: A critical review. *Resour. Conserv. Recycl.* **2021**, *164*, 105107. [CrossRef]
9. Calvo, G.; Mudd, G.; Valero, A.; Valero, A. Decreasing ore grades in global metallic mining: A theoretical issue or a global reality? *Resources* **2016**, *5*, 36. [CrossRef]

10. Magdalena, R.; Valero, A.; Valero, A. Mining energy consumption as a function of ore grade decline: The case of lead and zinc. *J. Sustain. Min.* **2021**, *20*, 5. [CrossRef]
11. Izatt, R.M.; Izatt, S.R.; Bruening, R.L.; Izatt, N.E.; Moyer, B.A. Challenges to achievement of metal sustainability in our high-tech society. *Chem. Soc. Rev.* **2014**, *43*, 2451–2475. [CrossRef]
12. Budhathoki, R.; Vaïsänen, A. Particle size based recovery of phosphorus from combined peat and wood fly ash for forest fertilization. *Fuel Process. Technol.* **2016**, *146*, 85–89. [CrossRef]
13. Ayobami, A.B. Performance of wood bottom ash in cement-based applications and comparison with other selected ashes: Overview. *Resour. Conserv. Recycl.* **2021**, *16*, 105351. [CrossRef]
14. Kirkelund, G.M.; Damoe, A.J.; Ottosen, L.M. Electrodialytic removal of Cd from biomass combustion fly ash suspensions. *J. Hazard. Mater.* **2013**, *250–251*, 212–219. [CrossRef] [PubMed]
15. Wolffers, M.; Weibel, G.; Eggenberger, U. Waste wood fly ash treatment in Switzerland—Effects of co-processing with fly ash from municipal solid waste on Cr(VI) reduction and heavy metal recovery. *Processes* **2021**, *9*, 146. [CrossRef]
16. Funari, V.; Bokhari, S.N.H.; Vigliotti, L.; Meisel, T.; Braga, R. The rare earth elements in municipal solid waste incinerators ash and promising tools for their prospecting. *J. Hazard. Mater.* **2016**, *301*, 471–479. [CrossRef]
17. Vassilev, S.V.; Vassileva, C.G. Contents and associations of rare earth elements and yttrium in biomass ashes. *Fuel* **2020**, *262*, 116525. [CrossRef]
18. Wang, J.; Guo, M.; Liu, M.; Wei, X. Long-term outlook for global rare earth production. *Resour. Policy* **2020**, *65*, 101569. [CrossRef]
19. Alonso, E.; Sherman, A.M.; Wallington, T.J.; Everson, M.P.; Field, F.R.; Roth, R.; Kirchain, R.E. Evaluating rare earth element availability a case with revolutionary demand from clean technologies supporting information. *Environ. Sci. Technol.* **2012**, *46*, 3406–3414. [CrossRef]
20. Binnemans, K.; Jones, P.T.; Blanpain, B.; Van Gerven, T.; Yang, Y.; Walton, A.; Buchert, M. Recycling of rare earths: A critical review. *J. Clean. Prod.* **2013**, *51*, 1–22. [CrossRef]
21. Ebbers, B.; Ottosen, L.M.; Jensen, P.E. Comparison of two different electrodialytic cells for separation of phosphorus and heavy metals from sewage sludge ash. *Chemosphere* **2015**, *125*, 122–129. [CrossRef]
22. Liang, S.; Chen, H.; Zeng, X.; Li, Z.; Yu, W.; Xiao, K.; Hu, J.; Hou, H.; Liu, B.; Tao, S.; et al. A comparison between sulfuric acid and oxalic acid leaching with subsequent purification and precipitation for phosphorus recovery from sewage sludge incineration ash. *Water Res.* **2019**, *159*, 242–251. [CrossRef]
23. Ottosen, L.M.; Kirkelund, G.M.; Jensen, P.E. Extracting phosphorous from incinerated sewage sludge ash rich in iron or aluminum. *Chemosphere* **2013**, *91*, 963–969. [CrossRef] [PubMed]
24. Wang, Q.; Li, J.-S.; Tang, P.; Fang, L.; Poon, C.S. Sustainable reclamation of phosphorus from incinerated sewage sludge ash as value-added struvite by chemical extraction, purification and crystallization. *J. Clean. Prod.* **2018**, *181*, 717–725. [CrossRef]
25. Li, J.; Chen, Z.; Wang, Q.; Fang, L.; Xue, Q.; Cheeseman, C.R.; Donatello, S.; Liu, L.; Poon, C.S. Change in re-use value of incinerated sewage sludge ash due to chemical extraction of phosphorus. *Waste Manag.* **2018**, *74*, 404–412. [CrossRef] [PubMed]
26. Hjelmar, O.; Hyks, J.; Korpisjärvi, K.; Wahlström, M.; Grönholm, R. BAT (Best Available Techniques) for Combustion and Incineration Residues in a Circular Economy. 2022. Available online: https://pub.norden.org/temanord2022-542/ (accessed on 30 September 2022).
27. Tang, J.; Su, M.; Wu, Q.; Wei, L.; Wang, N.; Xiao, E.; Zhang, H.; Wei, Y.; Liu, Y.; Ekberg, C.; et al. Highly efficient recovery and clean-up of four heavy metals from MSWI fly ash by integrating leaching, selective extraction and adsorption. *J. Clean. Prod.* **2019**, *234*, 139–149. [CrossRef]
28. Berra, M.; Mangialardi, T.; Paolini, A.E. Reuse of woody biomass fly ash in cement-based materials. *Constr. Build. Mater.* **2015**, *76*, 286–296. [CrossRef]
29. Maresca, A.; Hyks, J.; Astrup, T.F. Recirculation of biomass ashes onto forest soils: Ash composition, mineralogy and leaching properties. *Waste Manag.* **2017**, *70*, 127–138. [CrossRef]
30. Allegrini, E.; Maresca, A.; Emil, M.; Sommer, M.; Boldrin, A.; Fruergaard, T. Quantification of the resource recovery potential of municipal solid waste incineration bottom ashes. *Waste Manag.* **2014**, *34*, 1627–1636. [CrossRef]
31. Gupta, D.K.; Chatterjee, S.; Datta, S.; Veer, V.; Walther, C. Role of phosphate fertilizers in heavy metal uptake and detoxification of toxic metals. *Chemosphere* **2014**, *108*, 134–144. [CrossRef]
32. Swiss Federal Council. Ordinance on the Avoidance and the Disposal of Waste (Waste Ordinance, ADWO). 2022. Available online: https://www.fedlex.admin.ch/eli/cc/2015/891/en#chap_3/sec_3 (accessed on 13 October 2022).
33. Fertilizer Act 711/2022—Original Regulations. 2022. Available online: https://www.finlex.fi/fi/laki/alkup/2022/20220711 (accessed on 13 October 2022).
34. Zhai, J.; Burke, I.T.; Stewart, D.I. Potential reuse options for biomass combustion ash as affected by the persistent organic pollutants (POPs) content. *J. Hazard. Mater. Adv.* **2022**, *5*, 100038. [CrossRef]
35. Al-Ghouti, M.A.; Khan, M.; Nasser, M.S.; Al-Saad, K.; Heng, O.E. A novel method for metals extraction from municipal solid waste using a microwave-assisted acid extraction. *J. Clean. Prod.* **2021**, *287*, 125039. [CrossRef]
36. Zhu, Y.; Zhai, Y.; Li, S.; Liu, X.; Wang, B.; Liu, X.; Fan, Y.; Shi, H.; Li, C.; Zhu, Y. Thermal treatment of sewage sludge: A comparative review of the conversion principle, recovery methods and bioavailability-predicting of phosphorus. *Chemosphere* **2022**, *291*, 133053. [CrossRef] [PubMed]

37. Abis, M.; Calmano, W.; Kuchta, K. Innovative technologies for phosphorus recovery from sewage sludge ash. *Detritus* **2018**, *1*, 23–29. [CrossRef]
38. Fang, L.; Li, J.-S.; Guo, M.Z.; Cheeseman, C.R.; Tsang, D.C.W.; Donatello, S.; Poon, C.S. Phosphorus recovery and leaching of trace elements from incinerated sewage sludge ash (ISSA). *Chemosphere* **2018**, *193*, 278–287. [CrossRef] [PubMed]
39. Li, J.; Tsang, D.C.W.; Wang, Q.; Fang, L.; Xue, Q.; Poon, C.S. Fate of metals before and after chemical extraction of incinerated sewage sludge ash. *Chemosphere* **2017**, *186*, 350–359. [CrossRef]
40. Weibel, G.; Eggenberger, U.; Kulik, D.A.; Hummel, W.; Schlumberger, S.; Klink, W.; Fisch, M.; Mäder, U.K. Extraction of heavy metals from MSWI fly ash using hydrochloric acid and sodium chloride solution. *Waste Manag.* **2018**, *76*, 457–471. [CrossRef]
41. Xu, H.; He, P.; Gu, W.; Wang, G.; Shao, L. Recovery of phosphorus as struvite from sewage sludge ash. *J. Environ. Sci.* **2012**, *24*, 1533–1538. [CrossRef]
42. Rahmani, A.; Thapa, R.; Aalto, J.-M.; Turhanen, P.; Vepsäläinen, J.; Lehto, V.-P.; Riikonen, J. Functionalized nanoporous silicon for extraction of Sc from a leach solution. *Hydrometallurgy* **2022**, *211*, 105866. [CrossRef]
43. Fang, L.; Li, J.-S.; Donatello, S.; Cheeseman, C.R.; Wang, Q.; Poon, C.S.; Tsang, D.C. Recovery of phosphorus from incinerated sewage sludge ash by combined two-step extraction and selective precipitation. *Chem. Eng. J.* **2018**, *348*, 74–83. [CrossRef]
44. Tang, J.; Steenari, B.M. Solvent extraction separation of copper and zinc from MSWI fly ash leachates. *Waste Manag.* **2015**, *44*, 147–154. [CrossRef]
45. Turhanen, P.A.; Vepsäläinen, J.J.; Peräniemi, S. Advanced material and approach for metal ions removal from aqueous solutions. *Sci. Rep.* **2015**, *5*, 8992. [CrossRef]
46. Zalupski, P.R.; Chiarizia, R.; Jensen, M.P.; Herlinger, A. Metal extraction by sulfur-containing symmetrically-substituted bisphosphonic acids. Part I. P,P′-di(2-ethylhexyl) methylenebisthio-phosphonic acid. *Solvent Extr. Ion Exch.* **2006**, *24*, 331–346. [CrossRef]
47. Lynn, C.J.; Dhir, R.K.; Ghataora, G.S.; West, R.P. Sewage sludge ash characteristics and potential for use in concrete. *Constr. Build. Mater.* **2015**, *98*, 767–779. [CrossRef]
48. Krüger, O.; Grabner, A.; Adam, C. Complete survey of German sewage sludge ash. *Environ. Sci. Technol.* **2014**, *48*, 11811–11818. [CrossRef]
49. Tang, P.; Xuan, D.; Li, J.; Cheng, H.W.; Poon, C.S.; Tsang, D.C.W. Investigation of cold bonded lightweight aggregates produced with incineration sewage sludge ash (ISSA) and cementitious waste. *J. Clean. Prod.* **2020**, *251*, 119709. [CrossRef]
50. Tian, Y.; Wang, R.; Luo, Z.; Wang, R.; Yang, F.; Wang, Z.; Shu, J.; Chen, M. Heavy metals removing from municipal solid waste incineration fly ashes by electric field-enhanced washing. *Materials* **2020**, *13*, 793. [CrossRef]
51. Yen, C.-P.; Zhou, S.-Y.; Shen, Y.-H. The recovery of Ca and Zn from the municipal solid waste incinerator fly ash. *Sustainability* **2020**, *12*, 9086. [CrossRef]
52. Fang, L.; Wang, Q.; Li, J.-S.; Poon, C.S.; Cheeseman, C.R.; Donatello, S.; Tsang, D.C.W. Feasibility of wet-extraction of phosphorus from incinerated sewage sludge ash (ISSA) for phosphate fertilizer production: A critical review. *Crit. Rev. Environ. Sci. Technol.* **2021**, *51*, 939–971. [CrossRef]
53. Meng, X.; Huang, Q.; Xu, J.; Gao, H.; Yan, J. A review of phosphorus recovery from different thermal treatment products of sewage sludge. *Waste Dispos. Sustain. Energy* **2019**, *1*, 99–115. [CrossRef]
54. Havukainen, J.; Nguyen, M.T.; Hermann, L.; Horttanainen, M.; Mikkilä, M.; Deviatkin, I.; Linnanen, L. Potential of phosphorus recovery from sewage sludge and manure ash by thermochemical treatment. *Waste Manag.* **2016**, *49*, 221–229. [CrossRef]
55. Fraissler, G.; Joller, M.; Mattenberger, H.; Brunner, T.; Obernberger, I. Thermodynamic equilibrium calculations concerning the removal of heavy metals from sewage sludge ash by chlorination. *Chem. Eng. Process.* **2009**, *48*, 152–164. [CrossRef]
56. Mattenberger, H.; Fraissler, G.; Brunner, T.; Herk, P.; Hermann, L.; Obernberger, I. Sewage sludge ash to phosphorus fertiliser: Variables influencing heavy metal removal during thermochemical treatment. *Waste Manag.* **2008**, *28*, 2709–2722. [CrossRef]
57. Guedes, P.; Couto, N.; Ottosen, L.M.; Ribeiro, A.B. Phosphorus recovery from sewage sludge ash through an electrodialytic process. *Waste Manag.* **2014**, *34*, 886–892. [CrossRef]
58. Ottosen, L.M.; Jensen, P.E.; Kirkelund, G.M. Phosphorous recovery from sewage sludge ash suspended in water in a two-compartment electrodialytic cell. *Waste Manag.* **2016**, *51*, 142–148. [CrossRef]
59. Ottosen, L.M.; Bertelsen, I.M.G.; Jensen, P.E.; Kirkelund, G.M. Sewage sludge ash as resource for phosphorous and material for clay brick manufacturing. *Constr. Build. Mater.* **2020**, *249*, 118684. [CrossRef]
60. Villen-Guzman, M.; Guedes, P.; Couto, N.; Ottosen, L.M.; Ribeiro, A.B.; Rodriguez-Maroto, J.M. Electrodialytic phosphorus recovery from sewage sludge ash under kinetic control. *Electrochim. Acta* **2018**, *287*, 49–59. [CrossRef]
61. Righi, C.; Lancellotii, I.; Barbieri, L.; Kirkelund, G.M. Benefits of pre-treating MSWI fly ash before alkali-activation. *Sustain. Chem. Pharm.* **2022**, *27*, 100671. [CrossRef]
62. Oliveira, V.; Labrincha, J.; Dias-Ferreira, C. Extraction of phosphorus and struvite production from the anaerobically digested organic fraction of municipal solid waste. *J. Environ. Chem. Eng.* **2018**, *6*, 2837–2845. [CrossRef]
63. Ash2Phos. A Circular Solution for Phosphorus Fertiliser. 2022. Available online: https://www.easymining.se/technologies/ash2phos (accessed on 6 September 2022).
64. Ottosen, L.M.; Thornberg, D.; Cohen, Y.; Stiernström, S. Utilization of acid-washed sewage sludge ash as sand or cement replacement in concrete. *Resour. Conserv. Recycl.* **2022**, *176*, 105943. [CrossRef]

65. Quina, M.J.; Bontempi, E.; Bogush, A.; Schlumberger, S.; Weibel, G.; Braga, R.; Funari, V.; Hyks, J.; Rasmussen, E.; Lederer, J. Technologies for the management of MSW incineration ashes from gas cleaning: New perspectives on recovery of secondary raw materials and circular economy. *Sci. Total Environ.* **2018**, *635*, 526–542. [CrossRef]
66. Schlumberger, S.; Schuster, M.; Ringmann, S.; Koralewska, R. Recovery of high purity zinc from filter ash produced during the thermal treatment of waste and inerting of residual materials. *Waste Manag. Res.* **2007**, *25*, 547–555. [CrossRef]
67. HaloSep. Pure Separation. 2022. Available online: https://www.halosep.com/ (accessed on 6 September 2022).
68. Karlfeldt Fedje, K.; Andersson, S. Zinc recovery from Waste-to-Energy fly ash—A pilot test study. *Waste Manag.* **2020**, *118*, 90–98. [CrossRef]
69. Smart City Sweden. 2022. Available online: https://smartcitysweden.com/renova-is-building-for-the-worlds-unique-metal-recycling/ (accessed on 29 November 2022).
70. Ottosen, L.M.; Kirkelund, G.M.; Jensen, P.E.; Pedersen, K.B. Extraction of Phosphorus from Sewage Sludge Ash-Influence of Process Variables on the Electrodialytic Process. *Sustainability* **2023**, *15*, 13953. [CrossRef]
71. Donatello, S.; Tong, D.; Cheeseman, C. Production of technical grade phosphoric acid from incinerator sewage sludge ash (ISSA). *Waste Manag.* **2010**, *30*, 1634–1642. [CrossRef]
72. Xu, Y.; Zhang, L.; Chen, J.; Liu, T.; Li, N.; Xu, J.; Yin, W.; Li, D.; Zhang, Y.; Zhou, X. Phosphorus recovery from sewage sludge ash (SSA): An integrated technical, environmental and economic assessment of wet-chemical and thermochemical methods. *J. Environ. Manag.* **2023**, *355*, 118691. [CrossRef]
73. Eagle, L.; Rechberger, H.; Krampe, J.; Zessner, M. Phosphorus recovery from municipal wastewater: An integrated comparative technological, environmental and economic assessment of P recovery technologies. *Sci. Total Environ.* **2016**, *571*, 522–542. [CrossRef]
74. Abramov, S.; He, J.; Wimmer, D.; Lemloh, M.L.; Muehe, E.M.; Gann, B.; Roehm, E.; Kirchhof, R.; Babechuk, M.G.; Schoenberg, R.; et al. Heavy metal mobility and valuable contents of processed municipal solid waste incineration residues from Southwestern Germany. *Waste Manag.* **2018**, *79*, 735–743. [CrossRef] [PubMed]
75. Zhuge, Y.; Duan, W.; Liu, Y. Utilization of wood waste ash in green concrete production. In *Sustainable Concrete Made with Ashes and Dust from Different Sources: Materials, Properties and Applications*; Woodhead Publishing: Sawston, UK, 2022; pp. 419–450.
76. De Rossi, A.; Simão, L.; Ribeiro, M.J.; Hotza, D.; Moreira, R.F.P.M. Study of cure conditions effect on the properties of wood biomass fly ash geopolymers. *J. Mater. Res. Technol.* **2020**, *9*, 7518–7528. [CrossRef]
77. Luukkonen, T. Alkali activation of water and wastewater sludges: Solidification/stabilization and potential aluminosilicate precursors. In *Development in Waste Water Treatment Research and Processes*; Elsevier: Amsterdam, The Netherlands, 2022; pp. 161–174. [CrossRef]
78. Krejcirikova, B.; Ottosen, L.M.; Kirkelund, G.M.; Rode, C.; Peuhkuri, R. Characterization of sewage sludge ash and its effect on moisture physics of mortar. *J. Build. Eng.* **2019**, *21*, 396–403. [CrossRef]
79. Carević, I.; Štirmer, N.; Trkmić, M.; Jurić, K.K. Leaching characteristics of wood biomass fly ash cement composites. *Appl. Sci.* **2020**, *10*, 8704. [CrossRef]
80. Carević, I.; Baričević, A.; Štirmer, N.; Šantek Bajto, J. Correlation between physical and chemical properties of wood biomass ash and cement composites performances. *Constr. Build. Mater.* **2020**, *256*, 119450. [CrossRef]
81. Chowdhury, S.; Mishra, M.; Suganya, O. The incorporation of wood waste ash as a partial cement replacement material for making structural grade concrete: An overview. *Ain Shams Eng. J.* **2015**, *6*, 429–437. [CrossRef]
82. Gabrijel, I.; Rukavina, M.J.; Štirmer, N. Influence of wood fly ash on concrete properties through filling effect mechanism. *Materials* **2021**, *14*, 7164. [CrossRef] [PubMed]
83. Milovanović, B.; Štirmer, N.; Carević, I.; Baričević, A. Wood biomass ash as a raw material in concrete industry. *Gradjevinar* **2019**, *71*, 505–514. [CrossRef]
84. Lothenbach, B.; Scrivener, K.; Hooton, R.D. Supplementary cementitious materials. *Cem. Concr. Res.* **2011**, *41*, 1244–1256. [CrossRef]
85. Ashraf, W. Carbonation of cement-based materials: Challenges and opportunities. *Constr. Build. Mater.* **2016**, *120*, 558–570. [CrossRef]
86. Koch, R.; Sailer, G.; Paczkowski, S.; Pelz, S.; Poetsch, J.; Müller, J. Lab-scale carbonation of wood ash for CO_2-sequestration. *Energies* **2021**, *14*, 7371. [CrossRef]
87. Li, L.; Wu, M. An overview of utilizing CO_2 for accelerated carbonation treatment in the concrete industry. *J. CO2 Util.* **2022**, *60*, 102000. [CrossRef]
88. Tominc, S.; Ducman, V. Methodology for Evaluating the CO_2 Sequestration Capacity of Waste Ashes. *Materials* **2023**, *16*, 5284. [CrossRef]
89. Tripathi, N.; Hills, C.D.; Singh, R.S.; Singh, J.S. Offsetting anthropogenic carbon emissions from biomass waste and mineralised carbon dioxide. *Sci. Rep.* **2020**, *10*, 958. [CrossRef]
90. Winnefeld, F.; Leemann, A.; German, A.; Lothenbach, B. CO_2 storage in cement and concrete by mineral carbonation. *Curr. Opin. Green Sustain. Chem.* **2022**, *38*, 100672. [CrossRef]
91. Regulation (EU) 2019/1009 of the European Parliament and of the Council of 5 June 2019. Official Journal of the European Union. 2019. Available online: https://eur-lex.europa.eu/legal-content/EN/TXT/PDF/?uri=CELEX:32019R1009&from=EN (accessed on 6 September 2022).

92. Kappel, A.; Viader, R.P.; Kowalski, K.P.; Kirkelund, G.M.; Ottosen, L.M. Utilisation of electrodialytically treated sewage sludge ash in mortar. *Waste Biomass Valorization* **2018**, *9*, 2503–2515. [CrossRef]
93. Donatello, S.; Freeman-Pask, A.; Tyrer, M.; Cheeseman, C.R. Effect of milling and acid washing on the pozzolanic activity of incinerator sewage sludge ash. *Cem. Concr. Compos.* **2010**, *32*, 54–61. [CrossRef]
94. Istuque, D.B.; Soriano, L.; Akasaki, J.L.; Melges, J.L.P.; Borrachero, M.V.; Monzó, J.; Payá, J.; Tashima, M. Effect of sewage sludge ash on mechanical and microstructural properties of geopolymers based on metakaolin. *Constr. Build. Mater.* **2019**, *203*, 95–103. [CrossRef]
95. Castro-Gomes, J.; Sedira, N.; Grünhäuser Soares, E. Feasibility for Alkali-Activation of a Sewage Sludge Ash (SSA). 2022. Available online: https://www.researchgate.net/publication/359439591_Feasibility_for_alkali-activation_of_a_Sewage_Sludge_Ash_SSA (accessed on 10 January 2023).
96. Chen, W.; Klupsch, E.; Kirkelund, G.M.; Jensen, P.E.; Ottosen, L.M.; Dias-Ferreira, C. *WASTES—Solutions, Treatments and Opportunities II: Selected Papers from the 4th Edition of the International Conference on Wastes: Solutions Treatments and Opportunities;* Recycling of MSWI Fly Ash in Clay Bricks—Effect of Washing and Electrodialytic Treatment; CRC Press: Porto, Portugal, 2018; pp. 183–190.
97. Al-Ghouti, M.A.; Khan, M.; Nasser, M.S.; Al-Saad, K.; Heng, O.E. Recent advances and applications of municipal solid wastes bottom and fly ashes: Insights into sustainable management and conservation of resources. *Environ. Technol. Innov.* **2021**, *21*, 101267. [CrossRef]
98. Das, S.; Lee, S.H.; Kumar, P.; Kim, K.H.; Lee, S.S.; Bhattacharya, S.S. Solid waste management: Scope and the challenge of sustainability. *J. Clean. Prod.* **2019**, *228*, 658–678. [CrossRef]
99. Fontseré Obis, M.; Germain, P.; Bouzahzah, H.; Richioud, A.; Benbelkacem, H. The effect of the origin of MSWI bottom ash on the H2S elimination from landfill biogas. *Waste Manag.* **2017**, *70*, 158–169. [CrossRef]
100. Chen, B.; Perumal, P.; Illikainen, M.; Ye, G. A review on the utilization of municipal solid waste incineration (MSWI) bottom ash as a mineral resource for construction materials. *J. Build. Eng.* **2023**, *71*, 106386. [CrossRef]
101. Liu, J.; Hu, L.; Tang, L.; Ren, J. Utilisation of municipal solid waste incinerator (MSWI) fly ash with metakaolin for preparation of alkali-activated cementitious material. *J. Hazard. Mater.* **2021**, *402*, 123451. [CrossRef]
102. Pérez-Martínez, S.; Giro-Paloma, J.; Maldonado-Alameda, A.; Formosa, J.; Queralt, I.; Chimenos, J.M. Characterisation and partition of valuable metals from WEEE in weathered municipal solid waste incineration bottom ash, with a view to recovering. *J. Clean. Prod.* **2019**, *218*, 61–68. [CrossRef]
103. Balaguera, A.; Carvajal, G.I.; Albertí, J.; Fullana-i-Palmer, P. Life cycle assessment of road construction alternative materials: A literature review. *Resour. Conserv. Recycl.* **2018**, *132*, 37–48. [CrossRef]
104. Lam, C.H.K.; Ip, A.W.M.; Barford, J.P.; McKay, G. Use of incineration MSW ash: A review. *Sustainability* **2010**, *2*, 1943–1968. [CrossRef]
105. Li, Y.; Hao, L.; Chen, X. Analysis of MSWI Bottom Ash Reused as Alternative Material for Cement Production. *Procedia Environ. Sci.* **2016**, *31*, 549–553. [CrossRef]
106. Parés Viader, R.; Jensen, P.E.; Ottosen, L.M. Electrodialytic remediation of municipal solid waste incineration residues using different membranes. *Chemosphere* **2017**, *169*, 62–68. [CrossRef] [PubMed]
107. Sekito, T.; Dote, Y.; Onoue, K.; Sakanakura, H.; Nakamura, K. Characteristics of element distributions in an MSW ash melting treatment system. *Waste Manag.* **2014**, *34*, 1637–1643. [CrossRef]
108. Kaderides, K.; Papaoikonomou, L.; Serafim, M.; Goula, A.M. Microwave-assisted extraction of phenolics from pomegranate peels: Optimization, kinetics, and comparison with ultrasounds extraction. *Chem. Eng. Process. Process Intensif.* **2019**, *137*, 1–11. [CrossRef]
109. Rahmati, S.; Abdullah, A.; Kang, O.L. Effects of different microwave intensity on the extraction yield and physicochemical properties of pectin from dragon fruit (Hylocereus polyrhizus) peels. *Bioact. Carbohydr. Diet. Fibre* **2019**, *18*, 100186. [CrossRef]
110. Su, D.-L.; Li, P.-J.; Quek, S.Y.; Huang, Z.-Q.; Yuan, Y.-J.; Li, G.-Y.; Shan, Y. Efficient extraction and characterization of pectin from orange peel by a combined surfactant and microwave assisted process. *Food Chem.* **2019**, *286*, 1–7. [CrossRef]
111. Zhan, X.; Kirkelund, G.M. Electrodialytic remediation of municipal solid waste incineration fly ash as pre-treatment before geopolymerisation with coal fly ash. *J. Hazard. Mater.* **2021**, *412*, 125220. [CrossRef]
112. Łach, M.; Mierzwiński, D.; Korniejenko, K.; Mikuła, J.; Hebda, M. Geopolymers as a material suitable for immobilization of fly ash from municipal waste incineration plants. *J. Air Waste Manag. Assoc.* **2018**, *68*, 1190–1197. [CrossRef]
113. Pan, S.; Ding, J.; Peng, Y.; Lu, S.; Li, X. Investigation of mechanochemically treated municipal solid waste incineration fly ash as replacement for cement. *Energies* **2022**, *15*, 2013. [CrossRef]
114. Ebert, B.A.R.; Kirkelund, G.M. Effects of chlorides and sulphates on heavy metal leaching from mortar with raw and electrodialytically treated MSWI fly ash. *Waste Biomass Valorization* **2022**, *13*, 2673–2688. [CrossRef] [PubMed]
115. Kirkelund, G.M.; Skevi, L.; Ottosen, L.M. Electrodialytically treated MSWI fly ash use in clay bricks. *Constr. Build. Mater.* **2020**, *254*, 119286. [CrossRef]
116. Sun, J.; Zhou, H.; Jiang, H.; Zhang, W.; Mao, L. Recycling municipal solid waste incineration fly ash in fired bricks: An evaluation of physical-mechanical and environmental properties. *Constr. Build. Mater.* **2021**, *294*, 123476. [CrossRef]
117. Voisniene, V.; Kizinievic, O.; Kizinievic, V. Feasibility study of using clay bricks made from municipal solid waste incinerator (MSWI) fly ash. *IOP Conf. Ser. Mater. Sci. Eng.* **2019**, *603*, 022058. [CrossRef]

118. Carević, I.; Serdar, M.; Štirmer, N.; Ukrainczyk, N. Preliminary screening of wood biomass ashes for partial resources replacements in cementitious materials. *J. Clean. Prod.* **2019**, *229*, 1045–1064. [CrossRef]
119. Ukrainczyk, N.; Vrbos, N.; Koenders, E.A.B. Reuse of woody biomass ash waste in cementitious materials. *Chem. Biochem. Eng. Q.* **2016**, *30*, 137–148. [CrossRef]
120. Royo, J.; Canalís, P.; Quintana, D. Chemical study of bottom ash sintering in combustion of pelletized residual agricultural biomass. *Fuel* **2022**, *310*, 122145. [CrossRef]
121. Khan, A.A.; de Jong, W.; Jansens, P.J.; Spliethoff, H. Biomass combustion in fluidized bed boilers: Potential problems and remedies. *Fuel Process. Technol.* **2009**, *90*, 21–50. [CrossRef]
122. Cheah, C.B.; Ramli, M. The implementation of wood waste ash as a partial cement replacement material in the production of structural grade concrete and mortar: An overview. *Resour. Conserv. Recycl.* **2011**, *55*, 669–685. [CrossRef]
123. Mu, L.; Li, T.; Zuo, S.; Yin, H.; Dong, M. Effect of leaching pretreatment on the inhibition of slagging/sintering of aquatic biomass: Ash transformation behavior based on experimental and equilibrium evaluation. *Fuel* **2022**, *323*, 124391. [CrossRef]
124. Guo, Q.; Yan, B.; Hu, Y.; Guo, X.; Wu, W.; Cheng, Z.; Chen, G.; Hou, L. A novel reutilization of ash from biomass gasification process: Feasibility and products improvement analysis. *Fuel* **2023**, *339*, 127386. [CrossRef]
125. Drljaca, D.; Vukic, L.; Dragic, D.; Borkovic, A.; Botic, T.; Dugic, P.; Papuga, S.V.; Šolić, M.D.; Maletić, S.P.; Gvero, P.M.; et al. Leaching of heavy metals from wood biomass ash, before and after binding in cement composite. *J. Serbian Chem. Soc.* **2022**, *87*, 1091–1108. [CrossRef]
126. Akinyemi, B.A.; Dai, C. Development of banana fibers and wood bottom ash modified cement mortars. *Constr. Build. Mater.* **2020**, *241*, 118041. [CrossRef]
127. Elinwa, A.U. Effect of addition of sawdust ash to clay bricks. *Civ. Eng. Environ. Syst.* **2006**, *23*, 263–270. [CrossRef]
128. Hafez, R.D.A.; Tayeh, B.A.; Abd-Al Ftah, R.O. Development and evaluation of green fired clay bricks using industrial and agricultural wastes. *Case Stud. Constr. Mater.* **2022**, *17*, e01391. [CrossRef]
129. Gunning, P.; Hills, C.; Antemir, A.; Carey, P. Novel approaches to the valorisation of ashes using aggregation by carbonation. In Proceedings of the 2nd International Slag Valorisation Symposium 18-20 April 2011, Leuven, Belgium; pp. 103–116. Available online: https://www.researchgate.net/publication/228511247 (accessed on 13 October 2022).
130. Vassilev, S.V.; Vassileva, C.G.; Petrova, N.L. Mineral carbonation of biomass ashes in relation to their CO_2 capture and storage potential. *ACS Omega* **2021**, *6*, 14598–14611. [CrossRef]
131. Ke, X.; Baki, V.A.; Skevi, L. Mechanochemical activation for improving the direct mineral carbonation efficiency and capacity of a timber biomass ash. *J. CO2 Util.* **2023**, *68*, 102367. [CrossRef]
132. Abdulkareem, O.A.; Ramli, M.; Matthews, J.C. Production of geopolymer mortar system containing high calcium biomass wood ash as a partial substitution to fly ash: An early age evaluation. *Compos. Part B Eng.* **2019**, *174*, 106941. [CrossRef]
133. Zhang, Z.; Wang, H.; Provis, J.L. Quantitative study of the reactivity of fly ash in geopolymerization by FTIR. *J. Sustain. Cem. Based Mater.* **2012**, *1*, 154–166. [CrossRef]
134. da Costa, T.P.; Quinteiro, P.; Tarelho, L.A.C.; Arroja, L.; Dias, A.C. Life cycle assessment of woody biomass ash for soil amelioration. *Waste Manag.* **2020**, *101*, 126–140. [CrossRef]
135. Pasquali, M.; Zanoletti, A.; Benassi, L.; Federici, S.; Depero, L.E.; Bontempi, E. Stabilized biomass ash as a sustainable substitute for commercial P-fertilizers. *Land Degrad. Dev.* **2018**, *29*, 2199–2207. [CrossRef]

materials MDPI

Article

Experiment and Analysis of Variance for Stabilizing Fine-Grained Soils with Cement and Sawdust Ash as Liner Materials

Sadiq Iliyas [1,2], Ahmad Idris [3], Ibrahim Haruna Umar [1], Hang Lin [1,*], Ahmad Muhammad [4] and Linglin Xie [5,6,*]

1 School of Resources and Safety Engineering, Central South University, Changsha 410083, China; sadiqiliyasu11@gmail.com (S.I.); ibrahimharunaumar@yahoo.com (I.H.U.)
2 Department of Agriculture Engineering, Federal College of Land Resources Technology, Owerri 1518, Nigeria
3 Department of Civil Engineering, Bayero University, Kano 3011, Nigeria; aidris.civ@buk.edu.ng
4 Department of Civil Engineering, Kano State Polytechnic, Kano 3405, Nigeria; ahmad.civilengr@gmail.com
5 Key Laboratory of Natural Resources Monitoring and Supervision in Southern Hilly Region, Ministry of Natural Resources, Changsha 430071, China
6 Department of Investigation and Monitoring, The Second Surveying and Mapping Institute of Hunan Province, Changsha 430071, China
* Correspondence: hanglin@csu.edu.cn (H.L.); xielinglincs@126.com (L.X.)

Abstract: Due to volume change and low strength, fine-grained soils are problematic in construction. Stabilization with cement and sawdust ash (SDA) by-products can improve engineering properties. This study aimed to investigate the effectiveness of cement and sawdust ash (SDA) in stabilizing fine-grained soils for liner applications. Varying proportions of cement (0–9%) and SDA (0–10%) were added to soil samples (n = 24). Specimens were tested for unconfined compressive strength (UCS), hydraulic conductivity (HC), and volumetric shrinkage strain (VSS). Two-way ANOVA analyzed stabilization effects. Optimal stabilization occurred with 6% cement and 6% SDA, resulting in significant increases in UCS (51 to 375 kN/m^2) and decreases in HC (1.7×10^{-8} to 4.7×10^{-10} m/s) and VSS (12.8 to 3.51%) compared to untreated soil. ANOVA indicated that both cement and SDA had statistically significant ($p < 0.05$) effects on improving all three engineering properties. The addition of 6% cement and 6% SDA significantly improved the expansive soil's strength, hydraulic conductivity, and volume change properties. ANOVA confirmed the quantitative improvements and the significance of both stabilizers. Stabilization using the by-product SDA has the potential to be a sustainable soil improvement method.

Keywords: stabilization; fine-grained soils; cement; sawdust ash (SDA); analysis of variance (ANOVA)

1. Introduction

Clay liners are widely used in engineering applications such as landfill, waste containment, and groundwater protection. They act as barriers to prevent the migration of contaminants from the waste to the surrounding environment [1–4]. However, clay liners are susceptible to cracking and shrinking due to moisture content and temperature changes, which can compromise their performance and durability [5–7]. Therefore, improving the geotechnical properties of clay liners, such as strength, permeability, and shrinkage, is necessary to enhance their effectiveness and reliability [8–10]. One of the methods to improve the quality of clay liners is to stabilize them with additives, such as cement and sawdust ash (SDA) [11,12]. Cement is a common stabilizing agent that can increase the strength and reduce the permeability of clayey soils. However, cement also increases the brittleness and shrinkage of clay liners, which can lead to cracking and failure [13–15]. SDA is a waste material obtained from the combustion of sawdust, which is rich in silica and alumina. SDA can act as a pozzolanic material that reacts with cement to form additional cementitious products, improving the strength and durability of clay liners [16,17]. Moreover, SDA can reduce the shrinkage and cracking of clay liners by filling the pores and creating a more uniform structure [7]. SDA is also a low-cost and eco-friendly material that can utilize abundant and renewable sawdust resources and reduce the environmental impact of cement production [18].

There are still some research gaps and challenges in the use of cement and SDA for stabilizing fine-grained soils as clay liner materials, particularly in the context of unconfined compressive strength (UCS), hydraulic conductivity (HC), and volumetric shrinkage strain (VSS). Some research gaps are: the optimal proportion and dosage of cement and SDA for different fine-grained soils and clay liners [19]; the effect of curing time and temperature on the geotechnical properties and microstructure of cement and SDA-stabilized clay liners [20]; cement and SDA-stabilized clay liners' long-term performance and durability under various environmental and loading conditions, such as wetting–drying cycles, freezing–thawing cycles, and chemical attacks [21,22]; the comparison and evaluation of cement and SDA-stabilized clay liners with other types of clay liners, such as geosynthetic clay liners, bentonite-enhanced sand liners, and polymer-modified clay liners [23], and; the economic and environmental feasibility and benefits of cement and SDA-stabilized clay liners in engineering projects [24].

Previous research has shown that the amendment of cement stabilization with sawdust ash has led to significant strength benefits in expansive and soft clayey soils [12]. Studies have also focused on the environmental impact of using sawdust ash as a stabilizer for expansive soils, emphasizing the need for soil stabilization due to high plasticity [25]. Analysis of variance (ANOVA) is commonly used in geotechnical research, especially in studies evaluating and comparing different soil properties and liner materials. When testing soil samples under various conditions, ANOVA helps researchers to determine whether differences between group means are statistically significant or simply due to chance. Several studies have used ANOVA to compare the properties of natural and compacted clay liners for landfill and containment facilities. Benson and Trast performed a one-way ANOVA on the hydraulic conductivity values of three clay liners compacted to different moisture contents and densities. The analysis showed significant differences between liner types and preparation methods [26]. Daniel et al. used ANOVA to compare consolidation parameters, permeability, and shear strength of compacted kaolinite and bentonite liners with natural clays. Their ANOVA results indicated that the engineered clays had superior containment capacities. As a result, ANOVA can be used to analyze the factors affecting clay liners' integrity and performance. The technique confirms whether different additives, compositions, moisture levels, or compaction methods have a statistically significant effect on permeability, swelling potential, compressibility, shrinkage, and other properties that affect the effectiveness of clay liners for containment structures [27]. ANOVA provides quantitative validation of factors and trends affecting liner behavior that may not be apparent from the data alone [28].

This study investigates the effectiveness of cement and SDA combination for stabilizing fine-grained soils as clay liner materials. The experimental design consists of preparing clayey soil samples with different proportions of cement and SDA and testing them for unconfined compressive strength (UCS), hydraulic conductivity (HC), and volumetric shrinkage strain (VSS). The analysis of variance (ANOVA) method is used to evaluate the significance of the effects of cement and SDA on the geotechnical properties of clay liners. The results of this study can provide valuable insights into the optimal design and application of cement and SDA-stabilized clay liners in engineering projects. Figure 1 shows a liner system.

Figure 1. A liner system.

2. Materials and Method

2.1. Materials

Table 1 presents the protocol for preparing soil samples to study the effectiveness of combining cement and sawdust ash (SDA) for stabilizing fine-grained soils as clay liner materials. The table includes SDA content (ranging from 0% to 10%), cement content (ranging from 0% to 9%), and compaction water content (−2%, 0%, +2%, and +4% of the optimum moisture content (OMC)). These variables were varied to produce different soil specimens that were then tested for unconfined compressive strength (UCS), hydraulic conductivity (HC), and volumetric shrinkage strain (VSS). The results of these tests were used to evaluate the effectiveness of the cement and SDA combination in stabilizing fine-grained soils. Initially, soil specimens were prepared by thoroughly mixing the fine-grained soil with predetermined percentages of cement (ranging from 0% to 9%) and SDA (ranging from 0% to 10%). Each mixture needed to be homogenized to ensure consistent distribution of the stabilizing agents throughout the soil. Water was then added to achieve the desired moisture content (−2%, 0%, +2%, and +4% of the OMC). The thoroughly mixed material was compacted into molds to form specimens, which follows the standard procedure for soil compaction, using a mold and hammer at controlled weights and heights to achieve specified density and moisture levels.

Table 1. Test protocol.

S/No.	SDA Content (%)	Cement Content (%)	Soil Content (%)	Compaction Water Contents (%)	Tests	Tested Temperature (°C)
1	0	0	100	−2%, 0%, +2%, and +4% of the optimum moisture content (OMC)	Physical properties, unconfined compressive strength (UCS) test, hydraulic conductivity (HC) test, and volumetric shrinkage strain (VSS) test	~23
2		3	97			
3		6	94			
4		9	91			
5	2	0	98			
6		3	95			
7		6	92			
8		9	89			

Table 1. *Cont.*

S/No.	SDA Content (%)	Cement Content (%)	Soil Content (%)	Compaction Water Contents (%)	Tests	Tested Temperature (°C)
9	4	0	96	−2%, 0%, +2%, and +4% of the optimum moisture content (OMC)	Physical properties, unconfined compressive strength (UCS) test, hydraulic conductivity (HC) test, and volumetric shrinkage strain (VSS) test	~23
10		3	93			
11		6	90			
12		9	87			
13	6	0	94			
14		3	91			
15		6	88			
16		9	85			
17	8	0	92			
18		3	89			
19		6	86			
20		9	83			
21	10	0	90			
22		3	87			
23		6	84			
24		9	81			

2.1.1. Natural Soil

Fine-grained soil specimens were obtained from a burrow area situated on the Bauchi-Gombe Road in Inkil town, within the Bauchi Local Government District (positioned at 10°18′40″ N, 9°53′46″ E) of Bauchi state, Nigeria (Figure 2 shows the location on a map). The sampling was performed utilizing the disrupted sampling method at a depth range of 0.5 to 1.0 m. The samples were gathered into medium-to-large bags, sealed, and transported to the lab for testing. The fundamental attributes of the untreated soil specimen, cement, and sawdust ash are outlined in Tables 2–4. The particle size distribution graph of the natural soil, cement, and sawdust ash are displayed in Figures 3–5. Per the Unified Soil Classification System structure, the soil is categorized as CH.

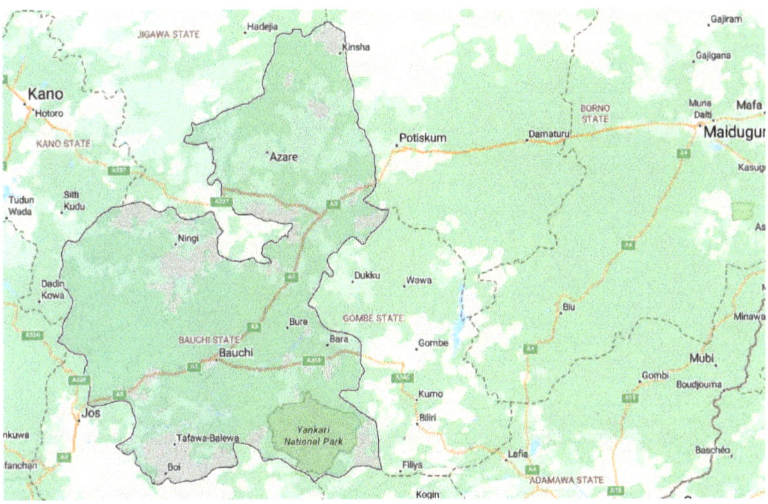

Figure 2. Bauchi-Gombe Road in Inkil Town location.

Table 2. Summary of physical properties of the natural soil.

Property	Soil
Natural moisture content, %	16.11
Specific gravity, %	2.68
Liquid limit, %	61.7
Plastic limit, %	26.8
Plasticity index, %	34.9
Linear shrinkage, %	17.4
% passing sieve No. 200	60
Free swell, %	64.8
USCS classification	CH
Optimum moisture content, %	23
Maximum dry density, Mg/m^3	1.63
Color	Reddish Brown
Dominant Soil Mineral	Kaolinite

Table 3. Summary of physical properties of the cement.

Property	Cement
Soundness, mm	0.38
Specific gravity, %	3.13
Fineness, m^2/kg	362
3 days UCS, Mpa	29.1
Bulk density, kg/m^3	1101
Initial Setting time, min	33.4
Final Setting time, min	242
Color	Ash

Table 4. Summary of physical properties of the sawdust ash.

Property	Sawdust Ash
Natural moisture content, %	26.07
Specific gravity, %	2.09
Liquid limit, %	22.5
Plastic limit, %	12.7
Plasticity index, %	10.2
Porosity, %	82.4
Fineness, micron	650
Rate of burning, g/hr	2224
Water absorption, %	18.23
Apparent	1.056
Linear shrinkage, %	5.05
Bulk density, kg/m^3	412
Modulus of rupture, Mg/m^3	1.249
Color	Ash

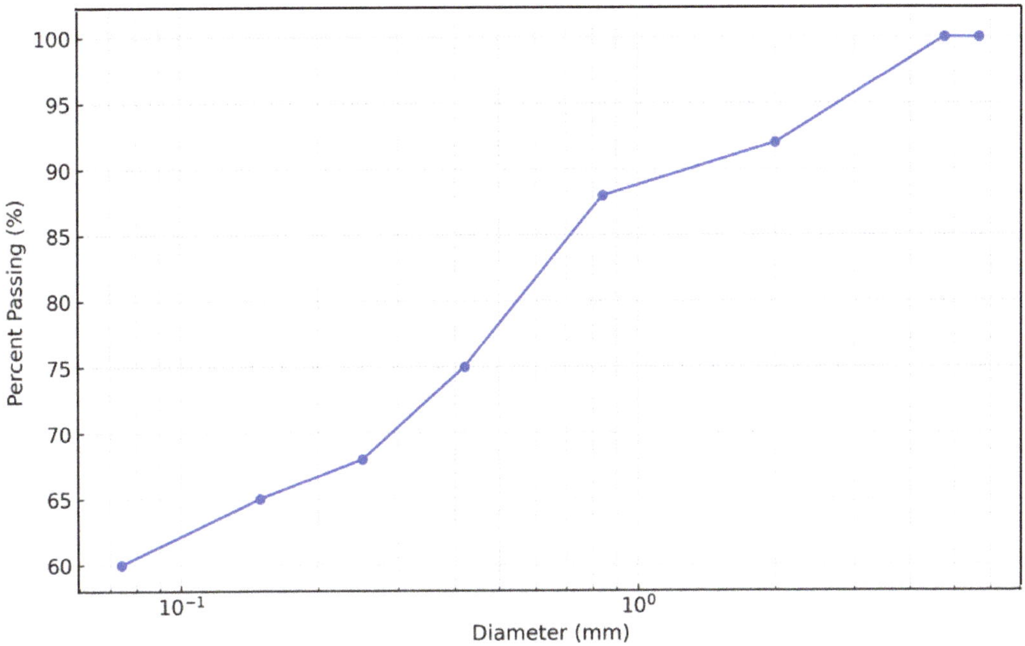

Figure 3. Particle size distribution curve.

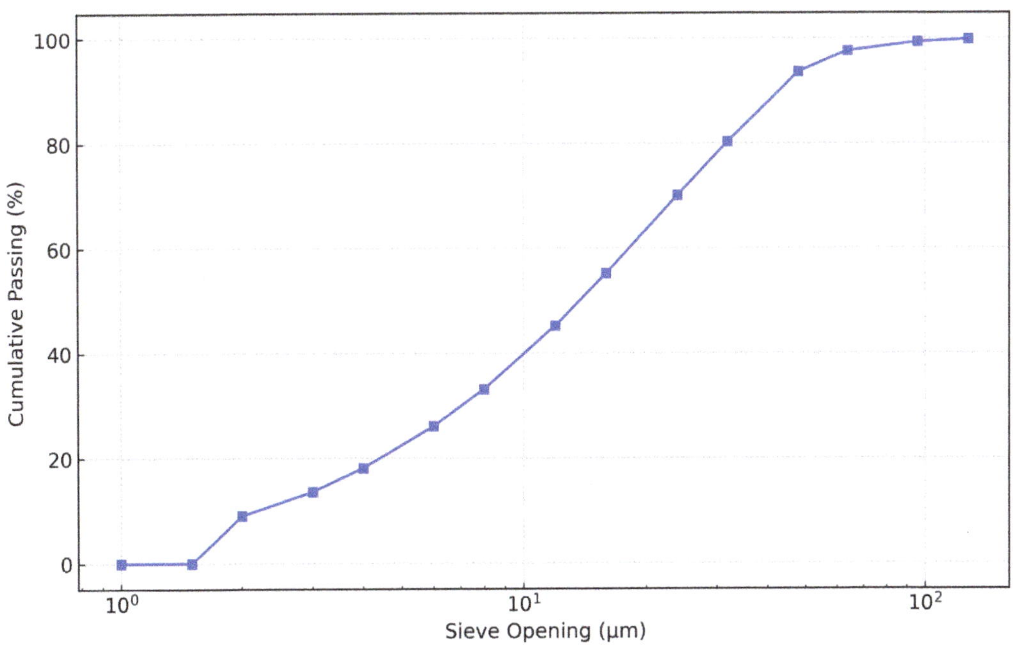

Figure 4. Particle size distribution curve of the cement.

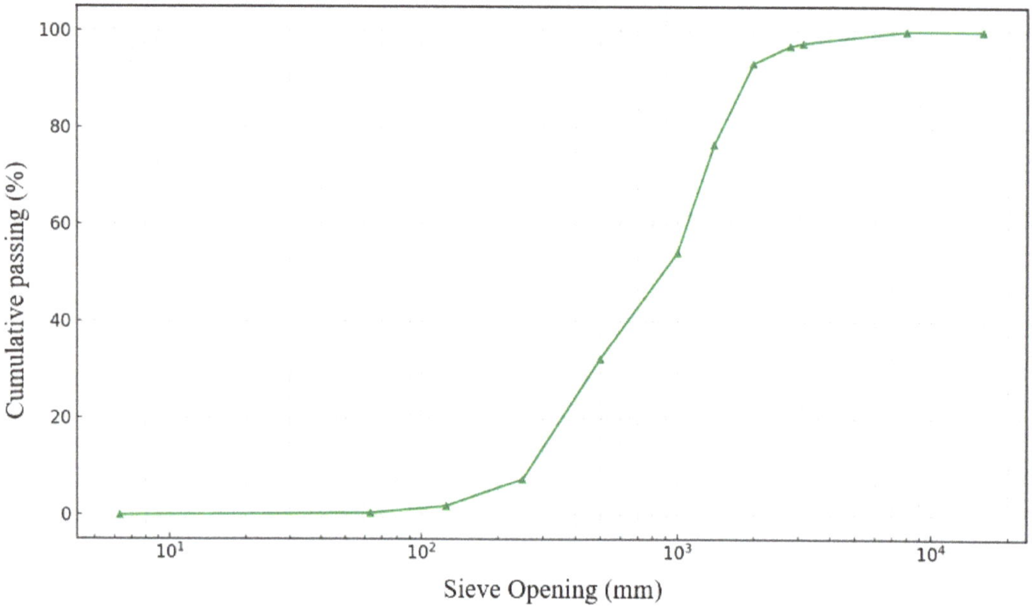

Figure 5. Particle size distribution curve of the sawdust ash.

2.1.2. Sawdust Ash

The sawdust used in the study was sourced from a local sawmill in the Gwamaja Yan Katako Dala Industrial District in Kano State (see Figure 6a). Subsequently, the sawdust was cleaned to eliminate excessive bark and organic matter. It was then air-dried and incinerated in a combustion furnace at 500 °C (see Figure 6b). The incinerated sawdust was stored in an air-tight container to avoid moisture exposure or contamination. The incinerated sawdust used in this study mainly comprised silica, alumina, and lime, with smaller amounts of other metal oxides, as shown in Table 3.

(**a**) Dried sawdust.

(**b**) Burning of sawdust.

Figure 6. *Cont.*

(**c**) Sawdust ash.

Figure 6. Sawdust process to ash.

2.1.3. Cement

The ordinary Portland cement (OPC) utilized in this study was commercially obtained from a local market in Sharada, Kano State. The chemical oxide compositions of both the cement and the sawdust ash are provided in Figure 7.

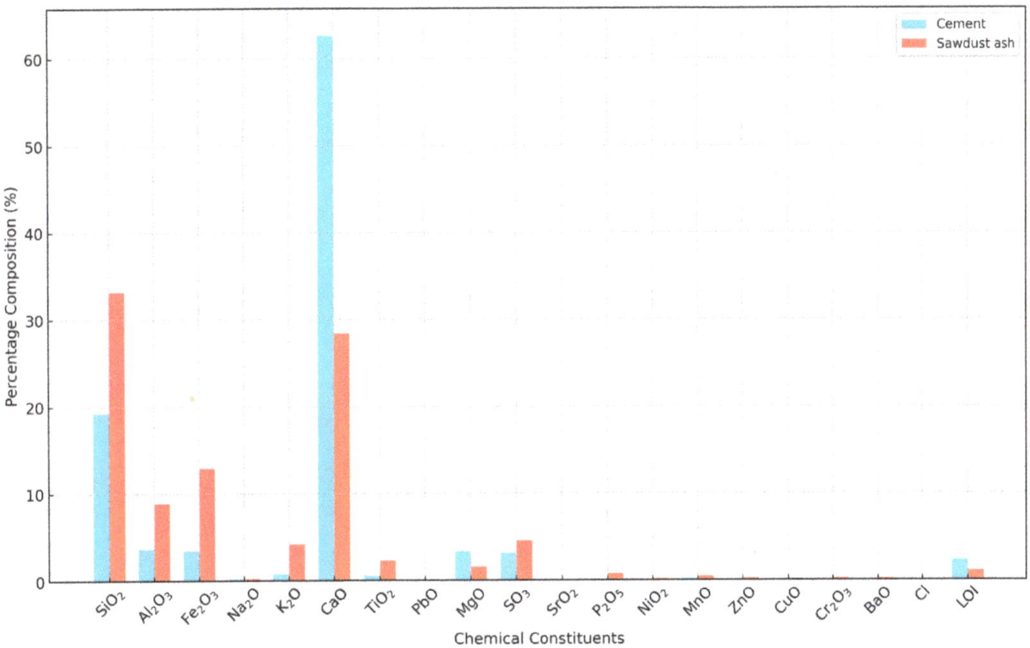

Figure 7. Percentage of oxide composition of cement and sawdust ash.

2.2. UCS Test

According to ASTM standards, the UCS test is a laboratory method for measuring chemicals' effect on improving fine-grained soils' engineering properties. The test involves preparing soil samples and applying a compressive force until they break, as in Figure 4. The standard procedure for this test is given in ASTM D2166 [29]. The test results provide valuable data on soil strength, which is critical in evaluating the feasibility of soil stabilization techniques. UCS test results are compared between stabilized and non-stabilized soils to determine the effectiveness of stabilizing agents. The test is critical to geotechnical studies and provides essential information for engineering design and construction projects. Figure 8 shows a triaxial testing machine. The same mixing and moisture conditioning is mentioned in the test protocol. The moist mixtures were compacted each in a 38 mm diameter × 76 mm height mold, applying 25 blows at each of the three layers with a 24.5 N rammer. After unmolding, sealing, and curing the specimen for 28 days in the humidity room, following ASTM D2166, the specimens were tested under unconfined axial compression at a strain rate of 0.5%/min using a loading machine with a spherically seated top plate, recording the peak compressive load at failure and documenting the complete stress–strain behavior until 20% strain.

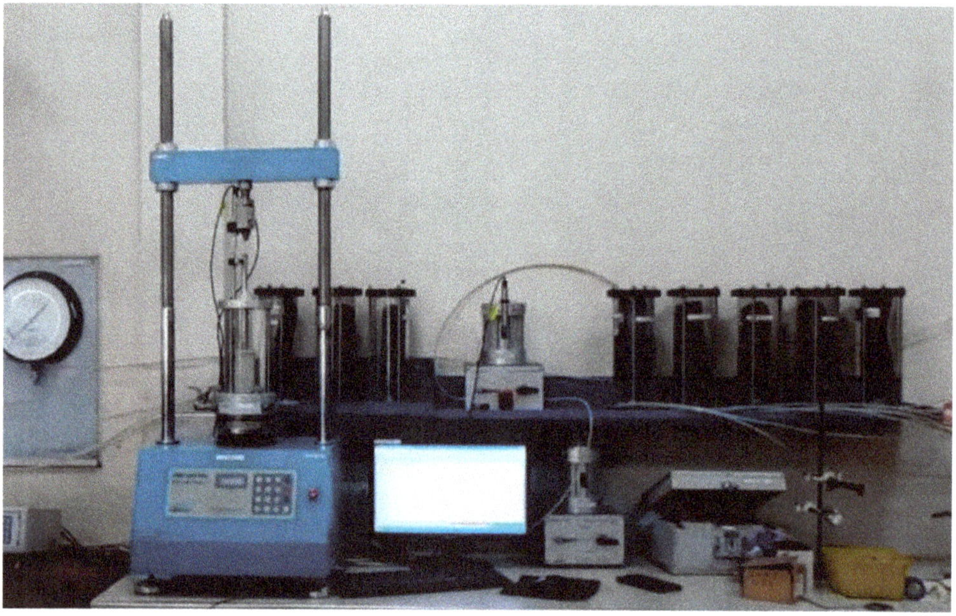

Figure 8. Triaxial machine.

2.3. Hydraulic Conductivity

The hydraulic conductivity test is a laboratory method for determining the rate at which water flows through soil [30]. The test follows standards that provide instructions for preparing soil samples and measuring hydraulic conductivity. The test involves applying a water pressure differential to the soil sample and recording the water flow rate. Hydraulic conductivity test results help to evaluate soil suitability for various uses, such as drainage, irrigation, and waste disposal [31]. The test results can also help to measure the effect of soil stabilization techniques. The falling head permeability test apparatus is shown in Figure 9.

For hydraulic conductivity testing following ASTM D5084 [32], 24 different soil–binder mixtures were prepared by thoroughly mixing the soil with varying proportions of cement (0–9% by dry weight) and SDA (0–10% by dry weight) in dry conditions, adding the desired moisture content until a uniform, moist mixture was obtained. A rigid-wall compaction mold of 71 mm diameter × 25 mm height was used, and three layers were compacted for each moist mixture, applying 25 blows per layer with a 24.5 N manual rammer. The compacted specimens were sealed and cured for 28 days in a humidity room maintained at >95% relative humidity and 23 ± 2 °C. Post-curing, each specimen was placed in a rigid-wall permeameter cell, confining pressure of 34.5 kN/m^2 was applied, and the hydraulic conductivity was measured by conducting a falling head test, recording water flow rates through the specimen.

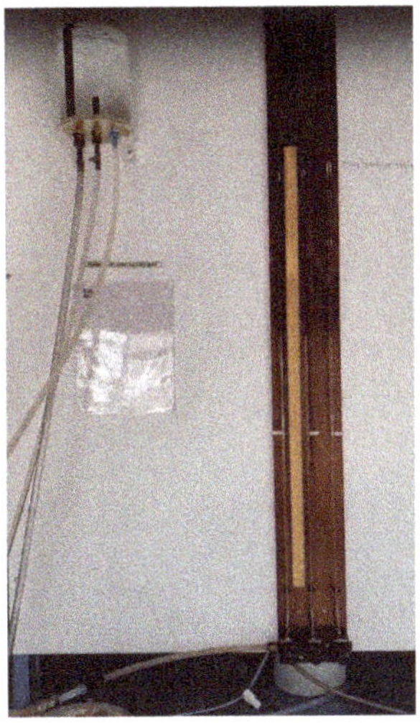

Figure 9. Falling head permeability test.

The hydraulic conductivity is evaluated from Equation (1).

$$k = 2.3\,\frac{aL}{At}\,\log_{10}\frac{h1}{h2} \quad (1)$$

k = hydraulic conductivity
a = cross-section area of standpipe
L = length of sample
A = cross-sectional area of the sample
h_1 = head at time, t_1
h_2 = head at time, t_2.

2.4. Volumetric Shrinkage

The volumetric shrinkage strain (VSS) test determines the extent of volume change when an intact soil sample can dry and shrink freely under controlled conditions [33]. The test involves taking an undisturbed soil sample, saturating it with water, determining its initial mass and volume characteristics, and then allowing it to dry entirely while carefully monitoring changes in mass and volume over time [34]. As the saturated soil sample dries, it contracts, and its volume decreases due to the evaporation of pore water held under tension in the delicate pores of the soil matrix. By determining the change in volume from the initial saturated state to the final oven-dried state, the volumetric shrinkage strain can be quantified as a percentage of the original sample volume. The test measures expansive soils' susceptibility to significant volume change as moisture content is reduced under structures or fill placement.

For volumetric shrinkage strain testing per ASTM D4943 [35], the same soil–binder mixtures prepared for hydraulic conductivity and UCS specimens. For each mixture, the moist soil–binder mixture was compacted in a 63.5 mm diameter × 127 mm height rigid-wall mold, applying 25 blows per layer with a 24.5 N rammer. Immediately after compaction, the initial specimen dimensions were measured with a caliper to 0.01 mm precision. The specimen was then placed in an oven maintained at 110 ± 5 °C to dry completely until constant mass, after which the final specimen dimensions was measured.

Then, the volumetric shrinkage strain was expressed as the percentage change in volume from the initial to the final state from Equation (2).

$$\text{VSS} = \frac{\text{Final volume} - \text{Initial volume}}{\text{Initial volume}} \times 100\% \quad (2)$$

2.5. Analysis of Variance (ANOVA)—Two-Way Analysis of Variance (ANOVA) without Replication

Analysis of variance (ANOVA) is a statistical test used to analyze the differences between group means in a sample. It compares the means of three or more independent groups to determine whether there is evidence that the associated population means are significantly different [36,37]. ANOVA tests the null hypothesis that all population group means are equal against the alternative hypothesis that at least one group mean is different from the others [28]. The test statistic used in ANOVA is the F-ratio, the ratio of the calculated between-group variance to the within-group variance. A large F-ratio indicates that the differences between group means are more significant than differences due to chance alone. Suppose the *p*-value associated with the F-test is less than the threshold (often 0.05) [38]. In that case, the null hypothesis of equal means is rejected at that level of significance, and it is concluded that there is a difference between the population means. ANOVA helps to determine whether a factor affects the mean of a dimension and allows an understanding of the differences between group means that raw data alone may not provide [39].

For this study, the ANOVA method is used to provide a more comprehensive understanding of the research. The use of sawdust ash as a stabilizer presents a sustainable alternative, as it involves utilizing a waste product from the timber industry. The research contributes to the development of effective and environmentally friendly methods for improving the engineering properties of fine-grained soils, which is of great significance in construction and environmental engineering. Also, the use of ANOVA in this study can help in identifying research gaps, comparing the findings with previous research, and establishing the novelty of the study by statistically evaluating the effectiveness of the cement and SDA combination for stabilizing fine-grained soils as clay liner materials.

2.5.1. *p*-Value (Probability Value)

The *p*-value is the probability of obtaining a result at least as extreme as the observed result, assuming that the null hypothesis is true. The null hypothesis typically states that there is no significant difference between the means of the groups being compared. A smaller *p*-value indicates more substantial evidence against the null hypothesis, while a more significant *p*-value indicates weaker evidence against the null hypothesis [40]. Typically, a significance level (α) is set, usually at 0.05 or 0.01. If the *p*-value is less than the significance level, the null hypothesis is rejected, suggesting that at least one of the groups is significantly different from the others. The *p*-value represents the probability of observing the obtained results or more extreme results, purely by chance if the null hypothesis is true. A small *p*-value indicates that the observed differences are unlikely to have occurred by chance alone, and there is likely a significant effect or difference between the groups [38]. The *p*-value is calculated from the F-distribution, a probability distribution that depends on the degrees of freedom for the numerator (dfN) and the denominator (dfD). The *p*-value is calculated from Equation (3).

$$p - \text{value} = (\text{F} \geq \text{F}_{\text{observed}} | \text{H}_0 \text{ is true}) \quad (3)$$

where:

$\text{F}_{\text{observed}}$ = calculated F-value from the data;
H_0 = null hypothesis, which states no significant difference between the group means;
p-value = the probability of observing an F-value as extreme or more extreme than the calculated F-value, assuming the null hypothesis is true.

2.5.2. F-Value (F-Statistic or F-Ratio)

The F-value is a ratio of two mean square values, i.e., the mean square due to the tested factor (treatment or effect) and the mean square due to the residual or error term. A larger F-value indicates that the variation between group means is much more considerable than the variation within groups, suggesting a significant effect of the factor being tested. The F-value is compared against a critical F-value from the F-distribution table, which depends on the degrees of freedom for the numerator and denominator and the chosen significance level (α). If the calculated F-value exceeds the critical F-value, the null hypothesis is rejected, indicating that at least one group mean significantly differs from the others [41].

The F-value represents the ratio of the variability between groups (due to the factor) to the variability within groups (due to random error). A larger F-value suggests that the tested factor significantly affects the response variable, as the variability between groups is much larger than the variability within groups. The F-value is calculated as the ratio of the mean square between the groups (treatment or effect) and the mean square within the groups. The $F - \text{value}$ is expressed from Equation (4).

$$F - \text{value} = \frac{MS_{\text{Between Groups}}}{MS_{\text{Within Groups}}} \tag{4}$$

where:

$$MS_{\text{Between Groups}} = \frac{\text{Sum of Squares between the groups}}{\text{Degrees of freedom between the groups}}$$

$$MS_{\text{Within Groups}} = \frac{\text{Sum of Squares within the groups}}{\text{Degrees of freedom within the groups}}$$

The sum of squares (SS) is calculated from Equations (5)–(7).

$$SS_{\text{total}} = \sum (y_i - y_{\text{mean}})^2 \tag{5}$$

$$SS_{\text{Between Groups}} = \sum \left(y_{\text{group mean}} - y_{\text{mean}}\right)^2 \tag{6}$$

$$SS_{\text{Within Groups}} = SS_{\text{total}} - SS_{\text{Between Groups}} \tag{7}$$

where:

y_i = individual observation;
y_{mean} = overall mean;
$y_{\text{group mean}}$ = mean of each group.

The degrees of freedom (df) are calculated as:

$df_{\text{between groups}}$ = number of groups − 1;
$df_{\text{within groups}}$ = total number of observations − number of groups;
df_{total} = total number of observations − 1

The F-value is then compared to the critical F-value from the F-distribution table, which depends on the degrees of freedom for the numerator ($dfN = df_{\text{between groups}}$) and the denominator ($dfD = df_{\text{within groups}}$), as well as the chosen significance level (α). If the calculated F-value exceeds the critical F-value, the null hypothesis is rejected, indicating that at least one group mean is significantly different from the others. These equations and calculations are used in ANOVA to determine the statistical significance of the factors being tested and their effects on the response variable.

3. Results and Discussion

Preliminary testing included the determination of the soil's specific gravity, natural moisture content, grain size distribution, free swelling properties, and Atterberg limits. In addition, engineering strength tests were performed, such as establishing the moisture–

density relationship, evaluating unconfined compressive strength, measuring hydraulic conductivity, and performing volumetric shrinkage tests on the soil samples.

3.1. Specific Gravity

Specific gravity, defined as the ratio of the unit weight of a material to the unit weight of water, is a quantitative measure of the density of soil particles, excluding the voids between them [42]. Figure 10 shows the specific gravity of stabilized soil mixtures containing varying sawdust ash (SDA) and cement proportions. The specific gravity ranges from 2.68 to 2.8 for all stabilized soil combinations, reflecting the additives' effects. The unamended soil has a baseline specific gravity of 2.68, which increases sequentially with the incremental additions of cement and SDA. This is consistent with the fact that both industrial by-products contain heavier and denser residuals that increase the overall particle density of the stabilized soil matrix [43]. Further observation shows that the specific gravity peaks at 6% SDA for a given percentage of cement, regardless of the amount used. However, beyond 6% SDA addition, excess unburned carbon in the ash tends to reduce the composite particle density again. The cement action exhibits similar initial increase and subsequent decrease behavior, with optimum densification occurring at 6% cement addition.

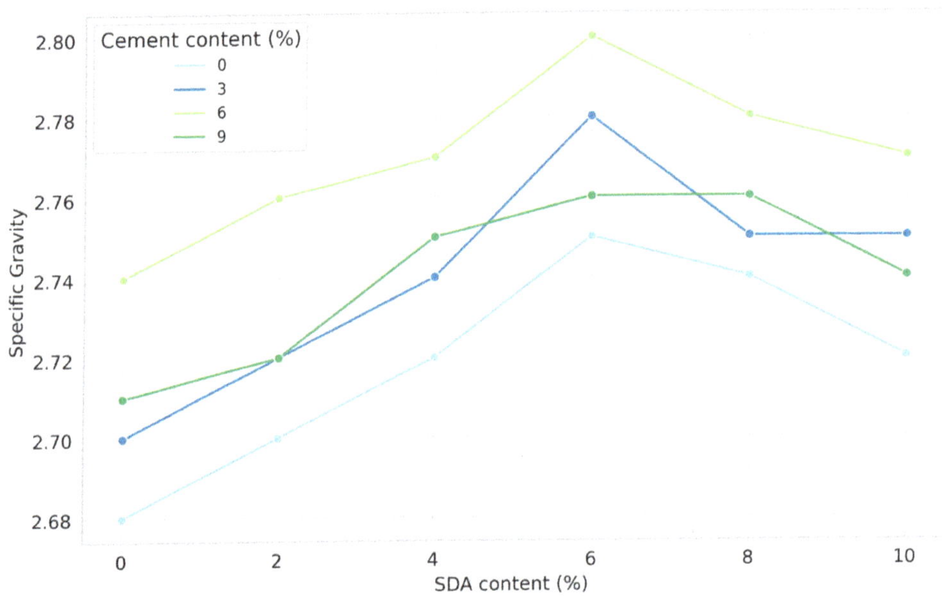

Figure 10. The amended and unamended soil samples' specific gravity characteristics.

The sawdust ash and cement increasing the soil's particle density until an optimal dosage is consistent with findings from studies on other industrial by-products like fly ash and slag [42]. These works also reported the existence of a threshold beyond which excessive additive amounts became detrimental to densification, corroborating our 6% optimum dosage. The initial increase in specific gravity can be attributed to the denser particle packing facilitated by the finer cement and ash particles filling the voids between soil grains. The subsequent decrease likely arises from excess unburnt carbon residues in the ash disrupting the stabilized matrix beyond the optimal proportion.

ANOVA of Specific Gravity

A two-factor ANOVA test without replication was performed to determine the statistical significance of the effects of varying cement and sawdust ash contents on the specific gravity of the stabilized soils under investigation (see Table A1 in Appendix A). The null hypothesis was that the means of the specific gravity data sets were equal for all cement–SDA proportions. At the 5% significance level, the analysis yielded F-statistic values of 66.2245 and 52.2653 for cement and SDA, respectively, which exceeded the critical F-value of 3.28738. Therefore, the null hypothesis was rejected. Thus, it can be concluded that cement and sawdust ash inclusions have significant measurable effects on the specific gravity, which is representative of the bulk density. The magnitude of the effect is also greater for cement, as indicated by its higher F_{CAL} vs. SDA. Cement dominance indicates that the compaction potency is primarily due to the fineness and pozzolanic properties of the cement particles as they blend into the soil matrix. Regarding statistical techniques, while ANOVA effectively established the significance, methods like multiple regression analysis could provide further insights into quantifying the individual parameter effects. The observed predominance of cement over sawdust ash aligns with studies that showed cement's superiority over lime in densification due to its higher pozzolanic reactivity [44].

Finally, the ANOVA testing provides substantial evidence that the additives produce notable and statistically significant improvements in stabilized soil density as quantified by specific gravity analysis. While the strength effects of cement/sawdust ash and various types of industrial waste have been extensively evaluated, the novelty of the current work lies in the use of statistical ANOVA tools to specifically establish significant causal relationships between low-cost sawdust ash alongside cement and its ultimate densification effects on treated native soil.

3.2. *Atterberg Limits*

3.2.1. Liquid Limit

The liquid limit (LL) test determines the water content during the transition between the liquid and plastic states in clay [45–48]. The liquid limit in Figure 11 shows a decreasing trend with increasing cement and sawdust ash (SDA) percentages in all mix combinations. This can be attributed to the replacement of clay with cementitious additives in the soil matrix, reducing its overall plasticity [42]. The decreasing pattern in LL is more pronounced for cement addition, with values consistently decreasing from 61.7% for untreated soil to as low as 32.9%. This highlights the dominant influence of cement over SDA in reducing the plastic zone of the soil through its superior coagulation effects, thereby lowering the optimum moisture requirements. The lowest LL of 30.1% occurs for the 6% cement and 6% SDA mix, representing the optimum additive composition for plasticity modification. In addition, some rebound is observed, perhaps due to the predominance of silt-sized SDA particles that interfere with soil–cement bonding. Nevertheless, cement couples more effectively with SDA up to 6% inclusion, ensuring significant LL reductions.

The decreasing liquid limit trend with increasing cement and sawdust ash dosages aligns with previous works by Nair et al. [49] and Harbottle et al. [50] that observed similar effects when using fly ash and ground granulated blast-furnace slag as soil additives, respectively. This consistency across different waste-based binders points to their common mechanism of reducing soil plasticity by progressively replacing the high-surface-area clay fraction with denser, less plastic particles. The rebound in liquid limit at higher additive doses is likely due to the disruption of inter-particle bonding by excess solids, as also reported in a study on lime stabilization [51]. The size range of the sawdust ash particles could also play a role, with finer fractions being more effective at filling voids as compared to the coarser silt-sized particles predominant beyond the optimal dosage. This aspect warrants further investigation via microstructural characterization techniques.

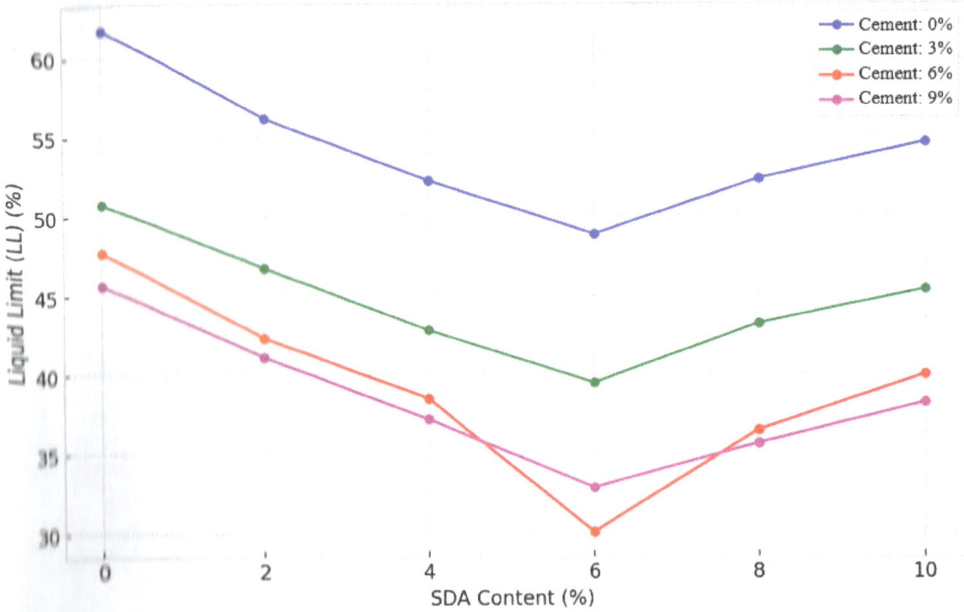

Figure 11. The amended and unamended soil samples' liquid limit characteristics.

3.2.2. ANOVA of Liquid Limit

The individual and interactive effects of different percentages of cement and sawdust ash (SDA) inclusions on the liquid limit (LL) of the stabilized expansive soils studied were quantified using the two-factor nonreplicated ANOVA test (see Table A2 in Appendix A). Following the standards of Andersen and Sivakugan, a 5% significance level was chosen to test the null hypothesis of no difference in mean LL values between the compositions [52]. The resulting F-statistic values were 278.907 and 73.77099 for cement and SDA, respectively, higher than the critical F-value of 3.28738. Therefore, the null hypothesis was conclusively rejected, indicating that both variables have statistically significant effects on the LL response. Furthermore, cement has a more dominant role between the two additives, as evidenced by its significantly higher F_{CAL}, in line with the discussions of Okyay and Dias [53]. This dominant influence can be attributed to cement's intense flocculation and moisture absorption effects. Overall, the ANOVA tests have demonstrated that the inclusion of cement and SDA, within acceptable limits, can significantly modify the consistency limits of expansive soils and that the changes can be quantitatively captured using statistical tools.

While previous studies by Faluyi and Akinmusuru focused on the effects of clay mineralogy on LL [54,55], the uniqueness here lies in the use of ANOVA to precisely determine the optimal waste binder formulations to regulate soil plasticity using locally available sawdust ash. Most prior studies, like Güllü and Kadioğlu, focused on characterizing liquid limit changes but did not employ statistical tools for compositional optimization. The ANOVA approach allows us to quantify the individual and interactive effects of cement and sawdust ash rigorously [53].

3.2.3. Plastic Limit

The plastic limit (PL) defines the moisture content at which clay transitions from a semi-solid state to a plastic solid that can be rolled into threads without crumbling [42]. The PL trends in the stabilized expansive soil mixtures provide important insights (see Figure 12). Eberemu noted that the PL shows an overall decreasing trend with increasing cement–SDA percentages in all the combinations tested [56]. This is due to the dense clustering of clay plates due to cation exchange and pozzolanic reactions induced by the additives, as explained by Osinubi and Eberemu [42]. In particular, cement lowers the PL response more than SDA due to its superior plasticity-modifying properties, consistent with Nwaiwu and Osinubi's discussions [57]. For a mixture of 6% cement and 6% SDA, the optimal decrease in PL is 15.2%. This synergistic effect is due to the integrated effects of cementitious bonding and SDA particle filling reaching a peak interaction at this composition, as previously hypothesized by Ratna et al. [58]. However, beyond the threshold proportions, the PL tends to increase slightly, indicating an overload of pozzolanic remnants that interfere with forming a coherent stabilized soil fabric. Overall, the PL test data prove that the calculated amounts of cement and SDA efficiently adjust the moisture–plasticity balance within acceptable limits.

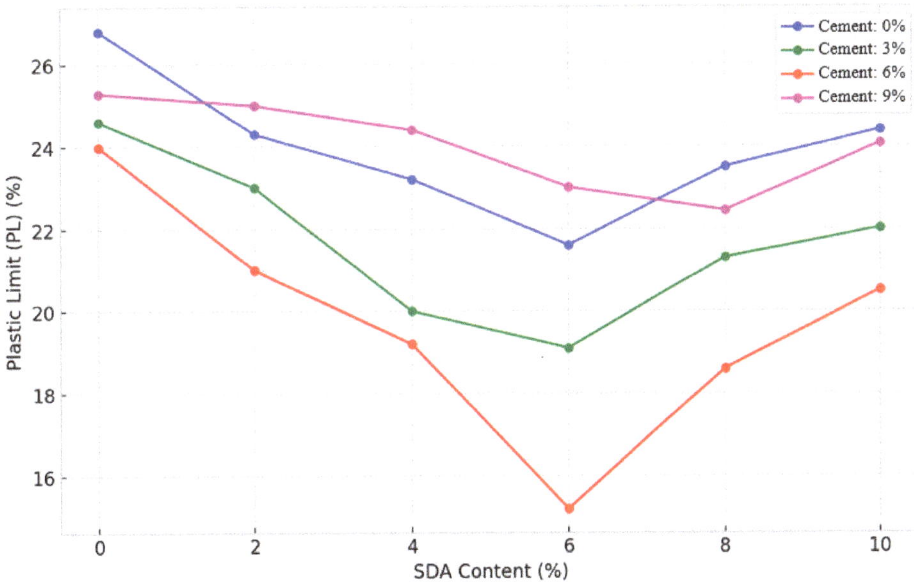

Figure 12. The amended and unamended soil samples' plastic limit characteristics.

The decreasing trend in plastic limit with increasing cement and sawdust ash dosages corroborates findings from several other researchers working with diverse stabilizing agents. Amu et al. reported similar reductions when using rice husk ash [59], while Chore and Dhole observed comparable effects with quarry dust additions [60]. This commonality stems from the fundamental mechanism of cementitious and pozzolanic additives inducing flocculation and agglomeration of clay particles. The superior performance of cement over sawdust ash aligns with the work of Al-Rawas et al., who found cement to be more effective than lime in reducing the plasticity of expansive soils. This dominance arises from cement's higher reactivity, finer particle size, and rapid cation exchange capacity compared to other supplementary additives [61]. The presence of an optimal additive dosage combination, beyond which plastic limit increases, has also been reported by other researchers such as Harichane et al. working with natural pozzolans. They attributed this effect to the potential development of dispersive forces at high additive concentrations that interfere with flocculation [16].

3.2.4. ANOVA of Plastic Limit

A two-factor ANOVA test without replication was conducted to statistically analyze the individual and collective effects of different percentages of cement and sawdust ash (SDA) on the plastic limit (PL) of the stabilized expansive soil (see Table A2 in Appendix A). At a 5% significance level, the F-test yielded F-statistic values of 25.74292 and 13.969 for cement and SDA, respectively, which exceeded the critical F-value of 3.28738. Thus, the null hypothesis of no difference between the means was rejected, which is consistent with similar conclusions by Safiuddin et al. [62]. It can be concluded that both cement and SDA induce measurable changes in the PL, proving that they are actively involved in altering soil plasticity. However, cement has a more dominant role than the two additives due to its intense flocculation effects, as Sariosseiri and Muhunthan explained [63]. Thus, the ANOVA technique has quantified and demonstrated the significance of the improvements provided by the binary additives in bridging the gap between liquid and plastic limits in the treated soil. Unlike previous ANOVA studies focused on strength properties, the uniqueness here is the use of ANOVA to precisely establish the statistical significance of cement–SDA mixtures in beneficially moderating the plasticity index of expansive soils. Indeed, most prior studies focused primarily on experimental characterization rather than rigorous quantification of statistical effects and optimization of binder proportions through techniques like ANOVA.

3.2.5. Plasticity Index

The plasticity index (PI), the numerical difference between the liquid and plastic limits, represents the moisture range over which the clay soil remains plastic [42]. Figure 13 shows that the inclusion of cement and sawdust ash (SDA) causes a downward curve in PI values across all compositional variants, as Oyelakin et al. noted [64]. This effect is related to the pozzolanic additives' replacement of clay mineral fractions, coupled with their intense water demand for hydration reactions, as Biricik et al. explained [65]. Comparatively, cement has a more pronounced decreasing effect on PI due to its excellent moisture-absorbing properties, as Kolias et al. hypothesized [12]. The addition of 6% cement and 6% SDA results in the lowest PI, proving the existence of an ideal clay–cement–SDA formulation blend where both additives work in interaction to modify the clay structure significantly. However, the slight PI rebound beyond threshold proportions mirrors the trends reported by Olutoge et al. [64] with cement-rice husk ash mixes, hypothesized to be due to excess ions hindering interparticle bonding. Similar non-linear effects on plasticity parameters have been noted in other studies evaluating cement–agriculture waste ash combinations [12]. The decreasing PI trends authenticate that incorporating calculated dosage of binder or binders helps to bridge the liquid and plastic limits, shrinking the moisture zone prone to volumetric instability, and corroborating past studies [65].

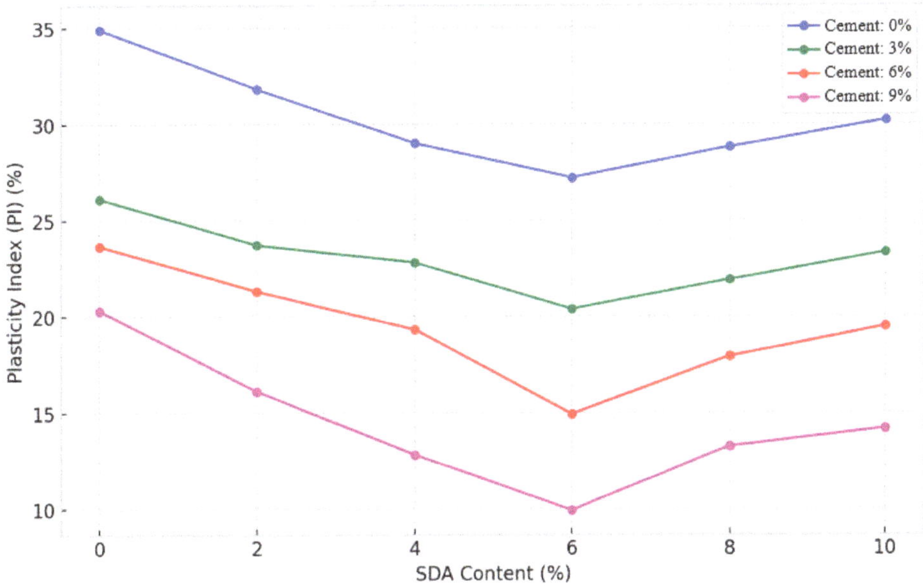

Figure 13. The amended and unamended soil samples' plasticity index characteristics.

3.2.6. ANOVA of Plasticity Index

A two-factor ANOVA test without replication was conducted to statistically verify the individual and collective effects of varying proportions of cement and sawdust ash (SDA) on the plasticity index (PI) of the stabilized expansive soil (Table A2 in Appendix A). At a 5% significance level, the resulting F-statistic was 417.5243 and 47.25471 for cement and SDA, respectively, exceeding the critical F of 3.28738. Thus, the null hypothesis was rejected, which is consistent with similar findings by Kolias et al. [12]. The analysis irrefutably proves that cement and SDA significantly modify the PI response of the stabilized soil matrix. Cement has a more pronounced influence between the two additives due to its intense flocculation-agglomeration capacities, as Horpibulsuk et al. hypothesized. As a result, cement causes more remarkable moisture migration and clay texture changes to reduce the gap between liquid and plastic states effectively [66]. The two-factor ANOVA confirmed the statistical significance of cement and SDA in reducing PI, analogous to statistical analyses by Olutoge and Saride et al. evaluating other cement–pozzolan formulations [62]. Overall, the ANOVA quantification and comparisons presented here highlight the statistically significant plasticity transitions induced in the treated expansive soil due to the synergistic interplay between cement and SDA inclusions.

3.2.7. Linear Shrinkage

The linear shrinkage (LS) test determines the one-dimensional volume reduction as a soil sample dries from the liquid limit to the oven-dry state [48]. Figure 14 shows increasing cement and sawdust ash (SDA) content significantly reduces shrinkage across all compositions, corroborating previous findings by Sabat and Muntohar with cement–ash amendments [67]. This shrinkage mitigation is attributed to two key factors supported by the literature: (1) reduced moisture affinity from cation exchange [68], and (2) limited swell–shrink capacity from pozzolanic bonding between clay and additives [48]. Cement exhibited a more pronounced shrinkage reduction effect compared to SDA, consistent with Muntohar's work on cement–rice husk ash stabilization [69]. This aligns with the excellent moisture-absorbing abilities of cement postulated by Kolias et al. and verified in other studies [67]. The minimum 4.14% LS occurred for 9% cement–10% SDA, suggesting an optimal formulation where cement hydration and SDA filler effects cooperatively alter the clay structure, analogous to findings with cement–fly ash blends [12]. The marginal LS fluctuation likely stems from excess ions impeding interparticle bonding, as hypothesized by Ingles and Metcalf and observed in other works evaluating high binder contents [69]. This non-linear behavior aligns with the generic volume change patterns reported for cement-stabilized clays [70]. The decreasing LS trends confirm cement chemically limits swelling potential, while SDA physically restricts moisture ingress, both mechanisms attenuating shrinkage appreciably within acceptable limits. This dual mechanism has been noted in prior studies on agricultural and industrial waste ashes [42].

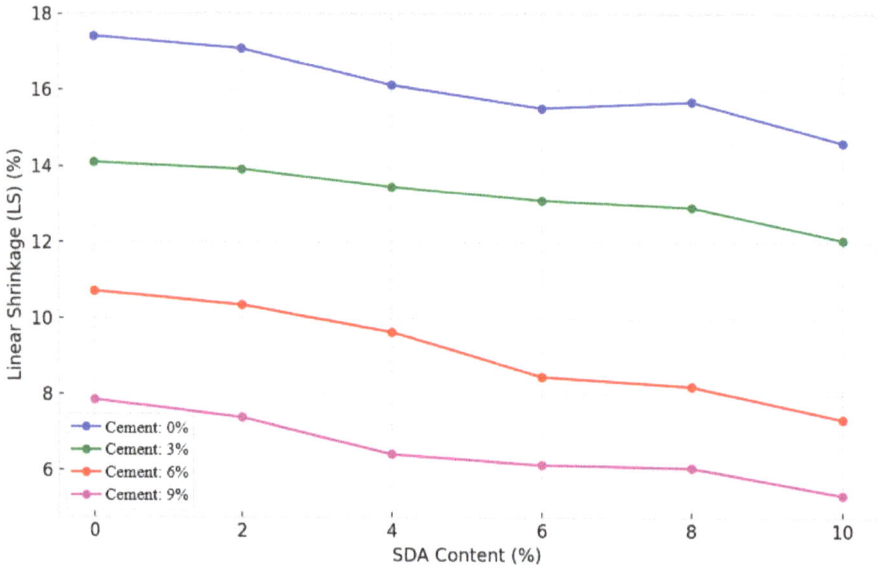

Figure 14. The amended and unamended soil samples' linear shrinkage characteristics.

3.2.8. ANOVA of Linear Shrinkage

A two-factor ANOVA test without replication was performed to statistically quantify the individual and collective effects of different doses of cement and sawdust ash (SDA) on the linear shrinkage (LS) response of the stabilized expansive soil as described in the standard guidelines of Daniel and Wu [12]. At 5% significance, the resulting F-statistic was 1289.157 for cement and 47.1054 for SDAs, exceeding the critical F-statistic of 3.28738 (Table A2 in Appendix A). Thus, the null hypothesis was conclusively rejected, which is consistent with similar studies by Sabat [67]. The analysis provides evidence that cement and SDA exert significant shrinkage modification effects, supporting their active participation in moisture affinity adjustments within the stabilized soil matrix. Cement plays a more dominant role between the two additives due to its prolific pore fluid absorption capacity, as hypothesized by Kolias et al. [12].

Consequently, cement causes more significant moisture migration from clay clusters to effectively reduce the volume change tendency. Overall, the ANOVA quantification and comparisons highlight the statistically substantial shrinkage changes induced in the expansive soil by the combined influences of cement and SDA inclusions. The ANOVA results statistically validate the considerable individual and combined influences of cement and SDA on shrinkage mitigation, corroborating similar statistical analyses by other researchers [70]. Cement exerted a more dominant role, ascribed to its superior fluid transport and moisture depletion capacity from clay surfaces [71].

3.2.9. Compacted Behavior

The standard Proctor compaction test determines the optimum moisture content (OMC) at which a soil's maximum dry density (MDD) is achieved under a given compaction effort. Analysis of the MDD and OMC trends with varying cement and sawdust ash (SDA) provides valuable insights (see Figure 15a–d). The OMC ranged from 9.7% to 34.85%, spanning a wide spectrum of moisture levels relative to the optimum moisture content (OMC), including molding points at −2%, 0%, +2%, and +4% of OMC. The dry density values exhibited a considerable range from 1.314 to 1.745 g/cm^3, while the specific gravity varied from 2.68 to 2.8, reflecting the changes in soil composition. Notably, Figure 15a–d includes the concept of the zero air-void line (ZAVL), which represents the theoretical maximum dry density achievable when the soil is fully saturated without any air voids. The ZAVL serves as a crucial reference baseline, as reliable compaction results should not surpass this line due to the inherent presence of air voids during practical compaction processes. For instance, a mixture composed of 100% soil and 0% cement and 0% SDA exhibited a maximum dry density of 1.63 g/cm^3 at an optimum water content of 23%.

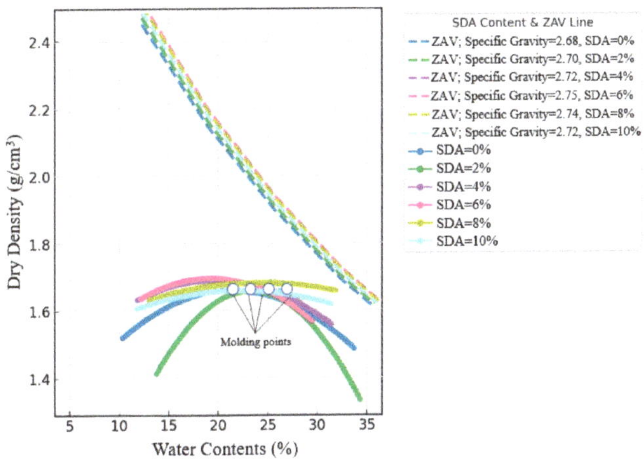

(**a**) Cement content: 0%.

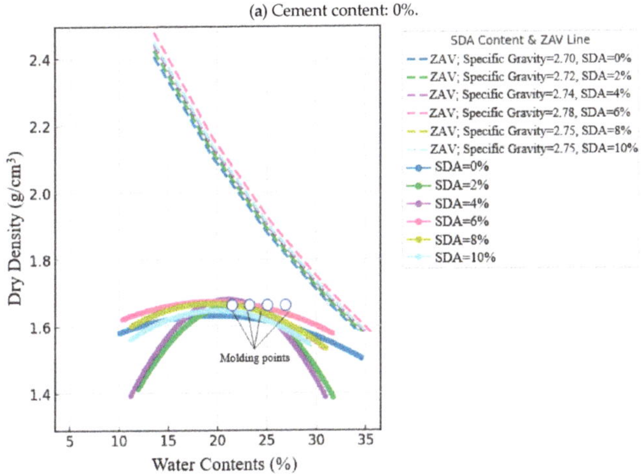

(**b**) Cement content: 3%.

Figure 15. *Cont.*

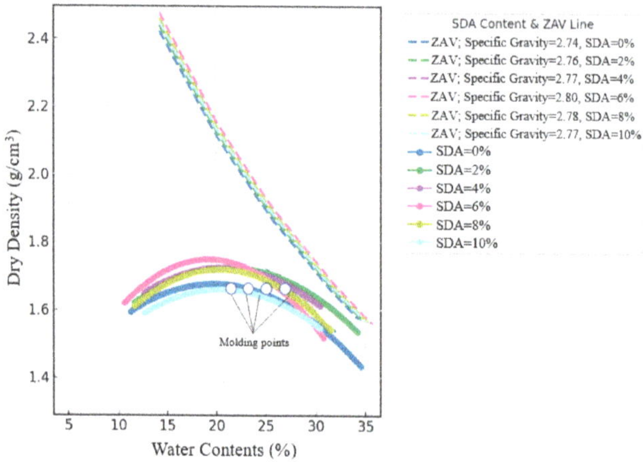

(c) Cement content: 6%.

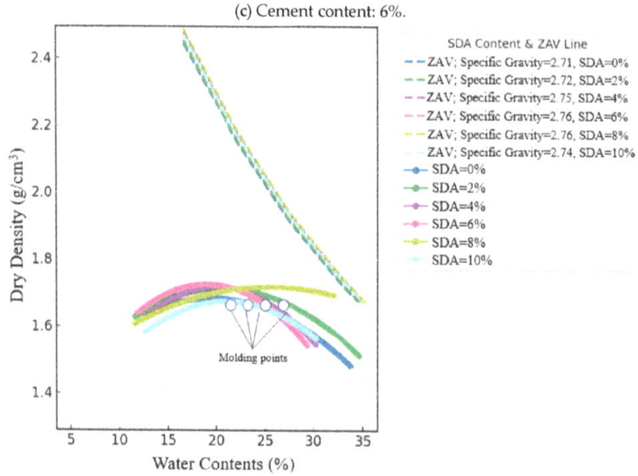

(d) Cement content: 9%.

Figure 15. The amended and unamended soil samples' compaction curves.

As noted by Agbede and Joel and corroborated by others, both MDD and OMC decrease progressively with increasing cement content at all SDA doses. This is consistent with the intense moisture demand due to accelerated pozzolanic reactions forming cementitious gels [72]. However, the descending MDD–OMC pattern is more pronounced for SDA augmentations, aligning with findings from studies on other agricultural waste ashes like rice husk ash [73]. This behavior is attributed to the dry, highly porous SDA particles significantly absorbing pore fluid while occupying larger volumes themselves, resulting in a relatively lower bulk density. The maximum MDD and minimum OMC occurred for the 6% cement–6% SDA blend, suggesting an optimal moisture adjustment by the complementary cement hydration and SDA filler mechanisms before excess residuals start disrupting the clay–water balance. Similar optimal dosage trends have been reported for cement–fly ash and cement–GGBS combinations [74].

The decreasing MDD and OMC trends confirm that adding cement and SDA beneficially tailors the compaction response through synergistic moisture regulation mechanisms, as observed in prior works on binary binder stabilization. This moisture buffering action improves the compacted fabric and density.

3.2.10. ANOVA of Compacted Behavior

A two-factor ANOVA without replication was conducted to determine the statistical significance of cement and sawdust ash (SDA) inclusion on the compaction characteristics of stabilized expansive soil. For MDD, at a 5% significance level, the F-statistics were 51.31148 and 20.90164 for cement and SDA, respectively, exceeding the critical F-value of 3.287382 (Table A3 in Appendix A). Thus, both additives were shown to have significant measurable effects on the maximum dry density response. Between the two, cement had a more pronounced influence, in line with the extensive discussions by Sabat [67]. This is due to the formation of stable cementation bonds that rigidify the composite soil structure. Similarly, for OMC, the F-statistics were 39.33962 and 23.18868 for cement and SDA, respectively, exceeding the critical F-value. Thus, significant effects were statistically observed on the moisture–density equilibrium imparted by both binary additives, especially on the moisture-requiring cement hydration reactions. The results indicate that both additives exerted statistically significant effects on MDD and OMC, corroborating similar ANOVA outcomes from related studies on cement–pozzolan-stabilized soils [71].

3.3. Key Engineering Parameters to Characterize Clay Liner Barrier

The suitability of the locally available clayey soil as a hydraulic barrier material in landfill liner systems was evaluated by conducting several critical engineering tests to determine if it met the established design guidelines. First, standard Proctor compaction tests were conducted to establish the clay's moisture–density relationship and determine the maximum dry density and optimum moisture content. These results provide the specified compaction criteria to ensure adequate liner integrity. Unconfined compression tests measured the compacted clay's compressive strength capacity, indicating its ability to withstand the applied liner pressure without failure. Hydraulic conductivity tests using flexible wall permeameter equipment quantified the permeability coefficients that define the fluid flow rates that will occur through the liner when used to contain leachate in the field over its lifetime. Volumetric shrinkage tests determined the propensity of the compacted clay to desiccation cracking and assessed whether swelling pressures posed a stability risk. Evaluation of the soil's moisture–density response aids in the selection of appropriate compaction specifications. In contrast, the strength, permeability, and shrinkage/swelling measurements directly aid in the assessment of whether landfill liner guidelines are being met. Engineering test data and subsequent statistical analysis more reliably validate whether locally available soils can provide technically and economically viable liner alternatives when appropriately amended with industrial by-products such as sawdust ash.

3.3.1. Hydraulic Conductivity

Hydraulic conductivity values of the natural and stabilized expansive soil samples were determined over a range of compaction moisture contents from 2% dry to 4% wet of optimum. Daniel and Benson mentioned that the moisture content is plotted relative to the optimum value obtained from standard Proctor compaction [44,75–77]. Analysis of the test data, as shown in Figure 16, indicates that the hydraulic conductivity (k) response is significantly affected by the moisture content of the form. The lowest k values, approaching 1×10^{-9} m/s, were achieved at moisture levels on the wet side of the optimum, particularly at 2% wet of optimum, for most stabilized mix combinations tested. This is consistent with the pore fluid viscosity–flow rate correlations for clays noted by Daniel and Wu [12].

Furthermore, soil samples modified with 3, 6, and 9% cement and 6% sawdust ash yielded satisfactory k values within 1.0×10^{-9} m/s when compacted between 12.5 and 17.0% moisture content. The optimum composition was 6% cement and 6% sawdust ash inclusion. In addition, the effects are likely to cause microcracking, which increases permeability. The uniqueness of the current work lies in delineating the specific moisture range for minimized conductivity using industrial sawdust ash to expand siting options for compacted clay liners through improved statistical interpretation of laboratory test data.

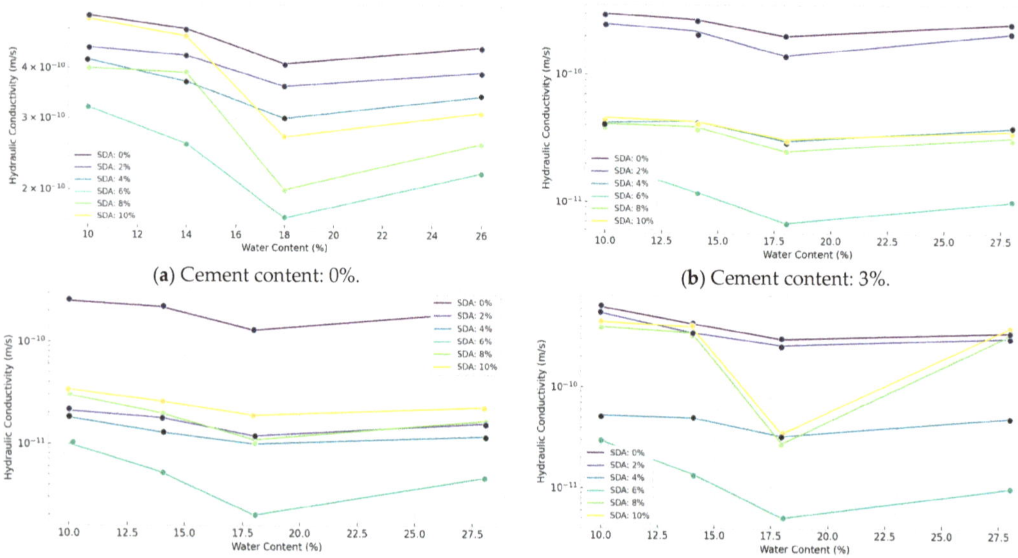

(**a**) Cement content: 0%. (**b**) Cement content: 3%. (**c**) Cement content: 6%. (**d**) Cement content: 9%.

Figure 16. Variation in hydraulic conductivity against water content.

The reduced hydraulic conductivity values at moisture contents slightly wet of optimum are consistent with findings reported by several researchers investigating stabilized soils and compacted clay liners. Kang and Shackelford noted a similar trend when working with zeolite-amended soil–bentonite backfills. They attributed this behavior to the increased soil fabric density and reduced void connectivity achieved under these near-optimal compaction conditions [12]. The existence of an ideal additive combination (6% cement + 6% sawdust ash in our case) yielding the lowest permeability values aligns with the work of Al-Tabbaa and Evans. They reported an optimum lime–GGBS proportion that maximized pozzolanic reactions and filler effects, thereby minimizing the flow potential in treated clays [78].

3.3.2. ANOVA of Hydraulic Conductivity against Water Content

ANOVA tests were performed on the hydraulic conductivity (HC) data of stabilized specimens compacted at four moisture conditions, namely, −2%, 0%, +2%, and +4% of the optimum value, to determine the statistical significance of cement and sawdust ash (SDA) additions across the compaction curve (Table A4 in Appendix A). At the 5% significance level, the F-statistics for both factors (cement and SDA) exceeded the critical F-value of 3.28738 at all moisture contents tested. This demonstrates that the pozzolanic admixtures significantly impact the permeability response regardless of the moisture content during compaction. However, between the two admixtures, cement had a consistently more pronounced influence at each moisture level, consistent with the detailed postulations of Ahmed et al. [79]. In contrast to previous statistical evaluations focusing on strength and consolidation factors, the current analysis uniquely applies ANOVA to confirm the versatility of the optimized waste binder formulation in significantly improving barrier efficiency from dry to wet of optimum moisture using locally available sawdust ash.

Most prior studies, like Shakri and Sobolev, employed ANOVA primarily to analyze strength or consolidation characteristics rather than hydraulic conductivity explicitly. Our approach quantifies how cement and sawdust ash impact barrier efficiency under varying moisture regimes. However, we acknowledge that a limitation of our study is the lack of microstructural investigations to elucidate the precise mechanisms behind the permeability trends observed. Future studies coupling hydraulic testing with techniques like SEM/XRD could provide deeper insights, as demonstrated by researchers like Mirzababaie et al. working on natural pozzolan-amended clays [78].

3.3.3. Effect of Sawdust Ash in Soil/Cement Mixture on Hydraulic Conductivity

As shown in Figure 17, the variation in hydraulic conductivity (HC) response of the stabilized soil mixtures with varying percentages of cement and sawdust ash (SDA) inclusions was determined for specimens compacted at the optimum moisture content. Analysis of the test data shows that HC values decrease progressively with increasing proportions of cement and SDA up to an optimum composition, beyond which the declining pattern recovers to some extent for higher additive doses. This is consistent with the coupled mechanisms of enhanced cementitious bonding and SDA particle filling effects, which initially reduce permeability more effectively through tailored moisture adjustments before thermal effects likely induce microcracking [79]. Specifically, the lowest conductivity coefficient of 4.7×10^{-10} m/s was obtained, corresponding to 6% cement–6% SDA content, consistent with the observations of Faluyi and Akinmusuru [48]. This confirmed the existence of an ideal soil–cement–SDA formulation mixture, in which the chemical and physical interactions between the two additives are maximized to enhance the permeability performance favorably.

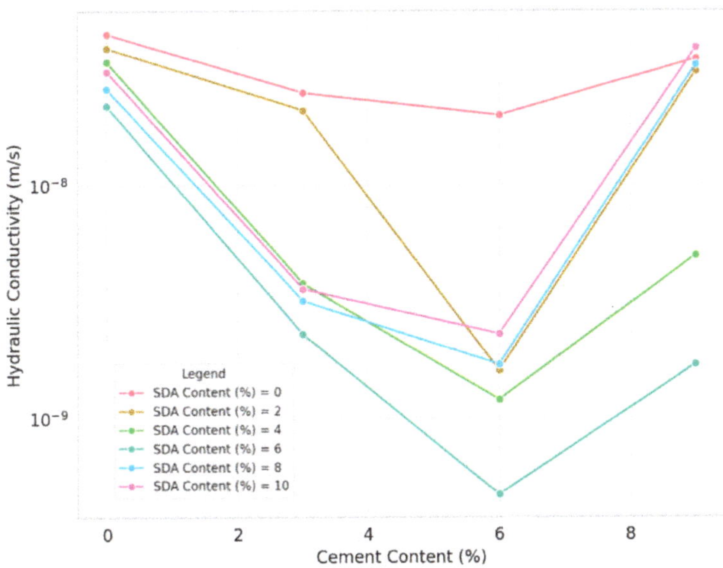

Figure 17. Variation in hydraulic conductivity of soil–cement mixture with sawdust ash content.

The decreasing trend in hydraulic conductivity with increasing cement and sawdust ash content up to an optimum point is consistent with observations reported by other researchers studying amended soils and clay liners. Moayedi et al. found a similar pattern when using coal ash as an additive, attributing it to the combined effects of pore filling, particle binding, and altered surface chemistry mechanisms [79]. The existence of an ideal admixture combination yielding minimum permeability aligns with the work of Tarefder et al., who identified optimal fly ash and lime contents for modifying the microstructure of clayey soils to restrict fluid migration. The rebound effect at higher dosages could be due to micro-cracking induced by excess unburnt/unreacted fractions, as Zhang et al. hypothesized in their study on lime-stabilized expansive clays.

3.3.4. ANOVA of Hydraulic Conductivity Results of Soil–Cement–Sawdust Ash Mixtures

A two-factor ANOVA test without replication was conducted to statistically quantify the individual and collective effects of varying proportions of cement and sawdust ash (SDA) on the hydraulic conductivity (HC) response of the treated expansive soil compacted under optimum moisture conditions (Table A5 in Appendix A). At the 5% significance level, the F-statistic values were 15.69096 and 4.788327 for cement and SDA, respectively, exceeding the critical F-value of 3.28738. Thus, statistical evidence was obtained that both pozzolanic admixtures caused significant changes in the permeability performance of the stabilized clay matrix. However, between the two additives, cement had a more predominant role in restricting moisture flow, consistent with the detailed discussions of Ingles and Metcalf [48]. Overall, the analysis provides quantitative evidence of the functional benefits of the binary blend inclusions in significantly improving the conductivity characteristics of the amended soil toward meeting the typical liner design criterion. Most prior studies primarily employed ANOVA for strength or compaction parameters rather than explicit barrier efficiency. Our work demonstrates the applicability of statistical tools in optimizing binder formulations to meet hydraulic conductivity criteria for critical applications like landfill liners. However, a limitation of the current study is the lack of microstructural characterization to elucidate the precise physicochemical mechanisms governing the observed permeability trends across the admixture range.

3.3.5. Volumetric Shrinkage Strain

The volumetric shrinkage strain (VSS) of the natural and stabilized expansive soil samples were determined using resin-impregnated soil block specimens over a range of forming moisture contents from 2% dry to 4% wet of the optimum. The moisture content is plotted relative to the optimum value obtained from standard Proctor compaction. Analysis of the test data, as shown in Figure 18, indicates that the VSS response is significantly affected by moisture content similarly for all amended and unamended soil combinations tested. The lowest VSS values approaching 4% were obtained at moisture levels on the wet side of optimum, especially at 2% wet of optimum, supporting the existence of a critical moisture threshold for minimized shrinkage potential. In addition, soil samples modified with 3, 6, and 9% cement and 6% sawdust ash yielded satisfactory VSS values within 4% when compacted between 15.0 and 23.5% moisture content. The optimum composition was 6% cement and 6% sawdust ash inclusion, consistent with the coupled effects of moisture-regulating mechanisms [48]. The uniqueness of the current work lies in delineating the specific compaction moisture range for minimized shrinkage potential using industrial sawdust ash to expand siting options for compacted clay liners through improved statistical interpretation of laboratory test data.

The existence of an optimal moisture content range (slightly wet of optimum in our case) for minimizing volumetric shrinkage strains aligns with observations reported by several researchers studying amended and compacted soil systems. Nagaraj et al. found a similar trend when investigating cement-treated expansive soils, attributing it to the formation of a denser and more stable soil fabric under these near-optimal compaction conditions [48]. The effectiveness of cement and supplementary cementitious materials like sawdust ash in reducing shrinkage potential is consistent with the work of Sabat, who studied lime–fly ash-stabilized expansive soil. He attributed the shrinkage mitigation to the pozzolanic reactions inducing pore fluid consumption and particle binding effects [79].

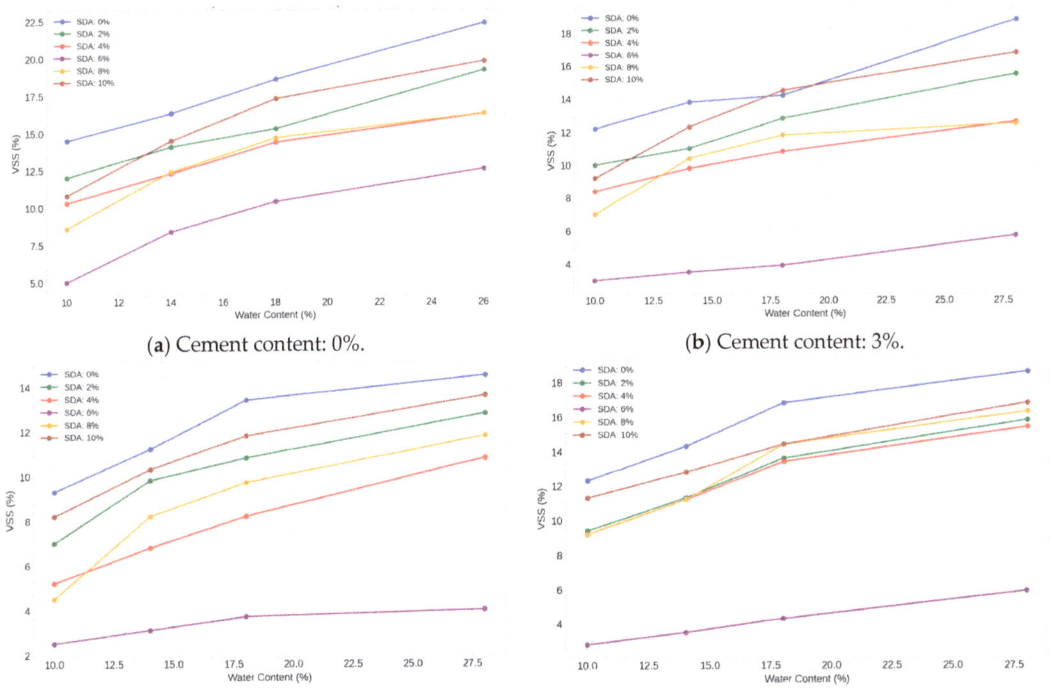

(**a**) Cement content: 0%. (**b**) Cement content: 3%. (**c**) Cement content: 6%. (**d**) Cement content: 9%.

Figure 18. Variation in volumetric shrinkage strain against molding water content.

3.3.6. ANOVA of Volumetric Shrinkage Strain against Molding Water Content

ANOVA tests were performed on the VSS data of stabilized specimens compacted at four moisture conditions, namely, −2%, 0%, +2%, and +4% of the optimum, to determine the statistical significance of cement and sawdust ash (SDA) additions across the compaction curve (Table A6 in Appendix A). At the 5% significance level, the F-statistics for both factors (cement and SDA) exceeded the critical F-value of 3.28738 at all moisture contents tested. This demonstrates that the pozzolanic admixtures significantly impact the shrinkage response regardless of the moisture content during compaction [48]. In addition, the effects of SDA were comparable to or even more significant than cement in some instances. Unlike previous shrinkage evaluations that focused on mineralogical factors, the current analysis uniquely applies ANOVA to confirm the versatility of the optimized waste binder formulation in significantly shrinking moisture-related dimensional instabilities using locally available sawdust ash.

Most prior studies, such as those by Nalbantoglu and Tuncer, primarily focused on experimental characterization rather than rigorous statistical analyses [79]. However, a limitation of the current study is the lack of detailed microstructural investigations to elucidate the fundamental mechanisms governing the observed shrinkage strain trends. Complementary techniques like mercury intrusion porosimetry (MIP) and environmental scanning electron microscopy (ESEM) could provide valuable insights into the pore structure evolution and moisture migration pathways, as demonstrated by researchers like Dang et al. studying shrinkage behavior in treated clays [78].

3.3.7. Effect of Sawdust Ash in Soil/Cement Mixture on Volumetric Shrinkage Strain

As shown in Figure 19, the variation in the VSS response of stabilized soil mixtures with varying percentages of cement and sawdust ash (SDA) inclusions was determined for specimens compacted at the optimum moisture content. Analysis of the test data indicates that the VSS values decrease progressively with increasing proportions of cement and SDA up to an optimum composition, beyond which the declining pattern recovers to some extent for higher additive doses. This is consistent with the coupled mechanisms of enhanced cementitious bonding and SDA particle filling effects that initially reduce shrinkage more effectively through tailored moisture adjustments before excess ions are likely to interfere with the stabilized clay fabric [48]. Specifically, a minimum VSS of 3.51% was obtained, corresponding to 6% cement–6% SDA content, consistent with the observations of Abdi et al. [80]. This confirmed the existence of an ideal soil–cement–SDA formulation mixture where the chemical and physical interactions between the two additives are maximized to enhance volume change performance favorably.

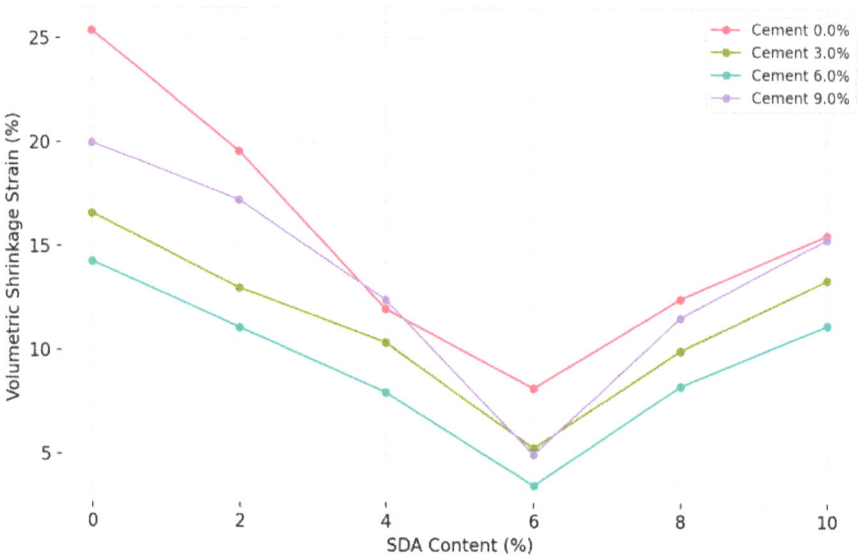

Figure 19. Variation in volumetric shrinkage of soil–cement mixture with sawdust ash content.

The optimal cement–sawdust ash combination minimizing volumetric shrinkage strains is consistent with findings reported by researchers studying other supplementary cementitious materials like fly ash and slag. Kalkan identified ideal fly ash dosages to reduce shrinkage in expansive soils, attributing it to the combined cementitious reactions and physical filler effects [80]. The dominant role of sawdust ash over cement in our study aligns with the work of Gow et al., who found pozzolanic additives like rice husk ash to be more effective than cement in reducing the surface area and moisture affinity of clays responsible for shrinkage phenomena [12].

3.3.8. ANOVA of Volumetric Shrinkage of Soil–Cement Mixture with Sawdust Ash Content

A two-factor ANOVA test without replication was conducted to statistically quantify the individual and collective effects of varying proportions of cement and sawdust ash (SDA) on the volumetric shrinkage strain (VSS) response of the treated expansive soil compacted under optimum moisture conditions (Table A7 in Appendix A). At the 5% significance level, the F-statistic values were 18.64996 and 38.157 for cement and SDA, respectively, exceeding the critical F-value of 3.28738. Thus, statistical evidence was obtained that both pozzolanic admixtures caused significant changes in the shrinkage behavior of the stabilized clay matrix. Between the two additives, SDA showed a more dominant role [80]. Overall, the analysis quantitatively demonstrates the functional benefits of the binary blend inclusions in significantly reducing the desiccation-related volume change risks of the amended soil to meet typical low-permeability liner design criteria. In contrast to previous statistical investigations focused on strength aspects, the uniqueness here lies in applying ANOVA to determine the importance of industrial sawdust ash alongside cement in tuning the moisture affinity of expansive soils to optimize barrier properties.

Most prior studies, such as those by Mehta and Mihradi, primarily focused on experimental characterization rather than rigorous statistical analyses and compositional optimization. Future studies focused on the effects of complementing shrinkage strain measurements with advanced microscopic techniques like environmental scanning electron microscopy (ESEM) could elucidate the microstructural evolution and moisture transport mechanisms governing volume change behavior across different admixture formulations, similar to the approach adopted by Dang et al. for lime-treated expansive soils [12].

3.3.9. Unconfined Compressive Strength (UCS)

A predominant finding was the significant decrease in UCS observed as moisture content increased from dry to optimum to wet (Figure 20). Excess moisture prevents complete interparticle contact, friction, and bonding, reducing the ability to withstand compaction and shear stresses. The study also demonstrates the potential of cement and SDA stabilization to produce high UCS values over a wide range of moisture contents, from dry to optimum to wet. The cement and SDA treatments appeared to counteract the weakening effect of increased mold moisture up to 4% wet of optimum, allowing UCS values over 200 kN/m^2. This highlights the effectiveness of these stabilizers in improving soil workability and stability under non-ideal moisture conditions. Mechanisms likely include flocculation and agglomeration effects that strengthen soil structure [42]. Satisfactory UCS values exceeding 200 kN/m^2 were achieved for samples prepared with 3–9% cement and 6% lime stabilization and compacted at 20.3–23.6% moisture content. The maximum UCS was achieved with a 6% cement and 6% SDA treatment combination. This highlights the benefits of using multiple types of stabilizers to take advantage of their synergistic effects. The cement provides early strength gains, while the SDA may induce long-term pozzolanic reactions [81]. This dual stabilization produced workable soils with sufficient moisture tolerance to achieve significant strength even at the wet end of the moisture range tested.

The maximum UCS achievement was with certain treatment combinations, highlighting the benefits of using multiple types of stabilizers to take advantage of their synergistic effects. This observation aligns with the findings of [42], who reported enhanced soil stabilization through the combined use of cement and supplementary cementitious materials like fly ash or ground granulated blast-furnace slag. The cement provides early strength gains, while the SDA may induce long-term pozzolanic reactions, resulting in improved overall soil performance [12].

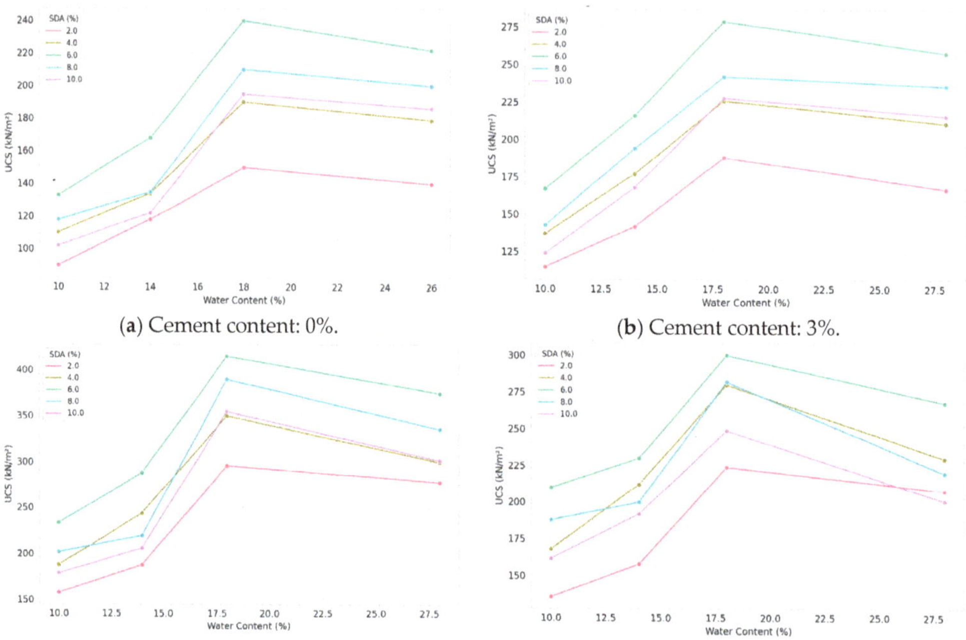

(**a**) Cement content: 0%. (**b**) Cement content: 3%.

(**c**) Cement content: 6%. (**d**) Cement content: 9%.

Figure 20. Effect of compaction water on unconfined compressive strength.

3.3.10. ANOVA of UCS Results of Soil–Cement–Sawdust Ash Mixtures

The influence of cement and slag-derived aggregate (SDA) stabilization on the unconfined compressive strength (UCS) of the modified soil was statistically analyzed at moisture contents of −2%, 0%, +2%, and +4% of the optimum moisture content (OMC) (Table A8 in Appendix A). At all moisture levels, the calculated F-values for cement and SDA effects significantly exceeded the critical F-values at the 5% confidence level. This indicates that the addition of cement and SDA resulted in significant improvements in UCS relative to the untreated soil over the entire moisture range tested, from dry to wet of optimum (F_{CRIT} = 3.287382 for cement and 2.901295 for SDA at 5% confidence). These results are consistent with previous studies showing the potential of cementitious stabilizers to increase strength and stiffness in fine-grained soils via flocculation, agglomeration, and pozzolanic reactions, even under non-ideal moisture conditions [82,83]. The SDA, composed of latent hydraulic compounds, offers a more sustainable alternative to traditional cement. By demonstrating its significant stabilization effects from dry to saturated conditions, this research expands SDA's known applicability and potential as a soil stabilizer. The achievement of significant UCS increases with SDA and cement from −2% to +4% OMC, which represents a novel contribution. Most stabilization work focuses on OMC conditions, but this study demonstrates that both additives can overcome moisture sensitivity and function over a wide moisture range. This highlights their versatility in improving marginal and moisture-prone soils. The dual SDA–cement stabilization produced optimal results, suggesting synergistic effects that could provide advantages over single stabilizer types.

Similar synergistic effects have been reported by [80], who observed enhanced soil stabilization through combined cement and supplementary cementitious materials like fly ash or ground granulated blast-furnace slag. Overall, the findings of this study are well supported by the existing literature and contribute novel insights into the efficacy of cement and SDA stabilization in improving soil strength and workability over a wide range of moisture conditions based on statistical approaches like linear regression.

3.3.11. Effect of Sawdust Ash in Soil/Cement Mixture on Unconfined Compressive Strength

The influence of cement and sawdust ash (SDA) stabilization on the unconfined compressive strength (UCS) of the soil at optimum moisture content (OMC) was evaluated. As shown in Figure 21, the UCS exhibited an increasing trend with incremental additions of cement and SDA up to optimum amendment dosages of 6% by dry weight. Further increases beyond 6% cement and 6% SDA resulted in a downward trend in UCS. This is consistent with previous findings that excessive amounts of stabilizer can cause soil aggregation and poor bonding, thereby reducing strength [84,85]. The peak UCS of 375 kN/m^2 was achieved at OMC with a 6% cement, 6% SDA combination. The complementary effects of cement hydration and SDA pozzolanic reactions appear responsible for this interaction, as the dual stabilization outperformed either admixture alone. Using SDA, an industrial by-product, as a supplementary cementitious material, represents a sustainable approach to soil stabilization. By verifying its effectiveness with cement, this research highlights the potential environmental and economic benefits of partially replacing conventional cement with SDA. It also expands the applicability of SDA to fine-grained soil improvement. While previous work has examined SDA and cement modification individually, the dual stabilization process maximized UCS while minimizing total cement requirements. The OMC moisture tolerance further validates the suitability of these stabilizers for soils under common field conditions. This study is an essential step toward more sustainable, high-performance soil stabilization techniques using industrial by-product admixtures.

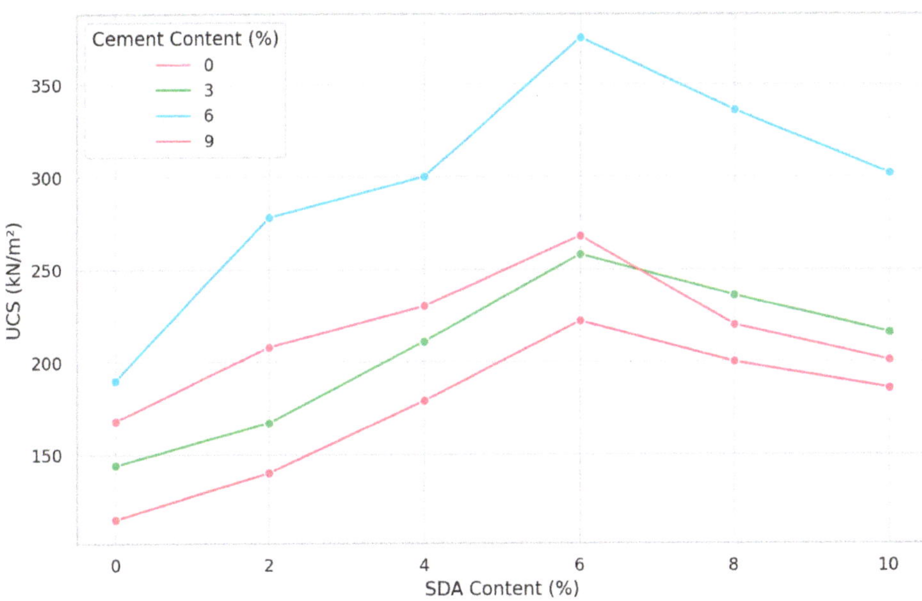

Figure 21. Sawdust ash in soil/cement mixture on unconfined compressive strength.

The increasing trend in UCS with incremental additions of cement and SDA up to optimum dosages of 6% by dry weight is consistent with observations reported in other studies using supplementary cementitious materials (SCMs) like fly ash, ground granulated blast-furnace slag (GGBS), and rice husk ash. The enhancement in soil strength is attributed to the pozzolanic reactions between the SCMs and the calcium hydroxide released during cement hydration, leading to the formation of additional cementitious compounds that densify the soil matrix [82,83].

3.3.12. ANOVA of Sawdust ash in Soil/Cement Mixture on Unconfined Compressive Strength

A two-factor ANOVA was conducted to evaluate the statistical significance of cement and sawdust ash (SDA) stabilization on the unconfined compressive strength (UCS) of the modified soil samples prepared at optimum moisture content (OMC) (Table A9 in Appendix A). The obtained F-values for cement (F_{CAL} = 56.24673) and SDA (F_{CAL} = 25.52665) both exceeded the critical F-value (F_{CRIT}) at the 5% confidence level (F_{CRIT} = 3.287382 for cement and 2.901295 for SDA). This indicates that the improvements in UCS resulting from the inclusion of cement and SDA were statistically significant compared to the untreated soil. These results confirm the ability of SDA to induce positive stabilization effects and complement the hydration reactions of cement in fine-grained soils, even at OMC conditions. Previous work has focused primarily on the individual impacts of cement or alternative supplementary cementitious materials such as fly ash [86,87]. This study provides a novel contribution by demonstrating for the first time the importance of dual cement–SDA stabilization. It verifies the synergistic potential of SDA as a more sustainable partial replacement for traditional cement content in soil improvement applications. Achieving significant strength gains using an industrial by-product such as SDA represents a promising approach to increasing cement-based soil stabilization's environmental and economic sustainability. This analysis demonstrates the feasibility and utility of SDA as a complementary cementitious soil stabilizer.

Alternative statistical techniques, such as response surface methodology (RSM) and multiple regression analysis, can model and optimize the effects of multiple stabilizers on soil properties. Incorporating such techniques could provide further insights into the interactions between cement, SDA, and other potential factors influencing soil stabilization.

While this initial study provides valuable insights into cement and sawdust ash stabilization of expansive soils, the research is limited in scope. Additional testing on more soil types, a more comprehensive range of stabilizer mixtures, simulated environmental conditions, long-term monitoring, and comprehensive geotechnical and leaching characterization would be required. The work evaluated a narrow range of materials, tests, and curing times on soils from one site. Significant further experimentation under diverse conditions is needed to optimize and implement the stabilization process for engineering applications. However, the work represents an essential first step in exploring a more sustainable approach to soil improvement using industrial by-product admixtures.

4. Conclusions

Based on the result of this investigation, it can be concluded that:

- There was a maximum dry density of 1.745 g/cm^3 and optimum moisture content of 23% for the untreated soil. At 6% cement and 6% SDA by dry weight of soil, this combination has the lowest liquid limit of 30.1%, indicating reduced plasticity, the lowest plasticity index of 15.2%, representing the smallest range of moisture content over which the soil exhibits plastic behavior, and the lowest linear shrinkage of 4.14%, implying reduced volumetric instability upon drying.
- The optimum proportions of cement and sawdust ash were found to be 6% cement and 6% SDA by dry weight of soil, resulting in maximum improvements in UCS (51 to 375 kN/m^2) and decreases in HC (1.7×10^{-8} to 4.7×10^{-10} m/s) and VSS (12.8 to 3.51%) compared to unamended soil.
- The ANOVA tests showed that cement and sawdust ash (SDA) inclusions significantly change the stabilized soil matrix's HC, UCS, and VSS. For example, at the 5% significance level, the F-statistic values were 15.69096 and 4.788327 for cement and SDA, respectively, exceeding the critical F-value of 3.28738 for hydraulic conductivity.
- Conducting a long-term field study to assess the durability and continued effectiveness of the stabilized soil under actual environmental conditions would be valuable. This could include monitoring the performance of the stabilized soil liner over time, including its resistance to cyclic wetting and drying and exposure to various envi-

ronmental factors. Such an evaluation would provide insight into the stabilization technique's long-term behavior and practical applicability.

Author Contributions: methodology, H.L., L.X., and A.I.; laboratory experiments, S.I., and A.M.; validation, I.H.U., S.I., and H.L.; formal analysis, A.M.; writing—original draft, L.X., A.I., and A.M.; writing—review and editing, I.H.U., S.I., and H.L. All authors have read and agreed to the published version of the manuscript.

Funding: This paper received funding from the Hunan Provincial Key Research and Development Program (2022SK2082), the Guizhou Provincial Major Scientific and Technological Program (2023-425), and the Project NRMSSHR-2022-Z08 supported by the Key Laboratory of Natural Resources Monitoring and Supervision in Southern Hilly Region, Ministry of Natural Resources.

Institutional Review Board Statement: On behalf of all authors, the corresponding author states that there are no conflicts of interest. This article contains no studies performed by authors with human participants or animals.

Informed Consent Statement: Not applicable.

Data Availability Statement: Some or all data, models, or codes that support the findings of this study are available from the corresponding author upon reasonable request.

Acknowledgments: The authors wish to acknowledge the support of the aforementioned funder from the Hunan Provincial Key Research and Development Program.

Conflicts of Interest: On behalf of all authors, the corresponding author states that there are no conflicts of interest.

Appendix A

Table A1. Two-way analysis of variance for specific gravity results of soil–cement–sawdust ash mixtures.

Property	Source of Variation	Degrees of Freedom	F_{CAL}	p-Value	F_{CRIT}	Remark
Specific gravity	Cement	3	66.22449	6.99×10^{-09}	3.287382	$F_{CAL} > F_{CRIT}$ Significant effect
	SDA	5	52.26531	5.85×10^{-09}	2.901295	$F_{CAL} > F_{CRIT}$ Significant effect

Table A2. Two-way analysis of variance for Atterberg limit results of soil–cement–sawdust ash mixtures.

Property	Source of Variation	Degrees of Freedom	F_{CAL}	p-Value	F_{CRIT}	Remark
Liquid limit	Cement	3	278.907	2.24×10^{-13}	3.287382	$F_{CAL} > F_{CRIT}$ Significant effect
	SDA	5	73.77099	5.08×10^{-10}	2.901295	$F_{CAL} > F_{CRIT}$ Significant effect
Plastic limit	Cement	3	25.74292	3.64×10^{-06}	3.287382	$F_{CAL} > F_{CRIT}$ Significant effect
	SDA	5	13.969	3.42×10^{-05}	2.901295	$F_{CAL} > F_{CRIT}$ Significant effect
Plasticity index	Cement	3	417.5243	1.14×10^{-14}	3.287382	$F_{CAL} > F_{CRIT}$ Significant effect
	SDA	5	47.25471	1.18×10^{-08}	2.901295	$F_{CAL} > F_{CRIT}$ Significant effect
Linear shrinkage	Cement	3	1289.157	2.58×10^{-18}	3.287382	$F_{CAL} > F_{CRIT}$ Significant effect
	SDA	5	47.10454	1.21×10^{-08}	2.901295	$F_{CAL} > F_{CRIT}$ Significant effect

Table A3. Two-way analysis of variance for compaction results of soil–cement–saw dust ash mixtures.

Property	Source of Variation	Degrees of Freedom	F_{CAL}	p-Value	F_{CRIT}	Remark
MDD	Cement	3	51.31148	4.04×10^{-08}	3.287382	$F_{CAL} > F_{CRIT}$ Significant effect
	SDA	5	20.90164	2.84×10^{-06}	2.901295	$F_{CAL} > F_{CRIT}$ Significant effect
OMC	Cement	3	39.33962	2.4×10^{-07}	3.287382	$F_{CAL} > F_{CRIT}$ Significant effect
	SDA	5	23.18868	1.45×10^{-06}	2.901295	$F_{CAL} > F_{CRIT}$ Significant effect

Table A4. Two-way analysis of variance for effect of compaction water on HC results of soil–cement–sawdust ash mixtures.

Property		Source of Variation	Degrees of Freedom	F_{CAL}	p-Value	F_{CRIT}	Remark
HC	−2%	Cement	3	17.47646	3.69×10^{-05}	3.287382	$F_{CAL} > F_{CRIT}$ Significant effect
		SDA	5	5.215878	5.68×10^{-03}	2.901295	$F_{CAL} > F_{CRIT}$ Significant effect
	0%	Cement	3	23.47024	6.43×10^{-06}	3.287382	$F_{CAL} > F_{CRIT}$ Significant effect
		SDA	5	6.148309	2.74×10^{-03}	2.901295	$F_{CAL} > F_{CRIT}$ Significant effect
	+2%	Cement	3	33.74422	6.53×10^{-07}	3.287382	$F_{CAL} > F_{CRIT}$ Significant effect
		SDA	5	12.55617	6.41×10^{-05}	2.901295	$F_{CAL} > F_{CRIT}$ Significant effect
	+4%	Cement	3	15.63979	6.93×10^{-05}	3.287382	$F_{CAL} > F_{CRIT}$ Significant effect
		SDA	5	4.894346	7.45×10^{-03}	2.901295	$F_{CAL} > F_{CRIT}$ Significant effect

Table A5. Two-way analysis of variance for HC results of soil–cement–sawdust ash mixtures.

Property	Source of Variation	Degrees of Freedom	F_{CAL}	p-Value	F_{CRIT}	Remark
HC	Cement	3	15.69096	6.8×10^{-05}	3.287382	$F_{CAL} > F_{CRIT}$ Significant effect
	SDA	5	4.788327	8.16×10^{-03}	2.901295	$F_{CAL} > F_{CRIT}$ Significant effect

Table A6. Two-way analysis of variance for effect of compaction water on VSS results of soil–cement–sawdust ash mixtures.

Property		Source of Variation	Degrees of Freedom	F_{CAL}	p-Value	F_{CRIT}	Remark
VSS	−2%	Cement	3	86.30367	1.67×10^{-10}	3.287382	$F_{CAL} > F_{CRIT}$ Significant effect
		SDA	5	112.3474	1.65×10^{-10}	2.901295	$F_{CAL} > F_{CRIT}$ Significant effect

Table A6. *Cont.*

Property		Source of Variation	Degrees of Freedom	F_{CAL}	p-Value	F_{CRIT}	Remark
VSS	0%	Cement	3	155.1971	1.62×10^{-11}	3.287382	$F_{CAL} > F_{CRIT}$ Significant effect
		SDA	5	155.3852	2.3×10^{-12}	2.901295	$F_{CAL} > F_{CRIT}$ Significant effect
	+2%	Cement	3	22.14195	9.17×10^{-06}	3.287382	$F_{CAL} > F_{CRIT}$ Significant effect
		SDA	5	37.69041	5.65×10^{-08}	2.901295	$F_{CAL} > F_{CRIT}$ Significant effect
	+4%	Cement	3	27.30304	2.52×10^{-06}	3.287382	$F_{CAL} > F_{CRIT}$ Significant effect
		SDA	5	49.79083	8.22×10^{-09}	2.901295	$F_{CAL} > F_{CRIT}$ Significant effect

Table A7. Two-way analysis of variance for VSS results of soil–cement–sawdust ash mixtures.

Property	Source of Variation	Degrees of Freedom	F_{CAL}	p-Value	F_{CRIT}	Remark
VSS	Cement	3	18.64996	2.54×10^{-05}	3.287382	$F_{CAL} > F_{CRIT}$ Significant effect
	SDA	5	38.157	5.2×10^{-08}	2.901295	$F_{CAL} > F_{CRIT}$ Significant effect

Table A8. Two-way analysis of variance for effect of compaction water on UCS results of soil–cement–sawdust ash mixtures.

Property		Source of Variation	Degrees of Freedom	F_{CAL}	p-Value	F_{CRIT}	Remark
UCS	−2%	Cement	3	121.1904	9.72×10^{-11}	3.287382	$F_{CAL} > F_{CRIT}$ Significant effect
		SDA	5	68.19628	8.9×10^{-10}	2.901295	$F_{CAL} > F_{CRIT}$ Significant effect
	0%	Cement	3	56.60906	2.06×10^{-08}	3.287382	$F_{CAL} > F_{CRIT}$ Significant effect
		SDA	5	37.6517	5.69×10^{-08}	2.901295	$F_{CAL} > F_{CRIT}$ Significant effect
	+2%	Cement	3	59.14261	1.53×10^{-08}	3.287382	$F_{CAL} > F_{CRIT}$ Significant effect
		SDA	5	23.92532	1.19×10^{-06}	2.901295	$F_{CAL} > F_{CRIT}$ Significant effect
	+4%	Cement	3	56.24673	2.16×10^{-08}	3.287382	$F_{CAL} > F_{CRIT}$ Significant effect
		SDA	5	25.52665	7.75×10^{-07}	2.901295	$F_{CAL} > F_{CRIT}$ Significant effect

Table A9. Two-way analysis of variance for UCS results of soil–cement–sawdust ash mixtures.

Property	Source of Variation	Degrees of Freedom	F_{CAL}	p-Value	F_{CRIT}	Remark
UCS	Cement	3	56.24673	2.16×10^{-08}	3.287382	$F_{CAL} > F_{CRIT}$ Significant effect
	SDA	5	25.52665	7.75×10^{-07}	2.901295	$F_{CAL} > F_{CRIT}$ Significant effect

References

1. Tian, K. Application of geosynthetic clay liner for coal combustion products disposal. In *Managing Mining and Minerals Processing Wastes*; Qi, C., Benson, C.H., Eds.; Elsevier: Amsterdam, The Netherlands, 2023; pp. 165–187.
2. Chen, J.; Peng, D. Management and disposal of alumina production wastes. In *Managing Mining and Minerals Processing Wastes*; Qi, C., Benson, C.H., Eds.; Elsevier: Amsterdam, The Netherlands, 2023; pp. 133–163.
3. Muralikrishna, I.V.; Manickam, V. Hazardous Waste Management. In *Environmental Management*; Muralikrishna, I.V., Manickam, V., Eds.; Butterworth-Heinemann: Oxford, UK, 2017; pp. 463–494.
4. Umar, I.H.; Lin, H.; Hassan, J.I. Transforming Landslide Prediction: A Novel Approach Combining Numerical Methods and Advanced Correlation Analysis in Slope Stability Investigation. *Appl. Sci.* **2024**, *14*, 3685. [CrossRef]
5. Safari, E.; Jalili Ghazizade, M.; Abduli, M.A.; Gatmiri, B. Variation of crack intensity factor in three compacted clay liners exposed to annual cycle of atmospheric conditions with and without geotextile cover. *Waste Manag.* **2014**, *34*, 1408–1415. [CrossRef] [PubMed]
6. Hewitt, P.J.; Philip, L.K. Problems of clay desiccation in composite lining systems. *Eng. Geol.* **1999**, *53*, 107–113. [CrossRef]
7. Chen, A.; Li, C.; Zhao, S.; Yang, B.; Ding, C. Study on the Dynamic Mechanism of the Desiccation Crack Initiation and Propagation in Red Clay. *Sustainability* **2023**, *15*, 11156. [CrossRef]
8. Kumar, R.; Kumari, S. Geotechnical properties of materials used in landfill clay liner: A critical review. *Sādhanā* **2023**, *48*, 64. [CrossRef]
9. Murthi, P.; Saravanan, R.; Poongodi, K. Studies on the impact of polypropylene and silica fume blended combination on the material behaviour of black cotton soil. *Mater. Today Proc.* **2021**, *39*, 621–626. [CrossRef]
10. Rasheed, R.M.; Moghal, A.A.; Jannepally, S.S.; Rehman, A.U.; Chittoori, B.C.S. Shrinkage and Consolidation Characteristics of Chitosan-Amended Soft Soil—A Sustainable Alternate Landfill Liner Material. *Buildings* **2023**, *13*, 2230. [CrossRef]
11. Jamaluddin, K.; Munirwan, R. Improvement of geotechnical properties of clayey soil with saw dust ash stabilization. *E3S Web Conf.* **2022**, *340*, 01009. [CrossRef]
12. Medina-Martinez, C.J.; Sandoval Herazo, L.C.; Zamora-Castro, S.A.; Vivar-Ocampo, R.; Reyes-Gonzalez, D. Use of Sawdust Fibers for Soil Reinforcement: A Review. *Fibers* **2023**, *11*, 58. [CrossRef]
13. Rachan, R.; Chinkulkijniwat, A.; Raksachon, Y.; Suddeepong, A. Analysis of Strength Development in Cement-stabilized Silty Clay from Microstructural Considerations. *Constr. Build. Mater.* **2010**, *24*, 2011–2021. [CrossRef]
14. Yin, Z.; Li, R.; Lin, H.; Chen, Y.; Wang, Y.; Zhao, Y. Analysis of influencing factors of cementitious material properties of lead-zinc tailings based on orthogonal tests. *Materials* **2023**, *16*, 361. [CrossRef] [PubMed]
15. Li, R.; Yin, Z.; Lin, H. Research status and prospects for the utilization of lead-zinc tailings as building materials. *Buildings* **2023**, *13*, 150. [CrossRef]
16. Yagüe, S.A.-O.; González Gaya, C.; Rosales Prieto, V.; Sánchez Lite, A.A.-O. Sustainable Ecocements: Chemical and Morphological Analysis of Granite Sawdust Waste as Pozzolan Material. *Materials* **2020**, *13*, 4941. [CrossRef] [PubMed]
17. Haruna, I.U.; Kabir, M.; Yalo, S.G.; Muhammad, A.; Ibrahim, A.S. Quantitative Analysis of Solid Waste Generation from Tanneries in Kano State. *J. Environ. Eng. Stud.* **2022**, *7*, 23–30. [CrossRef]
18. Olaiya, B.C.; Lawan, M.M.; Olonade, K.A. Utilization of sawdust composites in construction—A review. *SN Appl. Sci.* **2023**, *5*, 140. [CrossRef]
19. Sheikh, A.; Akbari, M.; Shafabakhsh, G. Laboratory Study of the Effect of Zeolite and Cement Compound on the Unconfined Compressive Strength of a Stabilized Base Layer of Road Pavement. *Materials* **2022**, *15*, 7981. [CrossRef] [PubMed]
20. Tran, T.; Ju, H.; Nguyen, T.-N.; Hoang, T.; Abdelaziz, S.; Brand, A. Effects of curing temperature on cement-stabilized soils. *IOP Conf. Ser. Mater. Sci. Eng.* **2023**, *1289*, 012097. [CrossRef]
21. Al-Mahbashi, A.M.; Dafalla, M.; Al-Shamrani, M. Long-Term Performance of Liners Subjected to Freeze-Thaw Cycles. *Water* **2022**, *14*, 3218. [CrossRef]
22. Jamshidi, R.; Lake, C.; Gunning, P.; Hills, C. Effect of Freeze/Thaw Cycles on the Performance and Microstructure of Cement-Treated Soils. *J. Mater. Civ. Eng.* **2016**, *28*, 04016162. [CrossRef]
23. Kong, D.-J.; Wu, H.-N.; Chai, J.-C.; Arulrajah, A. State-of-the-Art Review of Geosynthetic Clay Liners. *Sustainability* **2017**, *9*, 2110. [CrossRef]
24. Chen, J.; Dan, H.; Ding, Y.; Gao, Y.; Guo, M.; Guo, S.; Han, B.; Hong, B.; Hou, Y.; Hu, C.; et al. New innovations in pavement materials and engineering: A review on pavement engineering research 2021. *J. Traffic Transp. Eng. (Engl. Ed.)* **2021**, *8*, 815–999. [CrossRef]
25. Ikeagwuani, C.C.; Obeta, I.N.; Agunwamba, J.C. Stabilization of black cotton soil subgrade using sawdust ash and lime. *Soils Found.* **2019**, *59*, 162–175. [CrossRef]
26. Benson, C.; Zhai, H.; Wang, X. Estimating Hydraulic Conductivity of Compacted Clay Liners. *J. Geotech. Eng.* **1994**, *120*, 2. [CrossRef]
27. Di Emidio, G. Hydraulic and Chemico-Osmotic Performance of Polymer Treated Clays. Ph.D. Thesis, Ghent University, Ghent, Belgium, 2010.

28. Christensen, R. Analysis of Variance and Generalized Linear Models. In *International Encyclopedia of the Social & Behavioral Sciences*; Smelser, N.J., Baltes, P.B., Eds.; Pergamon: Oxford, UK, 2001; pp. 473–480.
29. *ASTM D2166-06*; Standard Test Method for Unconfined Compressive Strength of Cohesive Soil. ASTM: West Conshohocken, PA, USA, 2010.
30. Ilek, A.; Kucza, J. A laboratory method to determine the hydraulic conductivity of mountain forest soils using undisturbed soil samples. *J. Hydrol.* **2014**, *519*, 1649–1659. [CrossRef]
31. Gallage, C.; Kodikara, J.; Uchimura, T. Laboratory measurement of hydraulic conductivity functions of two unsaturated sandy soils during drying and wetting processes. *Soils Found.* **2013**, *53*, 417–430. [CrossRef]
32. *ASTM D5084-16a*; Standard Test Methods for Measurement of Hydraulic Conductivity of Saturated Porous Materials Using a Flexible Wall Permeameter. ASTM: West Conshohocken, PA, USA, 2016.
33. Julina, M.; Thyagaraj, T. Determination of volumetric shrinkage of an expansive soil using digital camera images. *Int. J. Geotech. Eng.* **2018**, *15*, 624–632. [CrossRef]
34. Tang, C.-S.; Cui, Y.J.; Tang, A.-M.; Shi, B. Volumetric shrinkage characteristics of soil during drying. *Yantu Gongcheng Xuebao/Chin. J. Geotech. Eng.* **2011**, *33*, 1271–1279.
35. *ASTM D4943-18*; Standard Test Method for Shrinkage Factors of Cohesive Soils by the Water Submersion Method. ASTM: West Conshohocken, PA, USA, 2018.
36. King, B.M. Analysis of Variance. In *International Encyclopedia of Education*, 3rd ed.; Peterson, P., Baker, E., McGaw, B., Eds.; Elsevier: Oxford, UK, 2010; pp. 32–36.
37. Smalheiser, N.R. ANOVA. In *Data Literacy*; Smalheiser, N.R., Ed.; Academic Press: Cambridge, MA, USA, 2017; pp. 149–155.
38. McIntosh, A.M.; Sharpe, M.; Lawrie, S.M. Research methods, statistics and evidence-based practice. In *Companion to Psychiatric Studies*, 8th ed.; Johnstone, E.C., Owens, D.C., Lawrie, S.M., McIntosh, A.M., Sharpe, M., Eds.; Churchill Livingstone: St. Louis, MO, USA, 2010; pp. 157–198.
39. Henson, R.N. Analysis of Variance (ANOVA). In *Brain Mapping*; Toga, A.W., Ed.; Academic Press: Waltham, MA, USA, 2015; pp. 477–481.
40. Dahiru, T. *p*-value, a true test of statistical significance? A cautionary note. *Ann. Ib. Postgrad. Med.* **2008**, *6*, 21–26. [CrossRef]
41. Kao, L.S.; Green, C.E. Analysis of variance: Is there a difference in means and what does it mean? *Natl. Inst. Health* **2008**, *144*, 158–170. [CrossRef]
42. Umar, I.H.; Lin, H.; Ibrahim, A.S. Laboratory Testing and Analysis of Clay Soil Stabilization Using Waste Marble Powder. *Appl. Sci.* **2023**, *13*, 9274. [CrossRef]
43. Muntohar, A.; Abidin, Z. A Comparative Study of Different Additive on the Index Properties of Expansive Soils. *J. Semesta Tek.* **2001**, *4*, 59–67.
44. Dąbska, A. Hydraulic Conductivity of Compacted Lime-Softening Sludge Used as Landfill Liners. *Water Air Soil Pollut.* **2019**, *230*, 280. [CrossRef]
45. Umar, I.H.; Orakoglu Firat, M.E. A Study on Uniaxial Compressive Strength and Ultrasonic Non-Destructive Analysis of Fine-Grained Soil in Seasonally Frozen Regions. *Turk. J. Sci. Technol.* **2022**, *17*, 267–277. [CrossRef]
46. Umar, I.H.; Orakoglu Firat, M.E. Investigation of Unconfined Compressive Strength of Soils Stabilized with Waste Elazig Cherry Marble Powder at Different Water Contents. In Proceedings of the 14th International Conference on Engineering & Natural Sciences, Sivas, Turkey, 18–19 July 2022.
47. Umar, I.H.; Muhammad, A.; Yusuf, A.; Muhammad, K.I. Study on the Geotechnical Properties of Road Pavement Failures "(A Case Study of Portion Of Malam Aminu Kano Way, Kano State From Tal-Udu Roundabout To Mambaya House Roundabout)". *J. Geotech. Stud.* **2020**, *5*, 8–15. [CrossRef]
48. Umar, I.H.; Muhammad, A.; Ahmad, A.; Yusif, M.A.; Yusuf, A. Suitability of Geotechnical Properties of Bentonite-Bagasse Ash Mixtures Stabilized Lateritic Soil as Barrier in Engineered Waste Landfills. In Proceedings of the 7th International Student Symposium, Ondokuz Mayiz University, Samsun, Turkey, 1–3 October 2021.
49. Darsi, B.P.; Molugaram, K.; Madiraju, S.V.H. Subgrade Black Cotton Soil Stabilization Using Ground Granulated Blast-Furnace Slag (GGBS) and Lime, an Inorganic Mineral. *Environ. Sci. Proc.* **2021**, *6*, 15. [CrossRef]
50. Firat, S.; Khatib, J.M.; Yilmaz, G.; Comert, A.T. Effect of curing time on selected properties of soil stabilized with fly ash, marble dust and waste sand for road sub-base materials. *Waste Manag. Res.* **2017**, *35*, 747–756. [CrossRef]
51. Ola, S.A. The geology and geotechnical properties of the black cotton soils of North Eastern Nigeria. *Eng. Geol.* **1978**, *12*, 375–391. [CrossRef]
52. Khatti, J.; Grover, K.S. Prediction of compaction parameters for fine-grained soil: Critical comparison of the deep learning and standalone models. *J. Rock Mech. Geotech. Eng.* **2023**, *15*, 3010–3038. [CrossRef]
53. Long, G.; Li, L.; Li, W.; Ma, K.; Dong, W.; Bai, C.; Zhou, J. Enhanced mechanical properties and durability of coal gangue reinforced cement-soil mixture for foundation treatments. *J. Clean. Prod.* **2019**, *231*, 468–482. [CrossRef]
54. Butt, W.; Gupta, K.; Jha, J. Strength behavior of clayey soil stabilized with saw dust ash. *Int. J. Geo-Eng.* **2016**, *7*, 18. [CrossRef]
55. Danish, A.; Totiç, E.; Bayram, M.A.-O.; Sütçü, M.A.-O.; Gencel, O.; Erdoğmuş, E.; Ozbakkaloglu, T.A.-O.X. Assessment of Mineralogical Characteristics of Clays and the Effect of Waste Materials on Their Index Properties for the Production of Bricks. *Materials* **2022**, *15*, 8908. [CrossRef]

56. Karakan, E. Comparative Analysis of Atterberg Limits, Liquidity Index, Flow Index and Undrained Shear Strength Behavior in Binary Clay Mixtures. *Appl. Sci.* **2022**, *12*, 8616. [CrossRef]
57. Oluremi, J.; Siddique, R.; Adeboje, E. Stabilization Potential of Cement Kiln Dust Treated Lateritic Soil. *Int. J. Eng. Res. Afr.* **2016**, *23*, 52–63. [CrossRef]
58. Nuaklong, P.; Jongvivatsakul, P.; Pothisiri, T.; Sata, V.; Chindaprasirt, P. Influence of rice husk ash on mechanical properties and fire resistance of recycled aggregate high-calcium fly ash geopolymer concrete. *J. Clean. Prod.* **2019**, *252*, 119797. [CrossRef]
59. Sani, J.E.; Yohanna, P.; Chukwujama, I.A. Effect of rice husk ash admixed with treated sisal fibre on properties of lateritic soil as a road construction material. *J. King Saud Univ.—Eng. Sci.* **2020**, *32*, 11–18. [CrossRef]
60. Nitin, M.; Guleria, H. To Study the Basic Geotechnical Characteristics of Fly Ash-Limegypsum Commix with Quarry Dust. *Int. J. Adv. Res. Eng. Technol.* **2020**, *11*, 921–930.
61. Al-Jabban, W.; Laue, J.; Knutsson, S.; Al-Ansari, N. A Comparative Evaluation of Cement and By-Product Petrit T in Soil Stabilization. *Appl. Sci.* **2019**, *9*, 5238. [CrossRef]
62. Abbey, S.; Olubanwo, A.; Ng'ambi, S.; Eyo, E.; Adeleke, B. Effect of Organic Matter on Swell and Undrained Shear Strength of Treated Soils. *J. Civ. Constr. Environ. Eng.* **2019**, *4*, 48–58. [CrossRef]
63. Sariosseiri, F.; Muhunthan, B. Effect of cement treatment on geotechnical properties of some Washington State soils. *Eng. Geol.* **2009**, *104*, 119–125. [CrossRef]
64. Moradi, S. Effects of CEC on Atterberg limits and Plastic Index in Different Soil Textures. *VictorQuest* **2013**, *4*, 2111–2118.
65. Yohanna, P. Evaluation of Compacted Black Cotton Soil–Sawdust Ash Mixtures as Road Construction Material. *Niger. J. Mater. Sci. Eng.* **2018**, *7*, 65–72.
66. Muhmed, A.; Mohamed, M. The Impact of Moisture and Clay Content on the Unconfined Compressive Strength of Lime Treated Highly Reactive Clays. *Geotech. Geol. Eng.* **2022**, *40*, 5869–5893. [CrossRef]
67. Sabat, A.K.; Nanda, R.P. Effect of marble dust on strength and durability of Rice husk ash stabilised expansive soil Effect of marble dust on strength and durability of Rice husk ash. *Int. J. Civ. Struct. Eng.* **2011**, *1*, 939–948.
68. Khan, S.; Khan, M.H. Improvement of Mechanical Properties by Waste Sawdust Ash Addition into Soil. *Electron. J. Geotech. Eng.* **2015**, *20*, Bundle 7. [CrossRef]
69. Soni, S.R.; Dahale, P.P.; Dobale, R.M. Disposal of solid waste for black cotton soil Stabilization. *Int. J. Adv. Eng. Sci. Technol.* **2011**, *8*, 113–120.
70. Farooq, K.; ur Rehman, Z.; Shahzadi, M.; Mujtaba, H.; Khalid, U. Optimization of Sand-Bentonite Mixture for the Stable Engineered Barriers using Desirability Optimization Methodology: A Macro-Micro-Evaluation. *KSCE J. Civ. Eng.* **2023**, *27*, 40–52. [CrossRef]
71. Mustafa, Y.M.H.; Zami, M.S.; Al-Amoudi, O.S.B.; Al-Osta, M.A.; Wudil, Y.S. Analysis of Unconfined Compressive Strength of Rammed Earth Mixes Based on Artificial Neural Network and Statistical Analysis. *Materials* **2022**, *15*, 9029. [CrossRef]
72. Mohamed, A.A.M.S.; Yuan, J.; Al-Ajamee, M.; Dong, Y.; Ren, Y.; Hakuzweyezu, T. Improvement of expansive soil characteristics stabilized with sawdust ash, high calcium fly ash and cement. *Case Stud. Constr. Mater.* **2023**, *18*, e01894. [CrossRef]
73. Das, B.M.; Sobhan, K.; Sobhan, K. *Principles of Geotechnical Engineering*; Cengage Learning: Boston, MA, USA, 2018.
74. Ikeagwuani, C.C.; Nwonu, D.C. Emerging trends in expansive soil stabilisation: A review. *J. Rock Mech. Geotech. Eng.* **2019**, *11*, 423–440. [CrossRef]
75. Daniel, D.; Benson, C. Water Content-Density Criteria for Compacted Soil Liners. *J. Geotech. Eng.* **1990**, *116*, 1811–1830. [CrossRef]
76. Osinubi, K.; Nwaiwu, C.M.O. Hydraulic Conductivity of Compacted Lateritic Soil. *J. Geotech. Geoenviron. Eng.* **2005**, *131*, 1034–1041. [CrossRef]
77. Hsiao, D.; Hsieh, C. Improving Mudstone Materials in Badland in Southwestern Taiwan by Increasing Density and Low-Cement Amount. *Appl. Sci.* **2022**, *12*, 2290. [CrossRef]
78. Umar, I.H.; Lin, H. Marble Powder as a Soil Stabilizer: An Experimental Investigation of the Geotechnical Properties and Unconfined Compressive Strength Analysis. *Materials* **2024**, *17*, 1208. [CrossRef]
79. Jorat, E.; Marto, A.; Namazi, E. Engineering Characteristics of Kaolin Mixed with Various Percentages of Bottom Ash. *Electron. J. Geotech. Eng.* **2011**, *16*, 841–850.
80. Mousavi, F.; Abdi, E.; Ghalandarayeshi, S.; Page-Dumroese, D.S. Modeling unconfined compressive strength of fine-grained soils: Application of pocket penetrometer for predicting soil strength. *Catena* **2021**, *196*, 104890. [CrossRef]
81. Agrawal, V.; Gupta, M. Expansive Soil Stabilization Using Marble Dust. *Int. J. Earth Sci. Eng.* **2011**, *4*, 59–62.
82. Archibong, G.A.; Sunday, E.U.; Akudike, J.C.; Okeke, O.C.; Amadi, C. A Review of the Principles and Methods of Soil Stabilization. *Int. J. Adv. Acad. Res. Sci. Technol. Eng.* **2020**, *6*, 89–115.
83. Cetin, B.; Aydilek, A.H.; Guney, Y. Stabilization of recycled base materials with high carbon fly ash. *Resour. Conserv. Recycl.* **2010**, *54*, 878–892. [CrossRef]
84. James, J.; Pandian, P.K. Strength and microstructure of micro ceramic dust admixed lime stabilized soil. *Rev. Constr.* **2018**, *17*, 5–22. [CrossRef]
85. Narnoli, V.; Suman, S.K.; Kumar, R. Stabilization of alluvial soilusing marble dust, lime and burnt brick dust for road construction. In Proceedings of the Indian Conference on Geotechnical and Geo-Environmental Engineering (ICGGE-2019), Prayagraj, India, 1–2 March 2019.

86. Osinubi Moses, G. Optimizing Soil-Cement-Ash Stabilization Mix for Maximum Compressive Strength. In Proceedings of the West Africa Built Environment Research Conference (WABER 2012) Conference, Abuja, Nigeria, 24–26 July 2012; pp. 1207–1218.
87. Khemissa, M.; Mahamedi, A. Cement and Lime Mixture Stabilization of an Expansive Overconsolidated Clay. *Appl. Clay Sci.* **2014**, *95*, 104–110. [CrossRef]

materials

MDPI

Article

New Eco-Cements Made with Marabou Weed Biomass Ash

Moisés Frías [1,*], Ana María Moreno de los Reyes [1,*], Ernesto Villar-Cociña [2], Rosario García [3], Raquel Vigil de la Villa [3] and Milica Vidak Vasić [4]

1 Eduardo Torroja Institute for Construction Sciences, IETcc-CSIC, 28033 Madrid, Spain
2 Department of Physics, Universidad Central de las Villas, Santa Clara 54830, Cuba; ernestovillarcocina@gmail.com
3 CSIC-Affiliated Geomaterials Unit, Department of Geology and Geochemistry, Autonomous University of Madrid, 28049 Madrid, Spain; rosario.garcia@uam.es (R.G.); vigildelavillaraquel@gmail.com (R.V.d.l.V.)
4 Institute for Testing of Materials IMS, 11000 Belgrade, Serbia; milica.vasic@institutims.rs
* Correspondence: mfrias@ietcc.csic.es (M.F.); ana.moreno@ietcc.csic.es (A.M.M.d.l.R.)

Abstract: Biomass ash is currently attracting the attention of science and industry as an inexhaustible eco-friendly alternative to pozzolans traditionally used in commercial cement manufacture (fly ash, silica fume, natural/calcined pozzolan). This paper explores a new line of research into Marabou weed ash (MA), an alternative to better-known conventional agro-industry waste materials (rice husk, bagasse cane, bamboo, forest waste, etc.) produced in Cuba from an invasive plant harvested as biomass for bioenergy production. The study entailed full characterization of MA using a variety of instrumental techniques, analysis of pozzolanic reactivity in the pozzolan/lime system, and, finally its influence on the physical and mechanical properties of binary pastes and mortars containing 10% and 20% MA replacement content. The results indicate that MA has a very low acid oxide content and a high loss on ignition (30%) and K_2O content (6.9%), which produces medium–low pozzolanic activity. Despite an observed increase in the blended mortars' total and capillary water absorption capacity and electrical resistivity and a loss in mechanical strength approximately equivalent to the replacement percentage, the 10% and 20% MA blended cements meet the regulatory chemical, physical, and mechanical requirements specified. Marabou weed ash is therefore a viable future supplementary cementitious material.

Keywords: Marabou weed biomass ash; mineralogical addition; binary eco-cement; electrical resistivity

1. Introduction

One of the five key elements of the cement industry's roadmap to achieving climate neutrality by 2050 is the 5C cement challenge to develop future commercial products with a lower clinker-to-cement ratio ($\leq$0.60) than those currently manufactured. According to the data available as of 2021, clinker content in the Spanish cement industry stands at around 0.8 (clinker/cement ratio), a high factor exacerbated by the economic crisis in the sector, COVID-19, lower demand for cement, and lower availability of traditional additives [1,2]. This combination makes it imperative to continue the search for new supplementary cementitious materials (SCMs) that offer an eco-friendly alternative to those traditionally used (fly ash, silica fume, and natural/calcined clay) [3]. In recent years, the scientific community has conducted ongoing research into the different industrial waste streams that, due to their chemical, mineralogical, and pozzolanic characteristics, offer scientifically, technically, and environmentally viable substitutes [4–10]. Within these emerging streams, ash from biomass combustion is attracting particular attention because of the vast volumes generated globally (170 Mt/yr) [11], mainly driven by the new international scenarios promoting the use of clean energy from renewable sources [12]. There is a wide variety of biomass ash rich in reactive silica and alumina (rice husk, bagasse cane, bamboo leaf, and paper sludge, among others) capable of providing cement with the improved properties required by the transition to a low-carbon economy [13–17]. However, biomass ash has several drawbacks, such as its heterogeneity (time of year generated, energy process, burning temperature, etc.) and its contribution of potentially negative elements/oxides to the cement matrix. The latter is the case with high-potash (K_2O) ash, which requires pre-washing to reduce its content and prevent reactivity and durability issues (alkali-aggregate reaction) in the new eco-cements [18–22].

Intending to expand the range of biomass ashes potentially viable as eco-pozzolans, this paper analyzes new ash obtained from the combustion of Marabou weed (harvested in Cuba). To this end, the study entails full characterization using a variety of instrumental techniques and analysis of pozzolanic reactivity and its influence on the physical and mechanical properties of binary eco-cements made with 10% and 20% replacement content used in the manufacture of CEM II-A cement.

2. Materials and Methods

2.1. Materials

Marabou (MA) weed (Dichrostachys cinerea) is a semi-deciduous shrub from the legume family. It is native to South Africa and can reach 7 m in height. In Cuba, the plant is estimated to grow on more than 1.2 million hectares at an average density of 37 t/ha [23–25]. It is considered an invasive species and an important alternative fuel for biopower stations. In 2016, Cuba produced 154 GWh of bioenergy from Marabou weed biomass [8]. Since Marabou weed harvesting is seasonal, the biopower plant did not use it as fuel at the sampling time. The ash (MA) was therefore produced at a laboratory scale, by calcination to 600 °C for 2 h in a muffle furnace, considered optimal conditions from a pozzolanic, energy, and economic point of view (Figure 1). In addition, carbonates will not degrade at this temperature, although organic matter will be completely decomposed.

Figure 1. The appearance of Marabou weed before and after thermal processing.

In this study, a commercial CEM I 52.5R-type cement (OPC) complying with European standard EN 197-1 [3] and supplied by Cementos Lemona, S.A. (Bilbao, Spain) was used. The binary cements were made by replacing part of the cement with 10% and 20% MA to obtain CEM II-A (6–20%) cements. Standardized commercial sand at a cement/sand ratio of 1:3 and a w/c ratio of 0.5 was used to manufacture the mortars.

2.2. Experimental Methodology

2.2.1. Pozzolanic Activity Method

To assess the pozzolanic activity of the MA, the accelerated chemical method was applied to the pozzolan/lime system [8]. The test consisted of adding 1 g of pozzolan to 75 mL of saturated lime solution kept in a laboratory stove at 40 °C for 7, 28, and 90 days of reaction. At the end of each period, the solid obtained was filtered and washed with ethanol and then dried in a stove at 60 °C for 24 h to halt the pozzolanic reaction. The fixed lime values were used to generate the diffusive kinetic model [26], as per Equation (1):

$$C_t = \frac{0.23 \cdot exp\left(\frac{-3t}{\tau}\right) \cdot \left(-1 + exp\left(\frac{t}{\tau}\right)\right) \cdot \frac{1}{\tau}}{D_e \cdot r_s} + \frac{0.23 \cdot exp\left(-\frac{1}{\tau}\right) \cdot \frac{1}{\tau}}{K \cdot r_s^2} + C_{corr} \quad (1)$$

where D_e is the effective diffusion coefficient, K is the reaction rate constant, τ is the time constant (the time interval during which the pozzolan radius diminishes to 37% of its initial radio r_s), C_t is the absolute loss of CH concentration with time for the pozzolan/lime system, and C_{corr} is the correction parameter that takes into account the concentration remainder of CH that is not consumed in the reaction. Depending on the pozzolanic reaction, the behavior is as follows: diffusive (described by the first term), kinetic (described by the second term), and kinetic–diffusive (both terms).

2.2.2. Rheological Properties

The water required for normal paste consistency (NPC), the initial setting time (IST), and the volume stability (S) of the fresh cement pastes were determined as per EN 196-3 [27].

2.2.3. Physical Properties

The capillary absorption capacity of the mortars was analyzed in prismatic specimens measuring 4 × 4 × 16 cm and previously cured for 28 days using the Fagerlund method, as described in UNE 83982 [28]. Upon completion of the curing phase, the specimens were conditioned as per Spanish standard UNE 83966 [29] to obtain homogeneous moisture distribution throughout them. After conditioning, the specimens were placed in a container and partially immersed in 5 mm of water. The capillary absorption coefficient (K) in $kg{\cdot}m^2{\cdot}min^{0.5}$, effective porosity ($\varepsilon$) in $cm^3{\cdot}cm^{-3}$, and resistance to water penetration by capillary absorption (m) in $min{\cdot}cm^{-2}$ were determined by applying Equations (2)–(4), respectively:

$$K = \delta^a \cdot \varepsilon_e / 10 \cdot \sqrt{m} \tag{2}$$

$$\varepsilon_e = Q_n - Q_0 / A \cdot h \cdot \delta_a \tag{3}$$

$$m = t_n / h^2 \tag{4}$$

where δ_a is the density of the water (considering the value of 1 $g{\cdot}cm^{-3}$), Q_n is the weight of the specimen at saturation ($t = t_n$), Q_0 is the weight of the specimen before the test ($t = 0$), A is the section of the specimen, h is the thickness of the specimen, and t_n is the period required to reach saturation.

Analysis of electrical resistivity in the mortars was performed on prismatic specimens measuring 4 × 4 × 16 cm saturated with water for up to 90 days of curing. To this end, the 4-electrode Wenner method, as described in Spanish standard UNE 83988-2 [30], was used. Resistivity (ρ) was calculated by applying Equation (5):

$$\rho = \rho_w \cdot F_f \tag{5}$$

where F_f is the form factor (which amounts to 0.172 for samples measuring 4 × 4 × 16 cm) and ρ_w is the Wenner resistivity. The age factor (q) [21,31] meanwhile describes the changes in resistivity over time and adjusts the resistivity curve over time through Equation (6):

$$\rho_t = \rho_0 (t/t_0)^q \tag{6}$$

where ρ_t represents the resistivity measured at time t, and ρ_0 represents the resistivity at time 0 (t_0).

2.2.4. Mechanical Properties

The mechanical flexural and compressive strength tests performed on the mortars were conducted at 2, 28, and 90 days of curing, as per European standard EN 196-1 [32]. The loading speed used for the flexural strength test was 50 $N{\cdot}s^{-1}$, and the loading speed used for the compressive strength test was 2400 $N{\cdot}s^{-1}$

2.3. Instrumental Techniques

Chemical quantification was performed using a Bruker S8 Tiger wavelength dispersive X-ray fluorescence (WDXRF) spectrometer. Particle size distribution was obtained using a Malvern Mastersizer 3000 laser diffraction device equipped with red and blue light sources (He-Ne and LED) operating in dry dispersion mode. Measurements were taken in the range of 0.01–3500 μm. Identification and quantification of crystalline mineralogical phases (XRD–Rietveld method) were carried out with a PANalytical X'Pert PRO diffractometer, utilizing the Crystallography Open Database (COD) collection of crystal structures, and quantified using the Match v.3 and Fullprof software v.23. The morphology and element microanalyses were performed using an Inspect (FEI Company, Hillsboro, OR, USA) scanning electron microscope (SEM) equipped with an energy-dispersive X-ray (EDX) analyzer and a Si/Li detector. The pore size distribution and total porosity in the mortars were analyzed using a mercury intrusion porosimeter (MIP; Micromeritics Model 9320) in microspecimens measuring approximately 1 cm^3.

The changes in heating curve and heat of hydration in the standardized mortars were obtained using the Langavant semi-adiabatic method set out in European standard EN 196-9 [33] employing an Ibertest IB32-101E and the WinLect32.06 software.

3. Results

3.1. Characterization of the Starting Materials

XRF Analysis

The chemical compositions of the OPC and the MA are shown in Table 1. The MA is calcic (45.42% calcium content), and the sum of its SiO_2 + Al_2O_3 + Fe_2O_3 (4.43%) oxides is far below the level (≥70%) specified in the ASTM 618 standard for artificial pozzolans (fly ash and calcined natural pozzolan). The loss on ignition (LOI) value for MA (30%) is also above the standard limit (≤10%) [34]. The 6.92% K_2O content is notable.

Table 1. Chemical composition of the OPC and MA, expressed in %.

	OPC	**MA**
SiO_2	20.06	2.63
Al_2O_3	4.96	1.02
Fe_2O_3	3.28	0.78
CaO	62.19	45.42
MgO	2.12	6.36
SO_3	3.14	4.14
Na_2O	0.37	0.48
K_2O	0.66	6.92
TiO_2	0.25	0.09
P_2O_5	0.29	1.13
Cl	0.07	0.77
ZnO	0.17	-
SrO	0.16	-
Cr_2O_3	0.04	-
LOI	2.11	30.00

The particle size distribution density curves for the two starting materials are shown in Figure 2. These materials show similar distribution densities, with the main peaks at 25.68 and 15.41 μm, respectively. This similarity is corroborated by the Dx values shown in Table 2.

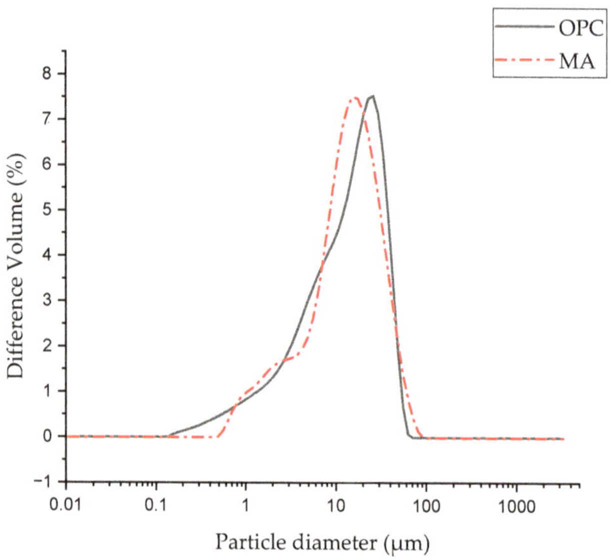

Figure 2. Density distribution curves of the starting materials.

Table 2. Dx values for the OPC and MA.

Raw Material	D_{10} (μm)	D_{50} (μm)	D_{90} (μm)
OPC	2.11	14.10	34.40
MA	2.40	13.60	34.70

Mineralogical quantification using the XRD–Rietveld method (Table 3) identifies anhydrous phases typical of OPC (C_3S, C_2S, C_3A, C_4AF, and $CaCO_3$), while the MA mostly comprises quartz, calcite, and dolomite.

Table 3. Mineralogical phases were quantified using the XRD–Rietveld method (n.d. = not detected; R_B and X^2 = adjustment factors).

Mineral (%)	OPC	MA
Quartz	n.d.	13
Calcite	4	34
Dolomite	n.d.	12
K-feldspar	n.d.	n.d.
C_3S	52	n.d.
C_2S	20	n.d.
C_4AF	6	n.d.
C_3A	9	n.d.
Amorphous matter	9	41
R_B	17.2	19.6
X^2	7.3	9.4

3.2. *Pozzolanic Activity*

3.2.1. Accelerated Chemical Method

The values obtained up to 90 days of reaction (Figure 3) show that the MA exhibits low–medium pozzolanic behavior since at 90 d it has only consumed 55% of the available lime. This fact is related to the low levels of silica and alumina in the ash (3.65%). Furthermore, the fixed lime values might be slightly overestimated if the possible effect of potash (K_2O = 6.92%) is taken into account, since it increases the pH of the solution and insolubilizes the portlandite, removing it from the medium [35].

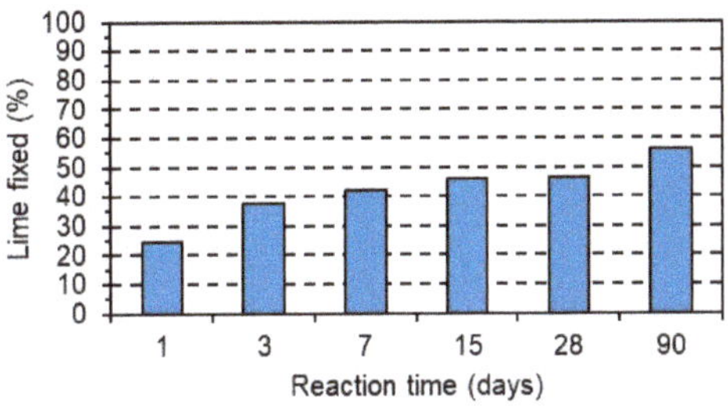

Figure 3. Changes in the amount of fixed lime with reaction time.

Applying the fixed lime values set out in Equation (1) quantifies the corresponding parameters shown in Table 4.

Table 4. Kinetic parameters of the pozzolanic reaction.

Parameters	$\tau(h)$	Rate K (h^{-1})	Free Energy of Activation AG#, kJ/mol	C_{corr}	Coefficient of Multiple Determination (R^2)
MA	108.9 ± 0.1	(2.57 ± 0.07)·10^{-3}	113.61	0.88 ± 0.007	0.9852

The K(10^{-3}) value obtained for the MA is of a lower order than other eco-pozzolans (silica fume, fly ash, natural pozzolan, and bagasse ash) [36].

3.2.2. Mineralogical Analysis of the Pozzolanic Reaction Using the XRD–Rietveld Method and SEM–EDX

After 28 days of reaction, the product presented no mineralogical differences compared to the starting MA (Figure 4), with quartz, calcite, and dolomite identified in both spectra (Table 5). The disappearance of the crystalline dolomite phase and the slight decrease in the amorphous phase due to the pozzolanic reaction are noteworthy. The calcite does not appear to intervene in the pozzolanic reaction, and its increase may be due to the carbonation of the sample during later testing, storage, and analysis.

Table 5. Analysis using the XRD–Rietveld method.

%	Quartz	Calcite	Dolomite	Amorphous Phase	R_B	X^2
Starting MA	13	34	12	41	19.6	9.4
28 d MA	16	47	traces	37	18.6	7.4

SEM/EDX observation at 28 days of curing (28 d MA) identifies the formation of CSH gels (C/S = 2.33) and ettringite as products of the pozzolanic reaction (Figure 5 and Table 6).

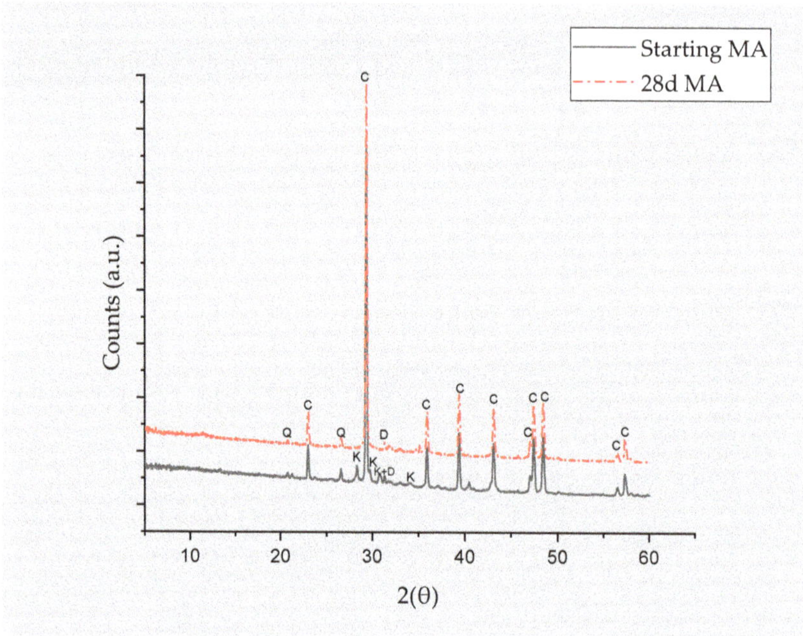

Figure 4. XRD diffractograms of the ash before and after 28 d. (C: calcite, Q: Quartz, K: Arkanite, and D: Dolomite).

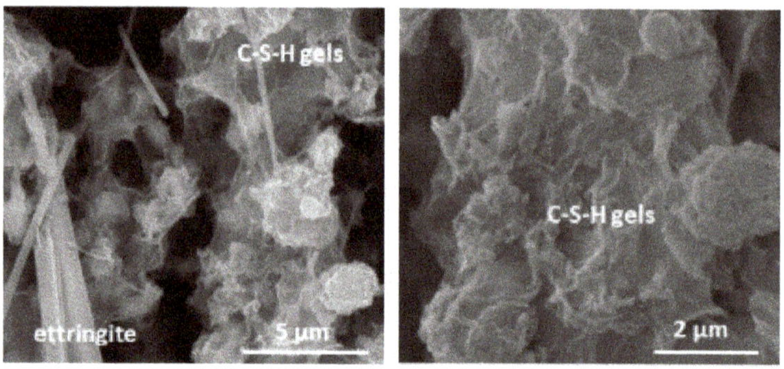

Figure 5. Ettringite and CSH gels (**left**) and detailed view of CSH gels (**right**).

Table 6. Chemical composition analysis using EDX (n.d. = not detected).

Oxides (%)	CSH Gels	Ettringite
MgO	12.74 ± 2.35	n.d.
Al_2O_3	6.65 ± 2.19	16.57 ± 2.25
SiO_2	20.91 ± 2.86	6.26 ± 2.87
SO_3	4.25 ± 1.63	22.71 ± 4.72
P_2O_5	3.34 ± 1.25	n.d.
K_2O	0.88 ± 0.32	n.d.
CaO	48.80 ± 4.26	54.43 ± 4.53
Fe_2O_3	2.44 ± 0.71	n.d.

3.3. Chemical Characterization of the Anhydrous Blended Cements

The addition of MA modifies the content of the major oxides in the blended cement (Table 7), decreasing the percentages of SiO_2, Al_2O_3, and CaO and increasing the share of potash and LOI. The blended cement containing 10% and 20% MA meets the chemical requirements (SO_3 and Cl^-) set out in EN 197-1 ($SO_3 \leq 3.5$–4.0% and $Cl \leq 0.1$%) [3]. Meanwhile, the EN 450 standard [37] on the addition of fly ash to concrete limits the equivalent Na_2O content ($Na_2O + 0.66\ K_2O$) to 5%, a value well above that obtained in this study (1.23% and 1.65%, respectively).

Table 7. Chemical composition (%) of the materials.

Oxides (%)	OPC	10% MA	20% MA	EN 197-1
SiO_2	20.06	18.32	16.57	-
Al_2O_3	4.96	4.57	4.17	-
Fe_2O_3	3.28	3.03	2.78	-
CaO	62.19	60.51	58.84	-
MgO	2.12	2.54	2.97	-
SO_3	3.14	3.24	3.34	≤3.5–4.0
Na_2O	0.37	0.38	0.39	-
K_2O	0.66	1.29	1.91	-
TiO_2	0.25	0.23	0.22	-
P_2O_5	0.29	0.37	0.46	-
Cl	0.06	0.05	0.05	≤0.1
ZnO	0.17	0.15	0.14	
SrO	0.16	0.14	0.13	
Cr_2O_3	0.04	0.04	0.03	
LOI	2.11	4.90	7.69	

3.4. Rheological Behavior of the Blended Cement Pastes

As per standard EN 196-3, normal consistency water (NCW), initial setting time (IST), and soundness (S) were analyzed (Table 8).

Table 8. Rheological behavior of the blended cement pastes.

	OPC	10% MA	20% MA	EN Standard
NCW (±1 g)	150	151	154	-
IST (±10 min)	138	129	126	≥60
S (mm)	0.15	0.10	0.10	≤10

The addition of MA to the cement causes a slight increase in the NCW values versus the OPC (of 0.7% and 2.6%, respectively) due to the greater specific surface area of the ash. It also slightly accelerates the IST [38], albeit within the deviation range of the method. The blended pastes do not experience any expansion effect or similar behavior to the OPC paste.

Based on these results, the types of cement analyzed meet the standardized physical specifications for the manufacture of future commercial cement.

3.5. Physical and Mechanical Behavior of the Blended Cement Mortars

3.5.1. Calorimetric Behavior of the Mortars

The blended mortars containing MA present a different heating curve to the OPC mortar (Figure 6), showing lower maximum values of 29.6 °C (10% MA) and 27.8 °C

(20% MA), respectively, versus 32 °C (OPC). In addition, this maximum value is recorded at greater reaction times—from 13 h (OPC) to 14 h—for the blended cement mortars [39,40]. These heating curve variations are reflected in the hydration heat values (Figure 6B), which decrease as the MA admixture percentage rises. Based on the data in Table 9 and the specifications in EN 197-1 [3], only the cement with 20% replacement content is considered a low heat-of-hydration cement since the value of 265.5 J/g is below the regulatory limit (≤270 J/g).

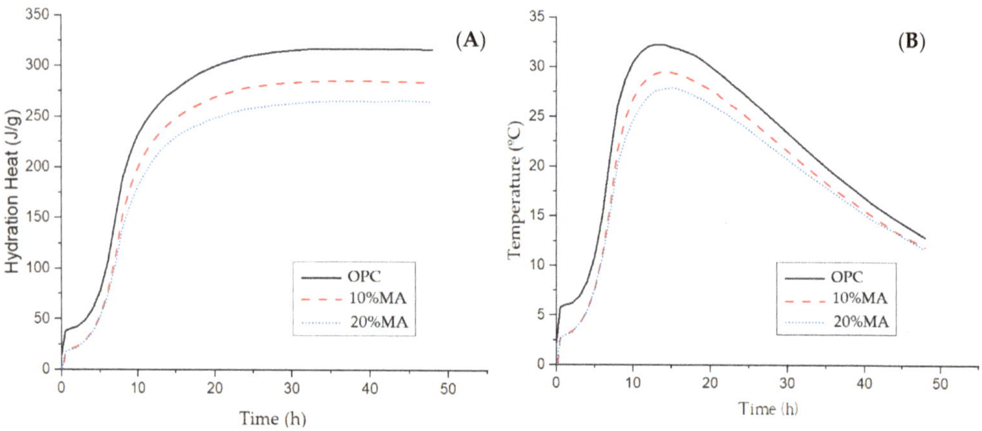

Figure 6. Langavant calorimetry: (**A**) heating curve and (**B**) hydration heat of the mortars.

Table 9. Heat of hydration (J/g) at 41 h of reaction.

Sample	Time (h)	Heat of Hydration (J/g)
OPC	41	317.4
10% MA	41	285.54
20% MA	41	265.54

3.5.2. Total and Capillary Water Absorption

Table 10 shows that the mortars containing MA experience an increase in water absorption capacity as the additive proportion grows, rising from 4.26% (OPC) to 5.63% for the 20% MA mortar, and an increase in absorption rate from 0.78 $g/min^{0.5}$ to 0.94 $g/min^{0.5}$. Similar results were obtained by other authors with ceramic waste, coal mining waste [41] and biomass ash [42]. Nevertheless, total absorption in MA cement mortars was below the 10% recommended for high-quality cement-based materials [43,44].

Table 10. Absorption coefficients in mortars.

Cement	Total Water Absorption (wt%)	Absorption Rate ($g/min^{0.5}$)	R^2
OPC	4.26	0.78	0.96
10% MA	4.83	0.86	0.97
20% MA	5.63	0.94	0.97

Table 11 shows the absorption rates obtained from the regression lines of Figure 7. An increase in the water absorption rate in the mortars containing MA is observed in the first and second absorption intervals (5 min–1 h) and (2–6 h), respectively, while a subsequent decrease versus the standard mortar is observed in the third interval (>6 h). The percentage of MA replacement content (10% and 20%) has little influence on the rate. This decrease in absorption rate is related to microstructural changes in the mortars studied due to the formation of hydrated phases, principally CSH gels [45].

Table 11. Water absorption rates (g/s).

Material	Rate 1	Rate 2	Rate 3
Intervals	2.24–7.75 min$^{0.5}$	10.95–18.97 min$^{0.5}$	37.95–14,400 min$^{0.5}$
OPC	0.682	0.058	0.006
10% MA	0.729	0.067	0.005
20% MA	0.788	0.092	0.007

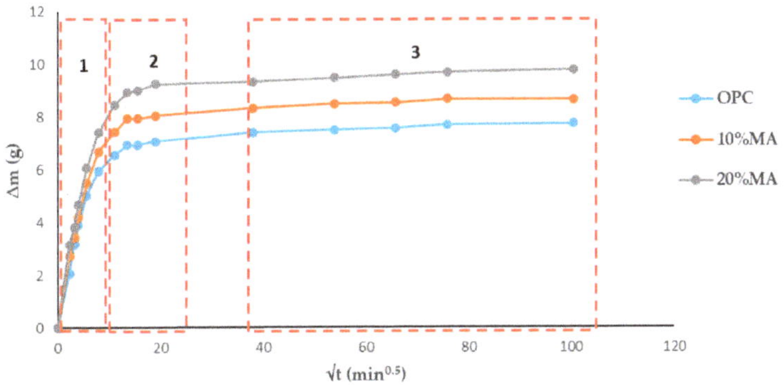

Figure 7. Total absorption curves in the mortars analyzed.

The capillary absorption tests (Figure 8) show the same trend as for total water absorption: the higher the MA percentage in the cement, the greater the capillary water absorption capacity [46]. Two behaviors are observed: the first (0–20 min$^{0.5}$) consists of water absorption via the capillary pore network, while the second (20–100 min$^{0.5}$) corresponds to the filling of air pores via the air diffusion and dissolution process until saturation state is reached [47].

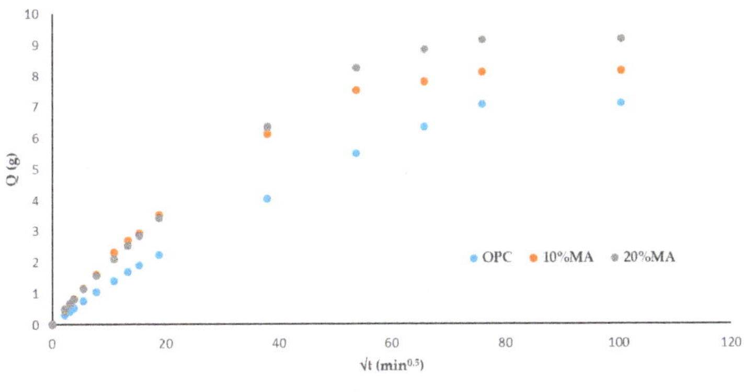

Figure 8. Capillary water absorption in the mortars.

Applying the Fagerlund method [8] and employing Equations (2)–(4) (Table 12) allow for the determination of the capillary absorption coefficient (K), effective porosity (ε_e), and resistance to water penetration by capillary absorption (m). The 10% and 20% MA mortars have a higher absorption coefficient than the OPC mortar. This is attributed to the higher density of the OPC mortar and to the sealing of the pore network, which contributes to a reduction in capillary porosity versus the blended cement mortars. These results are in line with those obtained previously (Table 11) and those obtained with other cements [48,49], but not with the findings reported in [50]. As regards the m-coefficient (Table 12), an increase in the effective porosity values is observed with the addition of MA and is very similar in all three cases to that of the OPC, indicating interconnection between capillary pores in the mortars [51]. In addition, the ε_e-coefficient (Table 12), classified according to the concrete's durability to penetration by aggressive agents as established by the CyTED RED DURAR [52,53] (where <10% indicates that the concrete is of good quality and compactness, 10–15% indicates moderate quality, and >15% indicates inadequate durability), shows that the values for the mortars studied are <10%. Taking into account the above, it can therefore be deduced that they are durable and of good quality [51].

Table 12. K, ε_e, and m-coefficients.

Material	K ($kg/m^2min^{0.5}$)	ε_e (cm^3/cm^3)	m (min/cm^2)
OPC	0.736	0.788	87.210
10% MA	1.112	1.190	87.316
20% MA	1.173	1.252	87.698

3.5.3. Electrical Resistivity

The electrical resistivity results show an increase with the curing time in all the mortars (Figure 9). At 90 d, the MA cement mortars present lower resistivity values than the OPC (59.5 Ω·m) which decreases as the additive percentage rises (55.9 and 51.9 Ω·m, respectively). This fact is likely related to the addition of MA lowering densification of the matrix, thus offering less resistance to the electrical current [54–56].

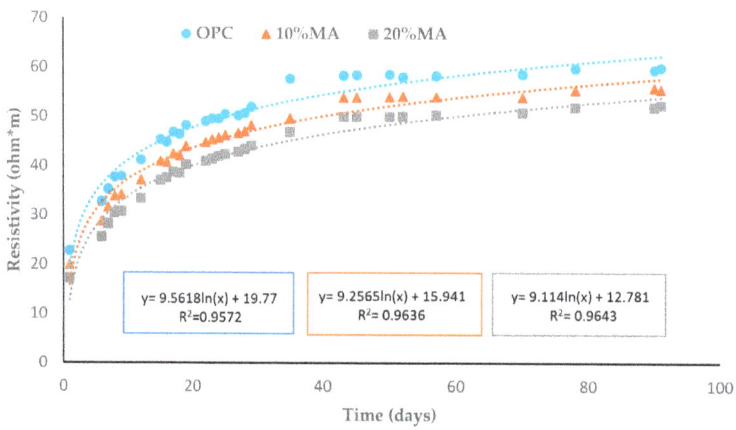

Figure 9. Changes in electrical resistivity as a function of hydration time.

The loss of resistivity observed in this paper is associated with content heterogeneity, especially as regards the presence of alkalis [57]—partly mirrors the findings of previous research into biomass ashes. The lower resistivity of the mortars containing MA is in line with the values obtained for the age factor (q) when applying Equation (5) (Section 2.2.3). The results corresponding to the age (q) and resistivity factors at time 0 (ρ_0) as per Equation (6), as well as the R^2 values of the regression, are shown in Table 13. The q-values increase significantly as the percentage of ash rises from 0.23 in the OPC to 0.27 in the 20% MA. These results are in line with those obtained for capillary absorption (Section 3.5.2) and all these q-values follow the reference values [58] for mortars prepared with CEM I and CEM II/A-P (0.22 and 0.37, respectively).

Table 13. Values for the q, ρ_0 and R^2 parameters of the mortars studied.

	OPC	10% MA	20% MA
q	0.231	0.246	0.265
ρ_0	19.716	17.566	15.179
R^2	0.964	0.969	0.971

3.5.4. Compressive Strength and Microporosity

Analysis of the relative losses in compressive strength versus the OPC mortar (Figure 10) reveals, in general, that the MA additive reduces strength in proportions greater than the percentage of replacement content, i.e., by approximately 17% and 29%, respectively, at 2 d of hydration. At 28 and 90 d, no noticeable mechanical differences are observed, with relative losses of around 11% and 23% for the 10% and 20% MA mortars, respectively.

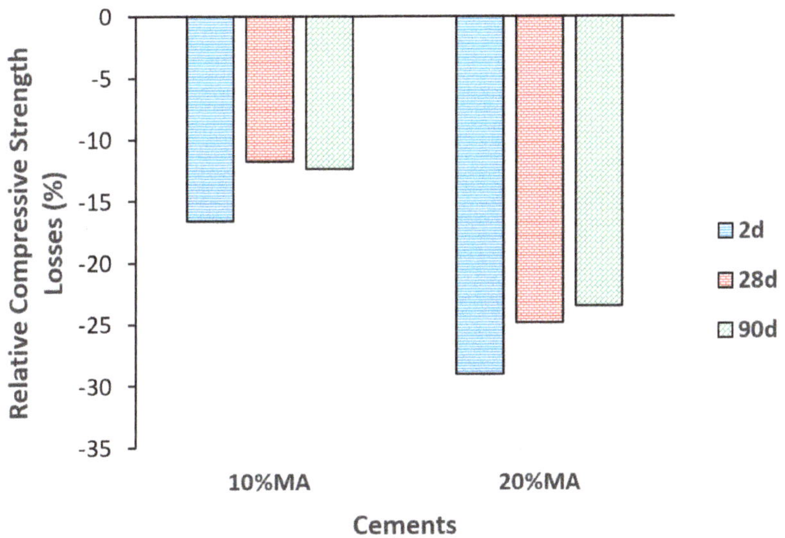

Figure 10. Changes in compressive strength losses versus the OPC mortar.

This decrease is related to the low pozzolanic activity, greater water absorption, lower density, and lower electrical resistivity in the MA mortars, as discussed above. Furthermore, the incorporation from the ash of additional K_2O, known for its negative effect on the previously discussed hydration and pozzolanic reactions, should be taken into account. Thus, adding MA would mostly have a filler effect on the cement matrix. According to Table 14 and the mechanical requirements set out in EN 197-1 [3], MA blended cement maintains the initial strength category at 2 days, while at 28 days only the 20% MA cement would drop a strength category to 42.5.

Table 14. Compressive strength (MPa) of the mortars and mechanical requirements.

Mortar	2 d	28 d	90 d	EN 197-1
OPC	41.23 ± 0.01	67.17 ± 0.18	69.03 ± 1.07	$\geq$20–30 (2 d) $\geq$52.5 (28 d)
10% MA	34.73 ± 0.24	59.29 ± 0.16	60.50 ± 0.08	
20% MA	29.27 ± 0.46	50.50 ± 0.61	52.81 ± 0.05	

An excellent correlation coefficient is found between the compressive strength and electrical resistivity values ($R^2 \geq 0.98$) (Figure 11) which, under standard conditions, makes it possible to predict one of these parameters based on the other [58].

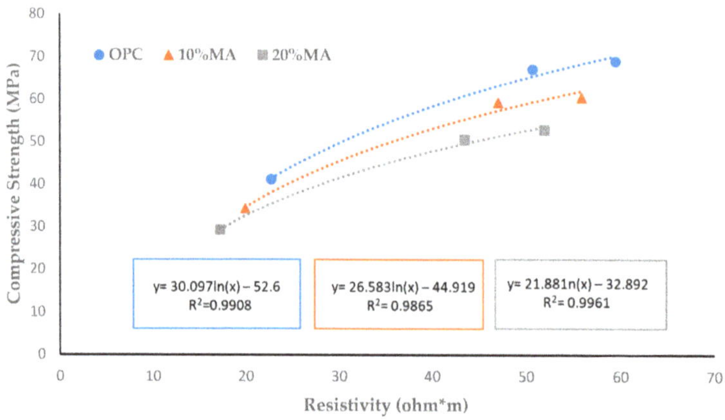

Figure 11. The relationship between compressive strength and resistivity.

The total porosity of the mortars (Table 15) shows that the addition of 10% and 20% MA produces an increase in porosity versus the OPC of between 21% and 23% and of between 11% and 21% at 2 and 90 d, respectively. However, the shapes of the pore size density curves at 90 d (Figure 12) are very similar, showing a single peak between 60 nm and 200 nm, albeit with a slight refinement of the pore sizes below 100 nm when MA is added. This phenomenon is related to the low pozzolanic activity of MA, which does not compensate for the cement replacement effect [59]. This behavior is in line with other industrial waste materials presenting low–medium pozzolanic activity [60].

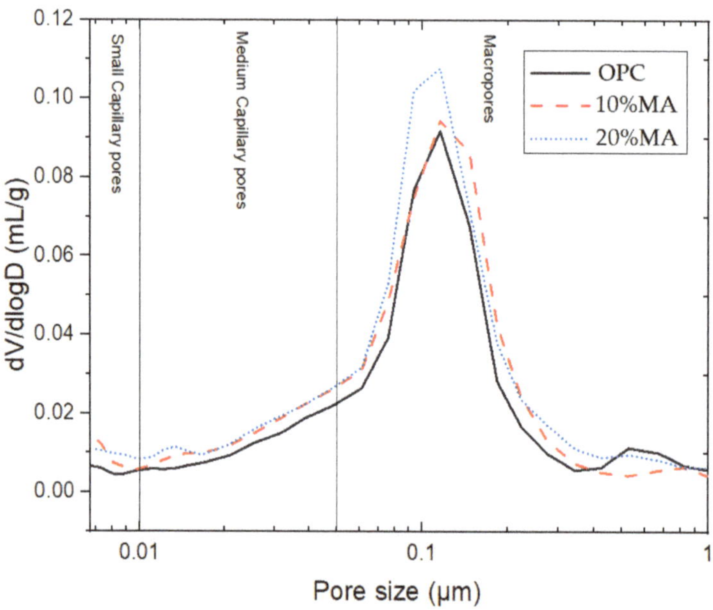

Figure 12. Pore distribution density curves at 90 d.

Table 15. Total mortar porosity (vol. %) at 2 and 90 d.

	Total Porosity (vol. %)	
	2 d	**90 d**
OPC	13.96	11.70
10% MA	16.83	13.01
20% MA	17.14	14.20

4. Conclusions

This work presents the quality of cement and mortar by replacing OPC with Marabou weed ash (MA) from Cuba. The following conclusions are drawn from the results obtained with MA:

1. Chemical analysis determined by the XRF of the MA shows that the sum of SiO_2 + Al_2O_3 + Fe_2O_3 (4.43%) oxides is far below the minimum value recommended by the regulations for active cement admixtures (≥70%). In addition, it presents 30% LOI due to its calcium and unburned material content, and 6.92% K_2O.
2. From a mineralogical point of view, MA comprises 46% calcareous minerals (calcite and dolomite) and 41% amorphous phase.
3. MA exhibits medium–low pozzolanic activity in the pozzolan/lime system due to its chemical and mineral composition. SEM-EDX analyses at 28 days of reaction identified CSH gels and ettringite as mineralogical phases produced by the pozzolanic reaction.
4. 10% and 20% MA blended cements meet the chemical (SO_3 and Cl^-) and physical (NCW, IST, and S) requirements set out in the regulations.
5. The heating curves and heat of hydration of the mortars containing 10% and 20% MA decrease in inverse proportion to the ash content. The types of cement made with 20% MA qualify as low heat-of-hydration cement (≤270 J/g).
6. In terms of intrinsic properties, the addition of MA results in an increase in water absorption (total and capillary), lower electrical resistivity (up to 13%), and classification of the blended mortars as durable from the point of view of penetration by aggressive agents.
7. Although, when compared to the OPC mortar, the MA cement mortars experience a decrease in compressive strength at 28 d of curing roughly equivalent to the replacement rate, the binary cement made with 10% and 20% MA meet the mechanical requirements for the manufacture of commercial cement. At 2 d of hydration, the materials maintain the initial strength category (52.5), but at 28 days the 20% MA cement drops to the lower strength category (42.5). A good coefficient of correlation (≥0.98) was found between strength and resistivity.
8. The porosimetry tests corroborate the findings of the intrinsic tests, highlighting that the MA cement admixture produces an increase in total porosity, although a slight refinement of less than 100 nm is observed in the pore density curves.

Based on the results obtained, Marabou weed biomass ash is viable as a mineralogical additive for use in the future manufacture of eco-friendly cement with a reduced carbon footprint. The findings of this pioneering study of MA will form the basis of future lines of research principally exploring higher replacement content percentages, synergies with other pozzolans, durability when subject to aggressive agents, etc.

Author Contributions: Conceptualization, M.F. and A.M.M.d.l.R.; methodology, M.F. and A.M.M.d.l.R.; validation, M.F.; formal analysis, A.M.M.d.l.R., R.G., R.V.d.l.V. and E.V.-C.; investigation, M.F., A.M.M.d.l.R., E.V.-C., R.G. and R.V.d.l.V.; data curation, A.M.M.d.l.R., R.G. and R.V.d.l.V.; writing—original draft preparation, M.F., A.M.M.d.l.R., R.G., R.V.d.l.V. and M.V.V.; writing—review and editing, M.F., A.M.M.d.l.R. and M.V.V.; supervision, M.F., A.M.M.d.l.R. and M.V.V.; project administration, M.F.; funding acquisition, M.F. All authors have read and agreed to the published version of the manuscript.

Funding: This research is part of the Cuban national project (10CB/PN223LH010-007) financed by the CITMA (Cuba). It is also supported by the Spanish national project PID2021-122390OB-C21/CIDECAR financed by the MICIU/AEI/10.13039/501100011033 and the ERDF/EU. The authors are thankful for the support by the Ministry of the Ministry of Science, Technological Development and Innovation of the Republic of Serbia (Contract No. 451-03-66/2024-03/200012), and the European Cooperation in Science and Technology through COST Action CA20133.

Institutional Review Board Statement: Not applicable.

Informed Consent Statement: Not applicable.

Data Availability Statement: The raw data supporting the conclusions of this article will be made available by the authors on request.

Conflicts of Interest: The authors declare that they have no known competing financial interests or personal relationships that could have appeared to influence the work reported in this paper.

References

1. Cement Roadmap. Available online: https://iea.blob.core.windows.net/assets/e3d8a122-455c-49f1-9347-635f46529826/Cement_Roadmap_Foldout_WEB.pdf (accessed on 29 May 2024).
2. Sanjuán, M.Á.; Martínez, R.; Alarcón Barrio, A.; Josa Garcia-Tornel, A. Análisis Prenormativo de Los Constituyentes Potenciales de Los Cementos Pórtland. *Cem. Hormig.* **2023**, *1015*, 27–39.
3. *AENOR UNE-EN 197-1*; Cemento. Parte 1: Composición, Especificaciones y Criterios de Conformidad de Los Cementos Comunes. Asociación Española de Normalización: Madrid, Spain, 2018.
4. Vashistha, P.; Park, S.; Pyo, S. A Review on Sustainable Fabrication of Futuristic Cementitious Binders Based on Application of Waste Concrete Powder, Steel Slags, and Coal Bottom Ash. *Int. J. Concr. Struct. Mater.* **2022**, *16*, 51. [CrossRef]
5. Meziani, M.; Leklou, N.; Chelouah, N.; Amiri, O. Mechanical and Thermal Modification of Mordenite-Rich Tuff and Its Effect on Cement Pastes. *Constr. Build. Mater.* **2022**, *318*, 126008. [CrossRef]
6. Pinheiro, V.D.; Alexandre, J.; Xavier, G.D.; Marvila, M.T.; Monteiro, S.N.; de Azevedo, A.R. Methods for Evaluating Pozzolanic Reactivity in Calcined Clays: A Review. *Materials* **2023**, *16*, 4778. [CrossRef]
7. Pillay, D.L.; Olalusi, O.B.; Mostafa, M.M.H. A Review of the Engineering Properties of Concrete with Paper Mill Waste Ash—Towards Sustainable Rigid Pavement Construction. *Silicon* **2021**, *13*, 3191–3207. [CrossRef]
8. Vegas, I.; Frías, M.; Urreta, J.; San José, J.T. Obtaining a Pozzolanic Addition from the Controlled Calcination of Paper Mill Sludge. Performance in Cement Matrices. *Mater. Constr.* **2006**, *56*, 49–60.
9. Frías, M. The Effect of Metakaolin on the Reaction Products and Microporosity in Blended Cement Pastes Submitted to Long Hydration Time and High Curing Temperature. *Adv. Cem. Res.* **2006**, *18*, 1–6. [CrossRef]
10. Velardo, P.; Sáez del Bosque, I.F.; Sánchez de Rojas, M.I.; De Belie, N.; Medina, C. Design and Evaluation of Physical, Mechanical and Micro-Structural Properties of Eco-Friendly Binary-Blended Mortars Using Biomass Bottom Ash or Construction and Demolition Waste Powder. *Cem. Concr. Compos.* **2023**, *143*, 105252. [CrossRef]
11. Zhai, J.; Burke, I.T.; Stewart, D.I. Beneficial Management of Biomass Combustion Ashes. *Renew. Sustain. Energy Rev.* **2021**, *151*, 111555. [CrossRef]
12. European Directive 2023/2413 to the Promotion of Energy from Renewable Source. Available online: https://eur-lex.europa.eu/legal-content/EN/TXT/?uri=CELEX:32023L2413&qid=1699364355105 (accessed on 29 May 2024).
13. Adhikary, S.K.; Ashish, D.K.; Rudžionis, Ž. A Review on Sustainable Use of Agricultural Straw and Husk Biomass Ashes: Transitioning towards Low Carbon Economy. *Sci. Total Environ.* **2022**, *838*, 156407. [CrossRef]
14. Silva, L.H.P.; de Paiva, F.F.G.; Tamashiro, J.R.; Kinoshita, A. Potential of Bamboo Leaf Ash as Supplementary Binder Materials—A Systematic Literature Review. *J. Build. Eng.* **2023**, *71*, 106547. [CrossRef]
15. Villar-Cociña, E.; Frías, M.; Savastano, H.; Rodier, L.; Sánchez de Rojas, M.I.; Sáez del Bosque, I.F.; Medina, C. Quantitative Comparison of Binary Mix of Agro-Industrial Pozzolanic Additions for Elaborating Ternary Cements: Kinetic Parameters. *Materials* **2021**, *14*, 2944. [CrossRef] [PubMed]
16. Moreno, S.; Rosales, M.; Rosales, J.; Agrela, F.; Díaz-López, J.L. Feasibility of Using New Sustainable Mineral Additions for the Manufacture of Eco-Cements. *Materials* **2024**, *17*, 777. [CrossRef] [PubMed]
17. Rithuparna, R.; Jittin, V.; Bahurudeen, A. Influence of Different Processing Methods on the Recycling Potential of Agro-Waste Ashes for Sustainable Cement Production: A Review. *J. Clean. Prod.* **2021**, *316*, 128242. [CrossRef]
18. Nakanishi, E.Y.; Frías, M.; Santos, S.F.; Rodrigues, M.S.; Vigil de la Villa, R.; Rodriguez, O.; Junior, H.S. Investigating the Possible Usage of Elephant Grass Ash to Manufacture the Eco-Friendly Binary Cements. *J. Clean. Prod.* **2016**, *116*, 236–243. [CrossRef]
19. Yin, K.; Ahamed, A.; Lisak, G. Environmental Perspectives of Recycling Various Combustion Ashes in Cement Production—A Review. *Waste Manag.* **2018**, *78*, 401–416. [CrossRef]
20. Kasaniya, M.; Thomas, M.D.A.; Moffatt, E.G. Efficiency of Natural Pozzolans, Ground Glasses and Coal Bottom Ashes in Mitigating Sulfate Attack and Alkali-Silica Reaction. *Cem. Concr. Res.* **2021**, *149*, 106551. [CrossRef]

21. Frías, M.; Caneda-Martínez, L.; Sánchez de Rojas, M.I.; Tenazoa, C.; Flores, E. Scientific and Technical Studies on Eco-Efficient Binary Cements Produced with Thermally Activated Ichu Grass: Behaviour and Properties. *Cem. Concr. Compos.* **2020**, *111*, 103613. [CrossRef]
22. Pedroso, D.T.; Kaltschmitt, M. Dichrostachys Cinerea as a Possible Energy Crop—Facts and Figures. *Biomass Convers. Biorefin.* **2012**, *2*, 41–51. [CrossRef]
23. Hernández Sardiñas, A.; González Morales, V.M.; Freide Orozco, M.L. Aprovechamiento de Las Posibles Fuentes de Biomasa Para Entregar Más Electricidad En La Fábrica de Azúcar Antonio Sánchez. *Centro. Azúcar.* **2017**, *44*, 88–97.
24. Rubio-González, A.; Iturria Quintero, P.J.; Rodríguez-Machín, L.; Palmero Marín, D. Propiedades del marabú (*Dichrostachys cinerea* L.) cosechado con máquinas, como combustible para la generación de electricidad (primera parte). *Centro. Azúcar.* **2021**, *48*, 93–104.
25. Sagastume Gutiérrez, A.; Cabello Eras, J.J.; Vandecasteele, C.; Hens, L. Data Supporting the Assessment of Biomass Based Electricity and Reduced GHG Emissions in Cuba. *Data Brief.* **2018**, *17*, 716–723. [CrossRef]
26. Villar-Cociña, E.; Valencia-Morales, E.; González-Rodríguez, R.; Hernández-Ruíz, J. Kinetics of the Pozzolanic Reaction between Lime and Sugar Cane Straw Ash by Electrical Conductivity Measurement: A Kinetic–Diffusive Model. *Cem. Concr. Res.* **2003**, *33*, 517–524. [CrossRef]
27. *AENOR UNE-EN 196-3*; Métodos de Ensayo de Cementos. Parte 3: Determinación del Tiempo de Fraguado y de La Estabilidad de Volumen. Asociación Española de Normalización: Madrid, Spain, 2017.
28. *Spanish Standard UNE 83982*; Determinación de La Absorción de Agua Por Capilaridad Del Hormigón Endurecido. Asociación Española de Normalización: Madrid, Spain, 2008.
29. *Spanish Standard IME 83966*; Acondicionamiento de Probetas de Hormigón Para Los Ensayos de Permeabilidad a Gases y Capilaridad. Asociación Española de Normalización: Madrid, Spain, 2008.
30. *Spanish Standard UNE 83988-2*; Durabilidad Del Hormigón. Métodos de Ensayo. Determinación de La Resistividad Eléctrica. Parte 2: Método de Las Cuatro Puntas o de Wenner. Asociación Española de Normalización: Madrid, Spain, 2014.
31. Attari, A.; McNally, C.; Richardson, M.G. A Probabilistic Assessment of the Influence of Age Factor on the Service Life of Concretes with Limestone Cement/GGBS Binders. *Constr. Build. Mater.* **2016**, *111*, 488–494. [CrossRef]
32. *AENOR UNE-EN 196-1*; Métodos de Ensayo de Cementos. Parte 1: Determinación de Resistencias Mecánicas. Asociación Española de Normalización: Madrid, Spain, 2018.
33. *European Standard EN 196-9*; Methods of Testing Cement—Part 9: Heat of Hydration-Semi-Adiabatic Method. European Committee for Standardization: Brussels, Belgium, 2011.
34. *American Standard ASTM C618-23e1*; Standard Specification for Coal Fly Ash and Raw or Calcined Natural Pozzolan for Use in Concrete. ASTM International: West Conshohocken, PA, USA, 2023.
35. Wang, H.; Liu, X.; Zhang, Z. Pozzolanic Activity Evaluation Methods of Solid Waste: A Review. *J. Clean. Prod.* **2023**, *402*, 136783. [CrossRef]
36. Villar-Cociña, E.; Frías, M.; Savastano, H. Quantitative Comparison of Mineral Ash from Agro-Industrial Waste for Use as Pozzolanic Additions in Cement: Kinetic Parameters. *Asp. Min. Miner. Sci.* **2020**, *5*, 626–634.
37. *European Standard EN 450-1*; Fly Ash for Concrete—Part 1: Definition, Specification and Conformity Criteria. European Committee for Standardization: Brussels, Belgium, 2013.
38. Jawed, I.; Skalny, J. Alkalies in Cement: A Review: II. Effects of Alkalies on Hydration and Performance of Portland Cement. *Cem. Concr. Res.* **1978**, *8*, 37–51. [CrossRef]
39. Mkahal, Z.; Maherzi, W.; Mamindy-Pajany, Y.; Bouzar, B.; Abriak, N.-E. Development of a Low-Carbon Binder Based on Raw, Ground, and Carbonated Waste Paper Fly Ash. *Sustain. Mater. Technol.* **2023**, *36*, e00650. [CrossRef]
40. Zeraoui, A.; Maherzi, W.; Benzerzour, M.; Abriak, N.E.; Aouad, G. Development of Flash-Calcined Sediment and Blast Furnace Slag Ternary Binders. *Buildings* **2023**, *13*, 333. [CrossRef]
41. Caneda-Martínez, L.; Medina, C.; Sánchez de Rojas, M.I.; Frías, M. Water Transport in Binary Eco-Cements Containing Coal Mining Waste. *Cem. Concr. Compos.* **2019**, *104*, 103373. [CrossRef]
42. Rosales, J.; Cabrera, M.; Beltrán, M.G.; López, M.; Agrela, F. Effects of Treatments on Biomass Bottom Ash Applied to the Manufacture of Cement Mortars. *J. Clean. Prod.* **2017**, *154*, 424–435. [CrossRef]
43. Senthamarai, R.M.; Manoharan, P.D.; Gobinath, D. Concrete Made from Ceramic Industry Waste: Durability Properties. *Constr. Build. Mater.* **2011**, *25*, 2413–2419. [CrossRef]
44. Neville, A.M. *Properties of Concrete*; Longman Scientific & Technical Jogn Wiley & Sons, Harlow: New York, NY, USA, 1995; Volume 4.
45. Naji Givi, A.; Abdul Rashid, S.; Aziz, F.N.A.; Salleh, M.A.M. The Effects of Lime Solution on the Properties of SiO_2 Nanoparticles Binary Blended Concrete. *Compos. Part B* **2011**, *42*, 562–569. [CrossRef]
46. Medina, J.M.; Sáez del Bosque, I.F.; Frías, M.; Sánchez de Rojas, M.I.; Medina, C. Durability of New Blended Cements Additioned with Recycled Biomass Bottom ASH from Electric Power Plants. *Constr. Build. Mater.* **2019**, *225*, 429–440. [CrossRef]
47. Lu, J.; Wang, K.; Qu, M.-L. Experimental Determination on the Capillary Water Absorption Coefficient of Porous Building Materials: A Comparison between the Intermittent and Continuous Absorption Tests. *J. Build. Eng.* **2020**, *28*, 101091. [CrossRef]
48. Torres-Castellanos, N.; Torres-Agredo, J.; Mejía-de-Gutierrez, R. Permeation Properties of Concrete Added with a Petrochemical Industry Waste. *Ing. Investig.* **2017**, *37*, 23–29. [CrossRef]

49. Silva, Y.F.; Lange, D.A.; Delvasto, S. Effect of Incorporation of Masonry Residue on the Properties of Self-Compacting Concretes. *Constr. Build. Mater.* **2019**, *196*, 277–283. [CrossRef]
50. Wygocka-Domagałło, A.; Garbalińska, H. The Effect of Pore Structure on the Water Sorption Coefficient of Cement Mortars Reinforced with 12 mm Polypropylene Fibres. *Constr. Build. Mater.* **2020**, *248*, 118606. [CrossRef]
51. Castañeda, A.; Albear, J.J.H.; Corvo, F.; Marrero, R. Concrete Quality Assessment before Building Structure s Submitting to Environment al Exposure Conditions. *Rev. Constr.* **2017**, *16*, 374–387.
52. Trocónis, O.; Romero, A.; Andrade, C.; Helene, P.; Díaz, I. *Manual de Inspección, Evaluación y Diagnóstico de Corrosión En Estructuras de Hormigón Armado*; CYTED Red. Temática XV. B. Durabilidad de la Armadura: Río de Janeiro, Brazil, 1997.
53. Mera, J.C.G.; Albear, J.J.H.; Valdés, A.C. Primeras Experiencias en el Desempeño por Durabilidad de un Hormigón Antes de Usarlo en el Perfil Costero de Manabí, Ecuador. *Rev. CENIC Cienc. Quím.* **2017**, *48*, 27–40.
54. Monasterio, M.; Caneda-Martínez, L.; Vegas, I.; Frías, M. Progress in the Influence of Recycled Construction and Demolition Mineral-Based Blends on the Physical–Mechanical Behaviour of Ternary Cementitious Matrices. *Constr. Build. Mater.* **2022**, *344*, 128169. [CrossRef]
55. Medina, C.; Sánchez de Rojas, M.I.; Thomas, C.; Polanco, J.A.; Frías, M. Durability of Recycled Concrete Made with Recycled Ceramic Sanitary Ware Aggregate. Inter-Indicator Relationships. *Constr. Build. Mater.* **2016**, *105*, 480–486. [CrossRef]
56. Hou, T.-C.; Nguyen, V.K.; Su, Y.-M.; Chen, Y.-R.; Chen, P.-J. Effects of Coarse Aggregates on the Electrical Resistivity of Portland Cement Concrete. *Constr. Build. Mater.* **2017**, *133*, 397–408. [CrossRef]
57. Rajamma, R.; Senff, L.; Ribeiro, M.J.; Labrincha, J.A.; Ball, R.J.; Allen, G.C.; Ferreira, V.M. Biomass Fly Ash Effect on Fresh and Hardened State Properties of Cement Based Materials. *Compos. Part B* **2015**, *77*, 1–9. [CrossRef]
58. Andrade, C. Resistivity Test Criteria for Durability Design and Quality Control of Concrete in Chloride Exposures. *Concr. Aust.* **2014**, *40*, 57–64.
59. Velardo, P.; Sáez del Bosque, I.F.; Sánchez de Rojas, M.I.; De Belie, N.; Medina, C. Effect of Incorporating Biomass Bottom Ash and Construction and Demolition Waste Powder on the Physical-Mechanical Properties and Micro-Structure of Ternary-Blended Mortars. *Constr. Build. Mater.* **2024**, *432*, 136628. [CrossRef]
60. Frías, M.; de Rojas, M.I.S.; Rodríguez, C. The Influence of SiMn Slag on Chemical Resistance of Blended Cement Pastes. *Constr. Build. Mater.* **2009**, *23*, 1472–1475. [CrossRef]

materials

MDPI

Article

Design, Characterization, and Incorporation of the Alkaline Aluminosilicate Binder in Temperature-Insulating Composites

Pavlo Kryvenko *, Igor Rudenko, Oleksandr Konstantynovskyi and Oleksandr Gelevera

Scientific Research Institute for Binders and Materials, Kyiv National University of Construction and Architecture, Povitroflotskyi Prospect 31, 03037 Kyiv, Ukraine; rudenko.ii@knuba.edu.ua (I.R.); konstantynovskyi.op@knuba.edu.ua (O.K.); gelevera.og@knuba.edu.ua (O.G.)

* Correspondence: kryvenko.pv@knuba.edu.ua

Abstract: This paper covers the design of binder formulations and technology for low-energy building materials based on alkaline aluminosilicate binders developed for special uses. The microstructure of the binders was investigated using scanning electron and atomic force microscopy examination techniques. The identification of phase compositions was performed by means of X-ray diffraction. The degree of binding of the alkali metal ions within the binder was determined with the help of chemical analysis of the pore fluid. Structure formation depending upon binder mix design and curing conditions was also studied. Some examples of the manufacture and application of binder-based glues and adhesives, including those developed for heat insulation and fire prevention, are discussed. The advantages of binder-based temperature-insulating composite materials compared with traditionally used materials are highlighted.

Keywords: alkaline aluminosilicate binder; composition design; binding; composite; curing conditions; durability; fire-proofness; heat insulation; structure formation; zeolite-like reaction products

1. Introduction

In view of the severe consequences of today's military actions, terrorism, and natural disasters, the demand for the reliable protection of critical military and civil infrastructure is increasingly growing. The requirements for such protective materials are the following: the use of easy-to-produce mineral binders as a binding agent, environmental friendliness, low energy consumption, and excellent performance properties under various severe conditions, e.g., high temperatures and fire. Aside from the above problems, nowadays, it is critical to reduce energy consumption through the use of highly efficient insulating composite materials.

One method of synthesizing alkaline aluminosilicate hydrates, the analogs of natural zeolites, using a mixture of natural (clays) and artificial aluminosilicates with alkali metal compounds under various curing conditions was established by Victor Glukhovsky [1]. This method is based on the modeling (simulating) of natural processes during the formation of alkaline zeolite-like aluminosilicates at different temperatures. An important conclusion was drawn: an increase in temperature promotes a smooth dehydration process and subsequent recrystallization of the hydration products with the formation of anhydrous aluminosilicates of alkali metals, i.e., Li, Na, K, Rb, and Cs [2]. Alkali metal compounds not only act as activators of hardening but are also responsible for the formation of the main structural elements of alkali-activated materials. These reaction products, being analogous to natural zeolites of the $Na_2O(K_2O)\cdot Al_2O_3\cdot(2\ldots4)SiO_2\cdot nH_2O$ type, have been discovered in ancient concretes (e.g., in Ancient Greece, Ancient Rome, Egypt, Syria) [3]. The durability of ancient concretes and the similarity of their structure to that of alkali-activated cement concretes allowed us to infer their high durability. The excellent performance properties of alkali-activated materials, such as their high strength and heat, corrosion, and freeze–thaw resistance [4–7], have been confirmed by over 65 years' experience of their use. The composition of zeolite-like products is determined by structural features of aluminosilicate components (precursor), alongside their composition, the applied curing conditions, the rate of chemical reactions, and the binder design [8–12]. The most appropriate materials for achieving the mentioned goals are natural clays in a dehydrated state, volcanic rocks, and industrial by-products like fuel ashes, slags of various origin, and red mud [9,13–19].

Alkali-activated materials can be divided into alkaline cements and geopolymers (or geocements) [20]. Depending on the precursor type, the following reaction products can be formed, $CaO\text{-}Al_2O_3\text{-}SiO_2\text{-}H_2O$, $R_2O\text{-}CaO\text{-}Al_2O_3\text{-}SiO_2\text{-}H_2O$, and $R_2O\text{-}Al_2O_3\text{-}SiO_2\text{-}H_2O$, along with R–Na, K, and Li [21]. In terms of composition, the binders can be divided into high-calcium, intermediate-calcium (hybrid), and low-calcium binders [22].

Several studies have been conducted on the production of composite materials based on zeolite-like matrices of the low-calcium $Na_2O(K_2O)\text{-}Al_2O_3\text{-}SiO_2\text{-}H_2O$ system, i.e., alkaline aluminosilicate binders [23–27].

An analysis of this research showed that the use of zeolite-like binder systems for the production of building materials for general and special applications presents a solution to the problems associated with durability, energy-saving ability, and environmental protection, where traditional binder systems fail [13,23].

Due to the polymerization of the aluminosilicate constituent, reaction products that are analogs to natural minerals (zeolites and feldspathoids) are formed. Considering their compositional make-up, zeolites may be referred to as inorganic polymers, which, on the one hand, are analogs to organic ones (in terms of such properties as elasticity, corrosion resistance, and adhesion to various materials) and, on the other hand, due to their mineral nature, are ecologically pure and heat- and fire-resistant [9,13]. As the theoretical premise for the creation of alkali-activated materials with a polymeric structure, this study aimed at the synthesis of analogs to natural minerals based on clays of various structural types and applied alkali metal compounds [3,8,26]. An extensive practical experiment of the application of these binding systems in different materials for construction and protection was conducted [28–38].

The purpose of this study was to test an approach to designing alkaline aluminosilicate binders (further, the binder) and developing a manufacturing technology for various curing conditions, especially for normal-temperature curing conditions, and for various applications.

2. Raw Materials and Testing Techniques

A binder of the system $Na_2O(K_2O)\text{-}Al_2O_3\text{-}SiO_2\text{-}H_2O$ with varying ratios of oxides ($Na_2O(K_2O)/Al_2O_3$ and SiO_2/Al_2O_3) was used as the object of this study.

The major components of the binder used in this study are as follows:

- Metakaolin (oxide content, % by mass: CaO—0.27; SiO_2—54.08; Al_2O_3—43.61; Fe_2O_3—0.77; MgO—0.52; K_2O + Na_2O—0.25; LOI—0.50) (specific surface area = 800 m^2/kg (by Blaine)) and kaolin (oxide content, % by mass: CaO—0.80; SiO_2—58.50; Al_2O_3—39.50; Fe_2O_3—0.50; TiO_2—0.70; LOI—0.19) (specific surface area = 800 m^2/kg (by Blaine)) were used as aluminosilicate components;
- Soluble sodium silicate (per CAS 1344-09-8, silicate modulus Ms = 2.8, density = 1430 Kg/m^3) was used as an alkaline component.

In order to adjust the composition of the binder with regard to the contents of main oxides, the following additives and alkali compounds were used:

- Tripoli powder (per CAS 1317-95-9) (oxide content, % by mass: CaO—0.83; SiO_2—85.59; Al_2O_3—6.20; Fe_2O_3—3.15; MgO—0.95; SO_3—0.39; TiO_2—2.03; K_2O + Na_2O—0.67), specific surface area = 800 m^2/kg (by Blaine), bulk density = 2300 kg/m^3, porosity $\geq$ 70%, degree of purity 98.0%;
- Solution of potassium hydroxide KOH (per CAS 1310-58-3), density= 1430 Kg/m^3;
- Solution of sodium hydroxide NaOH (per CAS 1310-73-2), density= 1430 Kg/m^3.

The liquid component of the binder was prepared by mixing the soluble sodium silicate with solutions of KOH and NaOH. The solid component of the binder was prepared by mixing metakaolin or kaolin with the Tripoli powder. The binder was prepared by mixing the alkaline component with the solid component in a Hobart mixer. The ratio of binder components was calculated in order to ensure the required oxide composition of the binder, $(Na_2O + K_2O)\cdot Al_2O_3\cdot mSiO_2\cdot nH_2O$.

The compressive strength of the alkaline aluminosilicate binding agent was determined per the EN 196-1:2016 standard with the following modifications: testing on 40 × 40 × 160 mm specimens (binder : CEN sand = 1:3) and hardening at t = 20 ± 2 °C and RH = 60 ± 5% for 28 days.

The water resistance was evaluated by considering the coefficient of water resistance (K_w) of the 40 × 40 × 40 mm specimens (binder: CEN sand = 1:1) hardened for 28 days at t = 20 ± 2 °C and RH= 60 ± 5%. After 28 days, part of the specimens was placed into the water until almost full water saturation was achieved (the changes in mass of the specimens over 24 h did not exceed 0.2%). The coefficient of water resistance is the ratio of the compressive strength of the water-saturated specimen to that of the specimen hardened for 28 days. A material can be considered water-resistant when $K_w \geq 0.8$.

The compressive and flexural strengths of the binder-based perlite-based materials (lightweight concrete) were determined per the EN 12390-3:2019 and EN 12390-5:2019 standards, respectively.

Adhesion was evaluated by considering the adhesion strength of the adhesive joints between the adhesive and various substrates and the characteristics of the connection of the adhesive with various materials, which were determined per the EN 24624:1992 standard.

The identification of the phase composition of the binders was achieved through X-ray diffraction (XRD), using a single-crystal X-ray diffractometer (DRON-3M, JSC "Bourevestnik", St. Petersburg, Russia) equipped with a copper anode and nickel filter at U = 30 kV, I = 10 to 20 mA, and a range of registered angles (2θ) (8 to 60) (the rotation speed of the counter was calculated as 2 degrees per minute using the technique reported in [39]). The X-ray spectra were identified according to references [40–42].

The microstructure of the binders was examined using a scanning electron microscope (SEM) with microanalyzer (REMMA 102-02, JSC "SELMI", Sumy, Ukraine). The settings of SEM were as follows: accelerating voltage—35 kV; SEI mode resolution—5 nm; and magnification—from 2500 to 23000. The settings of the energy-dispersive X-ray spectrometer were as follows: analyzed element range—Na; energy resolution—143 eV at MnKα; and energy range—up to 30 kV. The microstructure was also examined using an atomic force microscope (AFM) (NT–206, ALC «Microtestmachines», Gomel, Republic of Belarus) designed by the V.A. Belii Institute for Mechanics of Metal-polymer Systems of the National Academy of Sciences of The Republic of Belarus.

The degree of binding of the ions of alkali metals in the hardening binder was studied by the application of two methods of analysis (pH measurements and chemical analysis of the pore fluid) [43]. The measurement of pH values of the pore fluid of the binder was carried out using a pH meter, PL-700AL.

3. Results and Discussion

3.1. The Influence of Basic Binder Composition and Curing Conditions on Structure Formation of the Alkaline Aluminosilicate Binder

3.1.1. Influence of Binder Design

Figures 1 and 2 demonstrate the XRD patterns and SEM images of the reaction products represented by the alkaline aluminosilicate hydrates of the hardened binder of the composition with varying ratios of constituent oxides, namely SiO_2/Al_2O_3 = 2…7, $K_2O/(Na_2O + K_2O)$ = 0.15, and $(Na_2O+K_2O)/Al_2O_3$ = 1, after curing at t = 80 °C.

The phase composition of the reaction products of the binders with low ratios (SiO_2/Al_2O_3 = 2…3) was characterized by the following zeolite-like reaction products: analcime (d/n = 0.699; 0.365; 0.336; 0.293 nm), natrolite (d/n = 0.287; 0.243; 0.138 nm), and ussingite (d/n = 0.492; 0.347; 0.295 nm). The amorphous phases presented by the alkaline aluminosilicate hydrates and particles of non-reacted metakaolin were identified (Figure 2).

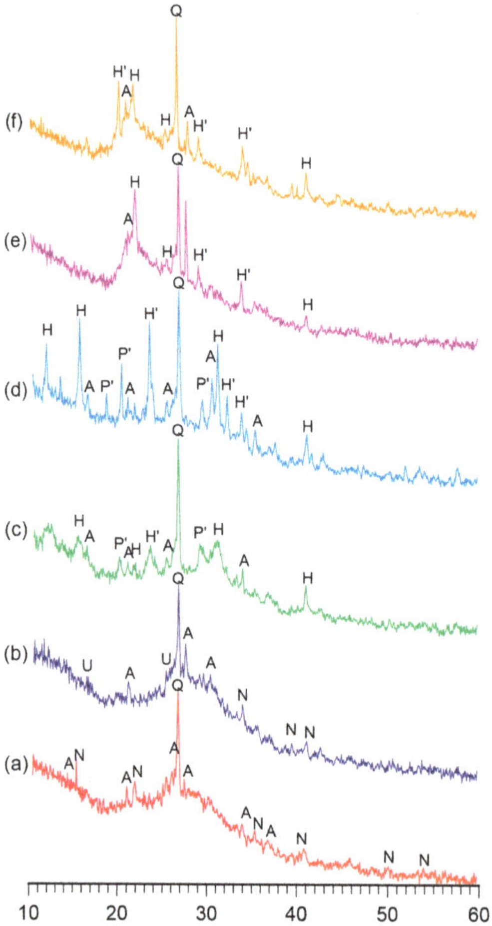

Figure 1. The XRD patterns of the binder with varying $K_2O/(Na_2O + K_2O) = 0.15$ and SiO_2/Al_2O_3 ratios, namely (**a**) 2, (**b**) 3, (**c**) 4, (**d**) 5, (**e**) 6, and (**f**) 7, after curing at t = 80 °C. Abbreviations: Q—quartz; N—natrolite; A—analcime; U—ussingite; P'—Na–K phillipsite; H—Na heulandite; H'—K heulandite.

The hardened binder with SiO_2/Al_2O_3 = 4. . .5 was characterized by the following zeolite-like reaction products: analcime (d/n = 0.699; 0.365; 0.336; 0.293 nm), sodium heulandite (d/n = 0.509; 0.392; 0.296 nm), potassium heulandite (d/n = 0.342; 0.281; 0.273 nm), and sodium-potassium phillipsite (d/n = 0.498; 0.408; 0.269 nm). The degree of crystallinity of the resulting structure was high as a result of intensive diffraction reflections (Figure 1) and due to the microstructure of the hardened binder (Figure 2).

As mentioned earlier, the modification of the binder with Ca-containing additives resulted in the formation of zeolite-like reaction products of the hybrid type, as follows: calcium–sodium aluminosilicate hydrates and Na- and K-heulandites (the latter products were found in small quantities) [44]. With such a modification, the strength gain of the binder under normal temperatures (t = 20 ± 2 °C) could be reached. The Ca-containing modifying additives used in this study were Portland cement, ground granulated blast furnace slag, $Ca(OH)_2$, and $CaCO_3$.

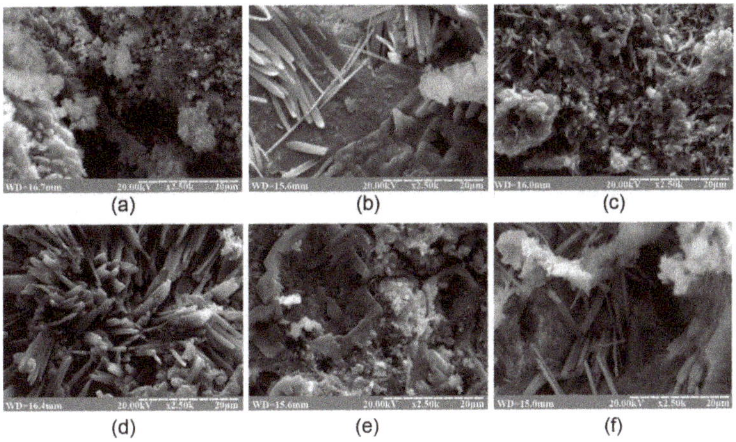

Figure 2. The SEM images of the cleaved surface of the binder with varying $K_2O/(Na_2O+K_2O) = 0.15$ and SiO_2/Al_2O_3 ratios, namely (**a**) 2, (**b**) 3, (**c**) 4, (**d**) 5, (**e**) 6, and (**f**) 7 (magnification ×2.5K), after curing at t = 80 °C.

3.1.2. Influence of Curing Conditions

The influence of curing temperature during the hardening of the binder on its phase composition was studied. It was found that the phase composition of the alkaline aluminosilicate hydrate with $K_2O/(Na_2O + K_2O) = 0.15$ and $SiO_2/Al_2O_3 = 5$ was characterized by the presence of the following zeolite-like reaction products, analcime, potassium, and sodium heulandite, as well as sodium–potassium phillipsite (Figure 3). With the temperature increase, the structure formation processes in the binder accelerated without affecting the phase composition, resulting in a higher degree of crystallinity of the reaction products (Figure 4).

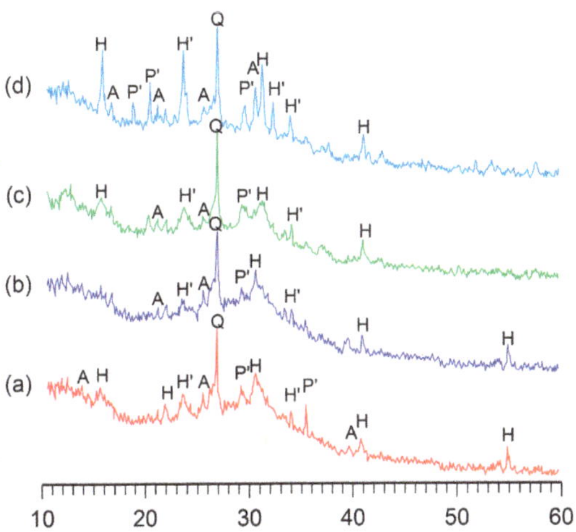

Figure 3. The XRD patterns of the binder with $K_2O/(Na_2O + K_2O) = 0.15$ and $SiO_2/Al_2O_3 = 5$ after curing at the following temperatures, °C: (**a**) 20; (**b**) 40; (**c**) 60; (**d**) 80. Abbreviations: Q—quartz; A—analcime; P′—Na–K phillipsite; H—Na heulandite; H′—K heulandite.

Figure 4. The SEM images of the cleaved surface of the binder with $K_2O/(Na_2O + K_2O) = 0.15$ and $SiO_2/Al_2O_3 = 5$ after curing at the following temperatures, °C: (**a**) 20; (**b**) 40; (**c**) 60; (**d**) 80 (magnification ×2.50K).

With $SiO_2/Al_2O_3 = 5$ and $K_2O/(Na_2O + K_2O) = 0.15$, an essential increase in strength tO + K place at t = 40. . .80 °C. The highest values of the coefficient of water resistance were reached after curing at t = 60. . .80 °C (Figure 5).

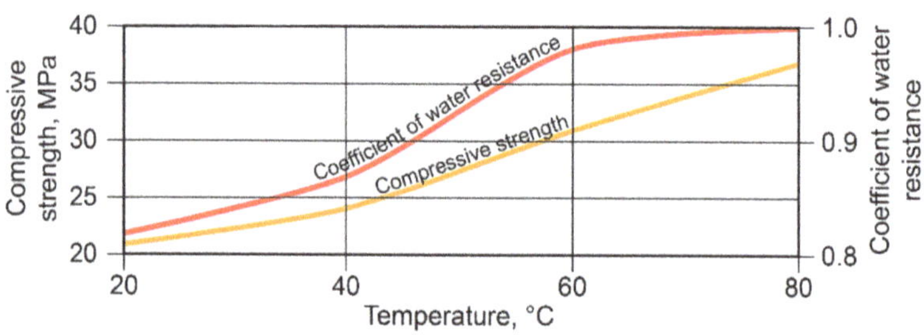

Figure 5. Compressive strength and coefficient of water resistance vs. curing temperature (binder with $SiO_2/Al_2O_3 = 5$ and $K_2O/(Na_2O + K_2O) = 0.15$).

A conclusion was drawn with regard to the binders of the general formula $(0.7\ldots1.0Na_2O + 0\ldots0.3K_2O)\cdot Al_2O_3\cdot 2\ldots7SiO_2\cdot nH_2O$ that the phase composition of the binder depended mainly on the ratio of oxides and that a curing temperature (in the range of 20. . .80 °C) accelerated the formation of zeolite-like aluminosilicate hydrates.

The SiO_2/Al_2O_3 ratio was found to be the main factor determining the type of reaction products that formed: the higher this ratio, the larger the quantity of silica formed in the crystal lattice in the zeolite-like phases. During the curing of these alkaline aluminosilicate hydrates under normal conditions, allowing them to reach the highest degree of structure crystallinity, the optimal ratio of SiO_2/Al_2O_3 was 4 to 5. The hardened binder with this ratio of oxides was characterized by the highest strength and water resistance.

The addition of potassium ions into the binder composition contributed to the formation of zeolite-like reaction products and an increase in the degree of crystallinity of the phases. To accelerate the formation of alkaline aluminosilicate hydrates under normal conditions, the following quantity of potassium oxide was required: $K_2O/(Na_2O + K_2O) = 0.15\ldots0.30$. The potassium ions also contributed to the higher water resistance and strength of the hardened binder, regardless of the curing temperature.

With the rise in temperature from 20 °C to 80 °C, the phase composition of the hardened binder hardly changed. However, this resulted in a higher rate of structure formation and degree of crystallinity.

The microscopic examination of the binders under study showed the presence of a sub-microcrystalline structure with microcrystals ≤3.5 microns in size alongside the amorphous phase. Continuously increasing the adhesion between the sub-microcrystalline and amorphous phases led to the higher strength and better durability of these structures, which, due to their nature, can be exposed to external deteriorating factors without significant resulting damage. The structure of the hardened binder was mainly represented by crystals of hexagonal form, which can be referred to as zeolite-like reaction products such as analcime, zeolite P (with the structure of garronite), and zeolite G (with the structure of chabazite). These reaction products were present in various quantities, depending on the initial binder composition, and differ in their degree of crystallinity (Figures 1–4). Changing the ratio of Na_2O to K_2O allowed us to obtain the required composition of the reaction products and to keep the structure formation in the binder and its resulting density under control. The density values of the zeolite-like reaction products were as follows: zeolite G (0.205–0.21 nm) < zeolite P (0.230 nm) < analcime (0.224 nm). The maximum density of the binder was obtained through the formation of equal quantities of aluminosilicate hydrates of the analcime and zeolite P types.

The atomic force microscopy (AFM) examination results coincide well with the results of AFM shown in Figure 6. The binder microstructure, prepared with the ratio of H_2O/Al_2O_3 = 12.5, was characterized by the presence of single microcrystals within the range of 0.1 to 0.5 microns that were chaotically distributed in the amorphous phase, testifying to the uneven crystallization process. A conclusion can be drawn about the absence of a clearly expressed stage in the formation of the sub-microcrystalline structure during the transformation of the binder from the amorphous state into the crystalline one. Nucleation of large crystals takes place in the amorphous phase, resulting in significantly slower crystallization and hardening processes.

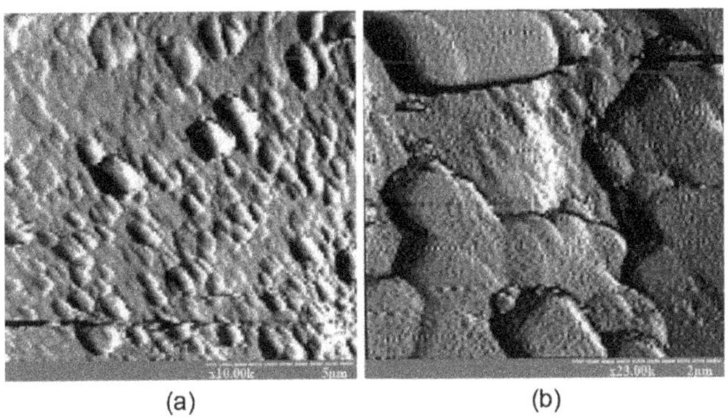

Figure 6. The AFM results for the binder with $Na_2O{\cdot}Al_2O_3{\cdot}4SiO_2{\cdot}10H_2O$ modified with the Ca-containing additive and cured for 28 d at t = 20 ± 2 °C: (**a**) magnification ×10 K; (**b**) magnification ×23 K.

The binding degree of the alkali metal ions in the binder was studied. The binding degree depends on the curing conditions and increases as the curing temperature rises (Figure 7). With the rise in temperature from 20 °C to 100 °C, the pH value of the pore fluid tended to decrease (from 11.2 to 9.7). The introduction of the Ca-containing additive reduced the pH value from 10.1 to 9.3. This phenomenon can be attributed to its participation in the binding process, with the formation of such reaction products as amicite (d = 0.564; 0.422; 0.314; 0.272 nm), garronite (d = 0.710; 0.501; 0.410; 0.316; 0.267 nm), and gismondine (d = 0.188; 0.191; 0.273; 0.274; 0.319; 0.324; 0.419; 0.467; 0.494 nm).

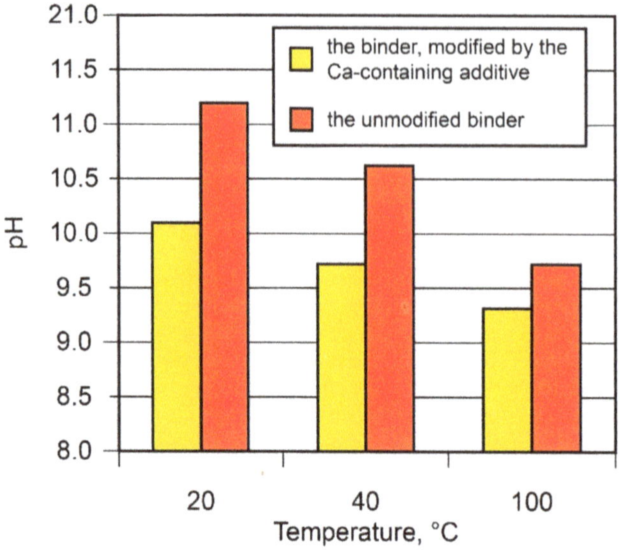

Figure 7. The pH values of the pore fluid vs. binder composition and curing temperature.

During the hardening of the binder, the binding process of the alkali metal ions tO + K place, with the most intense changes taking place during the first seven days (Figure 8). The Ca-containing additive accelerated the binding of the ions.

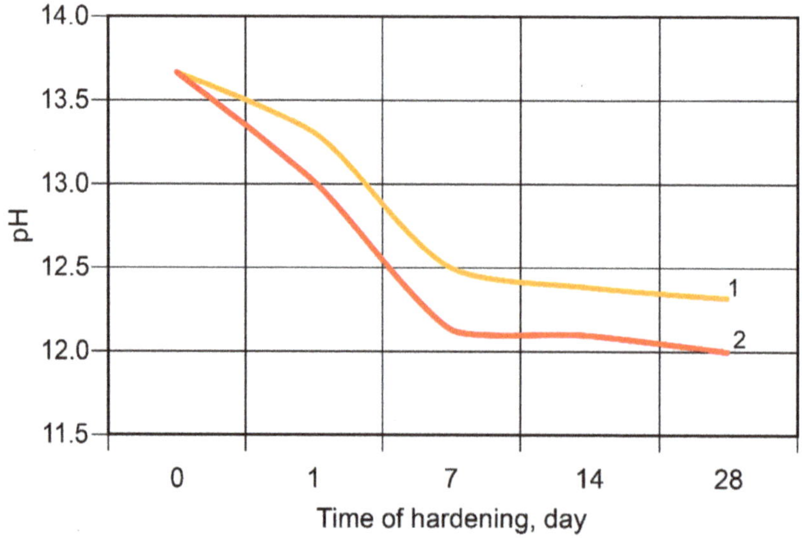

Figure 8. The pH values of the pore fluid vs. time of hardening at t = 20 °C: 1—the binder without modification; 2—the calcium-modified binder.

After 2 days, 84.99% of the free Na^+-ions were bound in the binder matrix without modification and 89.79% were bound in the Ca-modified binder (Figure 9). After 7 days, 91.26% and 92.36% of the ions, respectively, were bound in the binder matrix. Then, the rate of binding became considerably lower, especially in the case of the binder without modification; after 28 days, these values were 8.57% (binder without modification) and

1.20% (Ca-modified binder). The Ca-containing additive accelerated the binding of the Na^{+}-ions through the formation of zeolite-like reaction products.

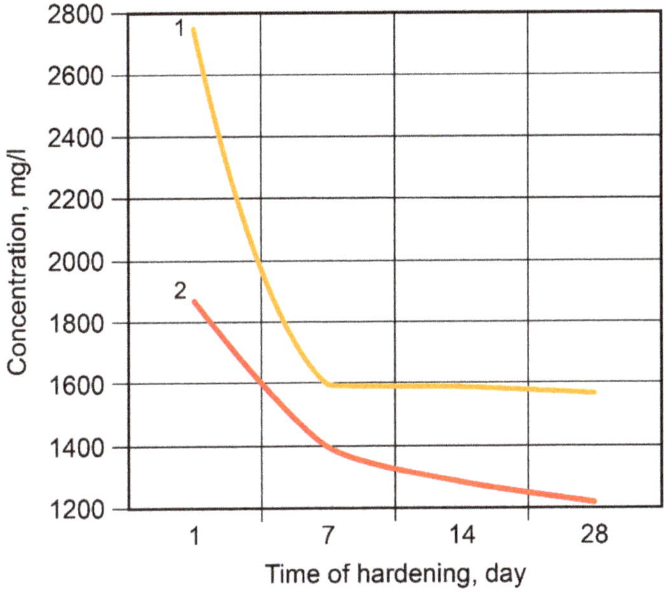

Figure 9. The concentration of the Na^{+}-ions in the pore fluid vs. time of hardening: 1—the binder without modification; 2—the Ca-modified binder.

Figure 10 presents the binding degree of the K^{+}-ions in the binder matrix. After 2 days, 95% of the free K^{+}-ions were bound in the binder without modification (compared with 96% of the Ca-modified binder).

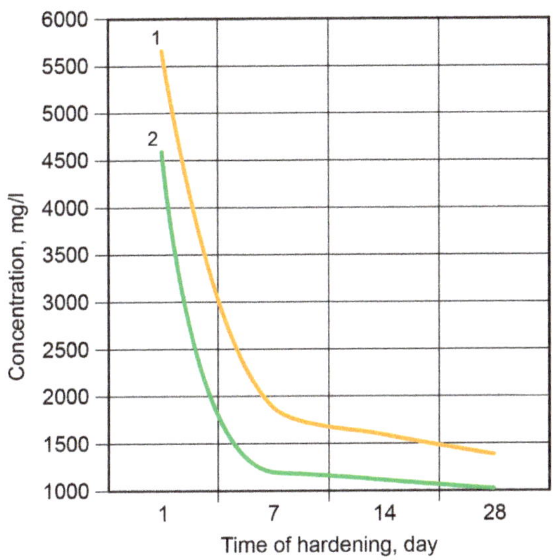

Figure 10. The concentration of K^{+}-ions in the pore fluid vs. time of hardening: 1—the binder without modification; 2—the Ca-modified binder.

After 7 days, approximately 99% of the K^+-ions were bound. Then, the binding rate became considerably lower, and after 28 days, it was 1.25% for the binder without modification and 0.85% for the Ca-modified binder.

3.2. Manufacturing Process and Examples of Application of the Alkaline Aluminosilicate Binder

In order to demonstrate the application of the binder in the manufacture of low-temperature-insulating composite materials, an alkaline aluminosilicate binder of the hydronepheline-analcime type ($Na_2O \cdot Al_2O_3 \cdot 3.5SiO_2$ (10.5; 12.5; 14.5)H_2O) was chosen as a binding agent. The influence of the number of water molecules contained in the binder produced using kaolin and metakaolin, after conditioning (maturing) of the aluminosilicate blends for 1.5 to 3.5 h and further high-temperature treatment, on the strength of the resulting ceramic matrices (low-temperature ceramics) is shown in Figures 11 and 12.

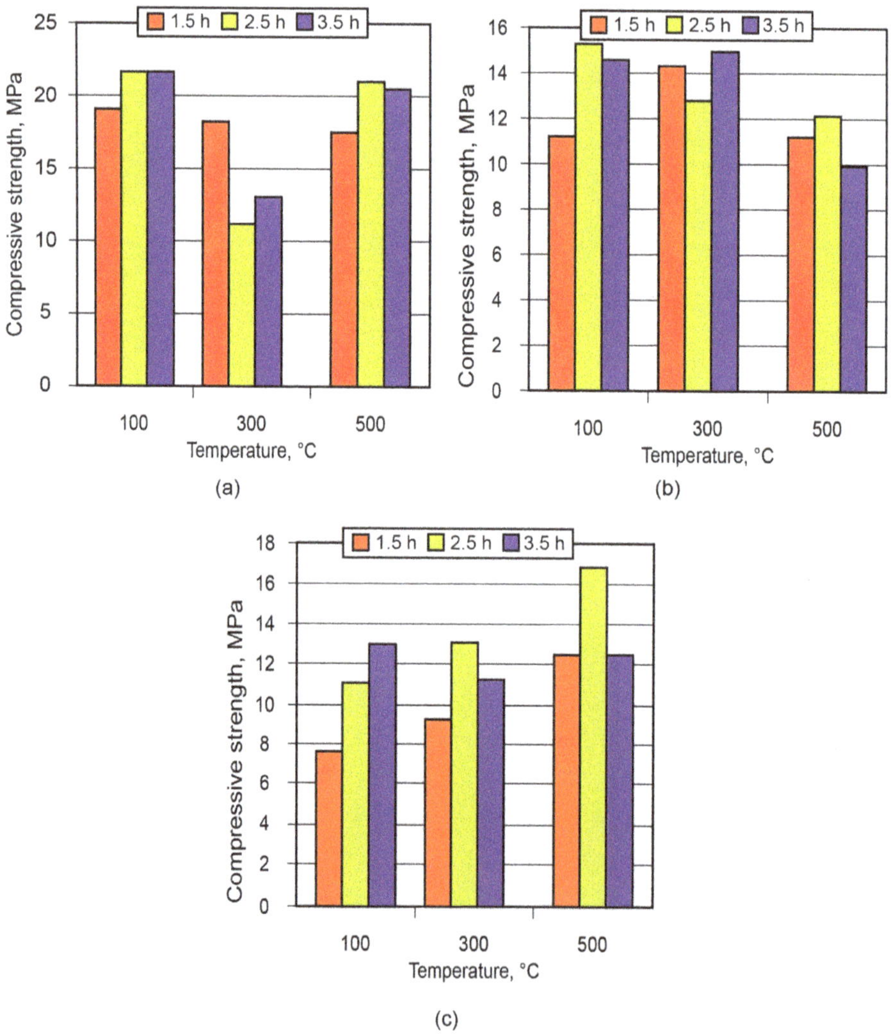

Figure 11. Strength of the ceramic matrix (aluminosilicate component—kaolin) vs. time of conditioning and treatment temperature: (**a**) the binder $Na_2OAl_2O_33.5SiO_210.5H_2O$; (**b**) the binder $Na_2OAl_2O_33.5SiO_212.5H_2O$; (**c**) the binder $Na_2OAl_2O_33.5SiO_214.5H_2O$.

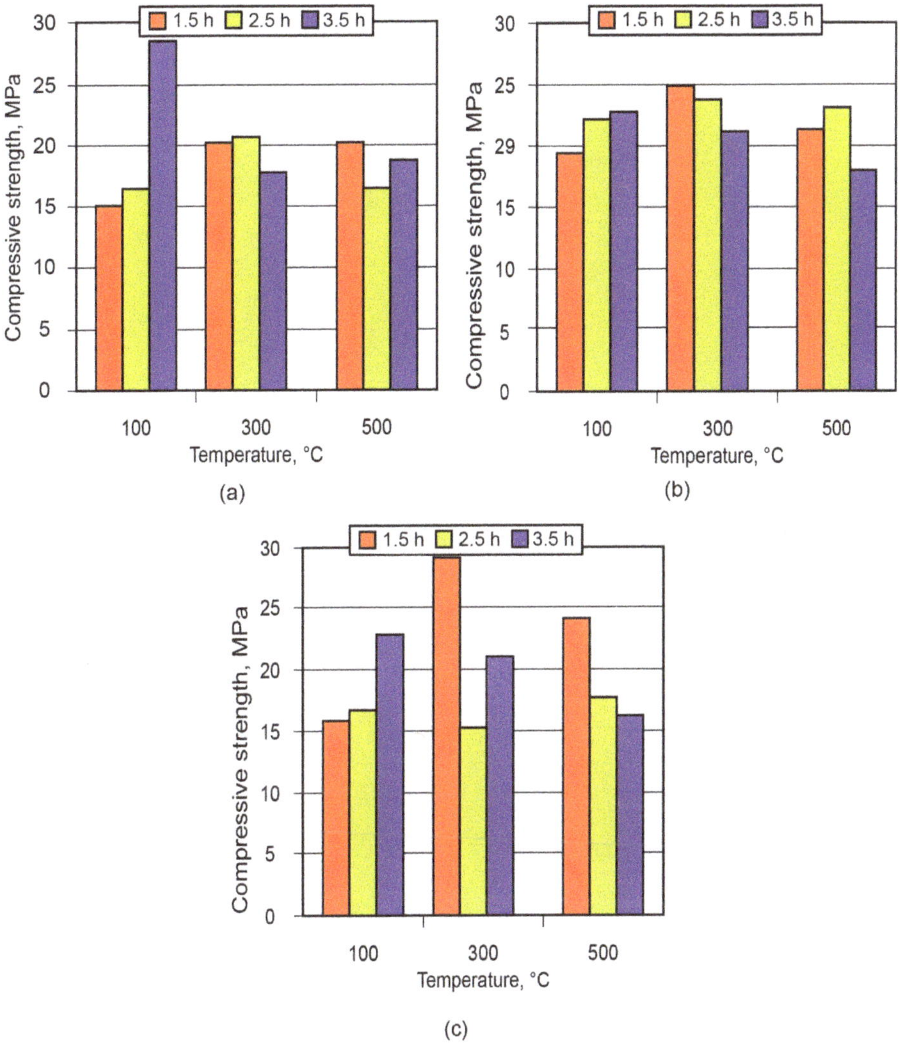

Figure 12. Strength of the ceramic matrix based on metakaolin vs. time of conditioning and treatment temperature: (**a**) binder $Na_2O{\cdot}Al_2O_3{\cdot}3.5SiO_2{\cdot}10.5H_2O$; (**b**) binder $Na_2O{\cdot}Al_2O_3{\cdot}3.5SiO_2{\cdot}12.5H_2O$; (**c**) binder $Na_2O{\cdot}Al_2O_3{\cdot}3.5SiO_2{\cdot}4.5H_2O$.

Analysis of the obtained results showed that the low-temperature ceramic matrix (the alkaline aluminosilicate binder, $Na_2OAl_2O_33.5SiO_214.5H_2O$, based on metakaolin) after conditioning for 1.5 h and treatment at t = 300 °C had the highest compressive strength among all the matrices.

The binder compositions used for the manufacture of perlite-based materials for heat insulation (lightweight concrete) were formulated and tested in pilot-scale production (Figure 13).

Figure 13. Pilot-scale production of the heat-insulating boards using expanded perlite.

The results of tests of the perlite-based materials are compared in Table 1. The composites prepared using the alkaline aluminosilicate binder were the best for high-temperature insulation applications.

Table 1. Basic physical–mechanical characteristics of the perlite-based materials (binding agents: the alkaline aluminosilicate binder and OPC (reference)).

Characteristic	**Density, kg/m^3**			
	150	**200**	**250**	**300**
Alkaline aluminosilicate binder				
Binder content (% by volume)	3.0	6.0	8.0	10.0
Compressive strength (MPa)	0.25	0.30	0.35	0.45
Flexural strength (MPa)	0.15	0.15	0.20	0.30
Heat conductivity at 25 °C (W/(m·°C))	0.062	0.068	0.076	0.082
Ordinary Portland cement (OPC)				
Binder content (kg/m^3)	–	–	80	110
Compressive strength (MPa)	–	–	0.15	0.25
Flexural strength (MPa)	–	–	0.16	0.20
Heat conductivity at 25 °C (W/(m·°C))	–	–	0.075	0.085

Considering the constantly increasing restrictions placed on energy consumption, the demand for heat-saving solutions has become a priority. Conserving heat in high-temperature installations is especially critical since the relative cost of energy produced at high temperatures (exceeding 500 °C) is considerably higher than that produced at lower temperatures. Binder formulations for high-temperature heat insulation for the insulation of boilers were developed. The guidelines on how to choose and arrange insulating materials were elaborated based on the study of the adhesion strength of adhesive joints through the design of alkaline-aluminosilicate-binder-based adhesives and using a numerical model for the design of high-temperature heat-insulating materials (Figure 14).

A solution to achieving the heat insulation of industrial boilers was developed. Among the many technical advantages of alkaline aluminosilicate binders as binding agents in the manufacture of adhesives are the following: ecological friendliness, incombustibility, absence of hazardous gases released under exposure to high temperatures, durability, and low cost. These materials are easy to prepare and apply, and they can be produced under almost any condition, in any quantity, with the use of primitive mixing equipment (e.g., electric drill).

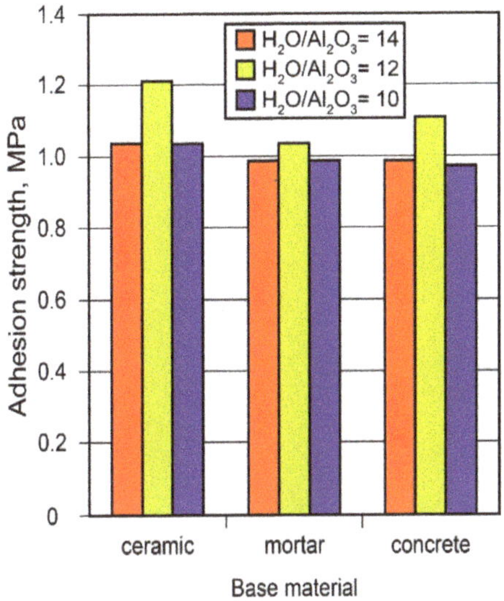

Figure 14. The adhesion strength of the adhesive joints between the adhesive and various other substrates (ceramics, mortar, concrete) vs. binder composition after 28 d.

Adhesives prepared using alkaline aluminosilicate binders as binding agents can be used for bonding layers of heat insulation products to each other for the heat insulation of structural elements with geometrically complicated shapes, impregnating basalt fiber cords intended for sealing holes, and for the direct sealing of holes. For these applications, a composition using ground sand as a filler was designed and applied with the help of an injection pistol into holes up to 4 mm in width and with a putty knife for holes ranging from 4 to 20 mm in width (Figure 15). The shelf life of this adhesive composition was between 1 h and 6 h, with a solidification time between 1 h and 8 h, and a strength gain period ranging from 24 h to 48 h, depending on the ambient temperature.

Figure 15. Work on sealing holes in the heat-insulating layer using an injection pistol.

Fire-resistant heat-insulating composite materials were developed from binder-based adhesives and basalt fibers.

One other important problem requiring a solution is how we ensure health and safety in the case of natural disasters and catastrophes in multistoried buildings with many people inside. One of the necessary tasks is to ensure quick localization of the fire source via a fire stop. Lifts and lift shafts should be protected by insulation using mineral materials. A solution to this problem was proposed, involving the use of binder-based adhesives with good fire-stopping ability and good heat insulation properties. These composite materials, due to their inorganic nature, do not release hazardous gaseous substances at high temperatures. A new design for fire-resistant doors was developed, since none of the previously used adhesives meet the requirements for fire-resistant lift shaft doors. Using an alkaline aluminosilicate binder in the manufacturing process of fire-resistant floor frames was also proposed. The manufacture and commercial-scale use of the mentioned products were initiated by the OTIS (Ukraine) (Figures 16 and 17).

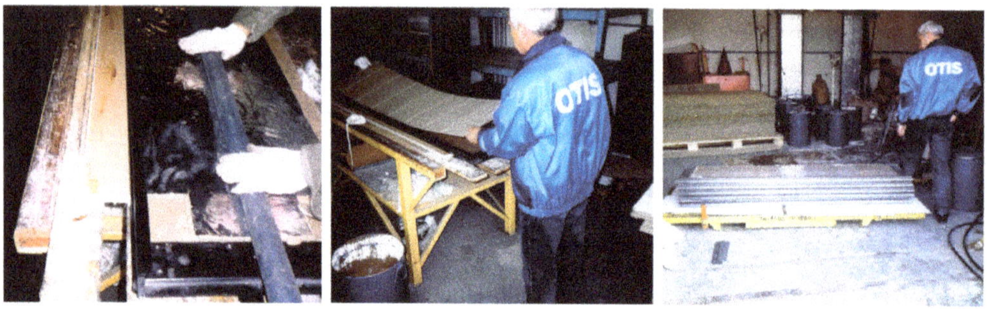

(a) (b) (c)

Figure 16. Technological operations: (**a**) adhesive application on aluminum foil; (**b**) gluing of basalt cardboard; (**c**) storage of the ready products (doors).

Figure 17. Production of floor frames of the lifts.

The alkaline-aluminosilicate-binder-based adhesives were tested on their ability to adhere to various materials: a glass crystalline plate, basalt cardboard, and steel and aluminum foil (Table 2).

Table 2. Characteristics of the adhesion.

Glued Materials	Correlation of the Total Areas at Uniform Tearing, %		
	Cohesive Destruction (in Glue), No More Than	Adhesive Destruction (Contact Zone Material: Glue), No More Than	Destruction in Fibrous Layer, Not Less Than
Plate–cardboard	5	5	90
Cardboard–steel	20	30	50
Plate–steel	20	30	50
Plate–aluminum foil	20	30	50

4. Conclusions

1. This study developed an approach to designing an alkaline aluminosilicate binder to be used in a variety of responsible-use applications, for example, for advanced heat- and high-temperature-insulating composite materials with higher efficiency compared with other traditionally used mineral and organic binding agents, including those intended for fire-stopping applications.
2. The structure formation processes in the alkaline aluminosilicate binders were characterized by the following oxide ratios: $(Na,K)_2O/Al_2O_3 = 1$, $SiO_2/Al_2O_3 = 2$ to 7, and $H_2O/Al_2O_3 = 10$ to 15. These caused the formation of zeolite-like reaction products, chiefly composed of aluminosilicate hydrates: analcime, zeolite P, and garronite. The introduction of a calcium-containing additive to the binder composition resulted in the formation of zeolite P and analcime as well as calcium silicate hydrates with the structure of ksonotlite and girolite. The additional quantities of SiO_2 added to the binder composition determined the predominance of zeolite-like reaction products with an increased content of SiO_2, i.e., minerals of the Na-shabasite- gmelenite, faujasite, and mordenite types.
3. The hardening process of the binder with $H_2O/Al_2O_3 < 10$ tO + K place with the formation of aluminosilicate hydrates through the following stages: amorphous, sub-microcrystalline, and crystalline. In the case of the binder with $H_2O/Al_2O_3 > 10$, the sub-microcrystalline structure was very poorly identified; as a result, the nucleation of large crystals tO + K place in the amorphous phase, leading to slower hardening and crystallization processes and, finally, to lower properties of the resulting cement matrix.
4. Hardening of the binder within 28 days was accompanied through the binding of 93 to 99% of Na^+- and K^+-ions in the binder matrix. The introduction of the Ca-containing additive allowed us to accelerate the binding of these ions due to the formation of zeolite-like reaction products of the amicite, garronite, and gismondine types.
5. The effectiveness of the designed binder as a binding agent for use in the manufacture of glues and adhesives is supported by the experience gained from pilot- and small-scale industrial manufacture and use. Such environmentally friendly mineral binder-based composites encompass a variety of insulating materials for heat insulation and fire stop applications.

Author Contributions: Project administration, funding acquisition, conceptualization, supervision, validation, and editing, P.K.; methodology, resources, data curation, formal analysis, visualization, and writing—original draft preparation, I.R. and O.K.; software and investigation, O.G. All authors have read and agreed to the published version of the manuscript.

Funding: This research work received financial support from the Ministry of Education and Science of Ukraine under the state budget, with registration no. 0123U101831 and no. 0123U101832 (implementation period: 2023–2025).

Institutional Review Board Statement: Not applicable.

Informed Consent Statement: Not applicable.

Data Availability Statement: The data are not publicly available, as this work was carried out within the framework of the mentioned projects, and its results can be made available on request after registration of a report in the state scientific institution of the Ukrainian Institute of Scientific and Technical Expertise and Information (http://www.uintei.kiev.ua/en) at https://nddkr.ukrintei.ua/.

Conflicts of Interest: The authors declare no conflicts of interest.

References

1. Glukhovsky, V.D. *Gruntosilikaty (Soil Silicates)*; Gosstroyizdat: Kyiv, Ukraine, 1959.
2. Glukhovsky, V.D. *Alkaline and Alkali-Alkali Earth Hydraulic Binders and Concretes*; Vyshcha Shkola Publ.: Kyiv, Ukraine, 1979.
3. Davidovits, J. Geopolymeric Reaction in Archaeological Cements and in Modern Blended Cement. *Geopolymer* **1988**, *1*, 93–105.
4. Kryvenko, P.; Rudenko, I.; Kovalchuk, O.; Gelevera, O.; Konstantynovskyi, O. Influence of Dosage and Modulus on Soluble Sodium Silicate for Early Strength Development of Alkali-Activated Slag Cements. *Minerals* **2023**, *13*, 1164. [CrossRef]
5. Fernández-Jiménez, A.; Pastor, J.Y.; Martín, A.; Palomo, A. High-Temperature Resistance in Alkali-Activated Cement: High Temperature Resistance in Alkali-Activated Cement. *J. Am. Ceram. Soc.* **2010**, *93*, 3411–3417. [CrossRef]
6. Krivenko, P.; Rudenko, I.; Konstantynovskyi, O.; Vaičiukynienė, D. Mitigation of Corrosion Initiated by Cl^- and SO_4^{2-}-Ions in Blast Furnace Cement Concrete Mixed with Sea Water. *Materials* **2022**, *15*, 3003. [CrossRef] [PubMed]
7. Krivenko, P.; Rudenko, I.; Konstantynovskyi, O. Effect of Technological Factors on Freeze-Thaw Resistance of Alkali-Activated Slag Cement Concrete in NaCl Solution; Kharkiv, Ukraine. *AIP Conf. Proc.* **2023**, *2684*, 040011. [CrossRef]
8. Skurchinskaya, Z. Progress in Alkaline Cements. In *Alkaline Cements and Concretes*; VIPOL Stock Company: Kyiv, Ukraine, 1994; Volume 1, pp. 271–298.
9. Krivenko, P.; KyrychO + K, V. Genesis of Structure and Properties of the Zeolite-Like Cement Matrices of the System Na(K)-Al_2O_3 -SiO_2-H_2O within a Temperature Range of 20–1200 °C. In *Advances in Geopolymer-Zeolite Composites—Synthesis and Characterization*; Vizureanu, P., Krivenko, P., Eds.; IntechOpen: London, UK, 2021. [CrossRef]
10. He, P.; Wang, M.; Fu, S.; Jia, D.; Yan, S.; Yuan, J.; Xu, J.; Wang, P.; Zhou, Y. Effects of Si/Al Ratio on the Structure and Properties of Metakaolin Based Geopolymer. *Ceram. Int.* **2016**, *42*, 14416–14422. [CrossRef]
11. Garcia-Lodeiro, I.; Palomo, A.; Fernández-Jiménez, A.; Macphee, D.E. Compatibility Studies between N-A-S-H and C-A-S-H Gels. Study in the Ternary Diagram Na_2O–CaO–Al_2O_3–SiO_2–H_2O. *Cem. Concr. Res.* **2011**, *41*, 923–931. [CrossRef]
12. Longhi, M.A.; Rodríguez, E.D.; Walkley, B.; Zhang, Z.; Kirchheim, A.P. Metakaolin-Based Geopolymers: Relation between Formulation, Physicochemical Properties and Efflorescence Formation. *Compos. Part B Eng.* **2020**, *182*, 107671. [CrossRef]
13. Shi, C.; Krivenko, P.V.; Roy, D. *Alkali-Activated Cements and Concretes*; Taylor & Francis: London, UK, 2006.
14. Wang, J.; Li, F.; Zhou, Z.; Du, P.; Xu, D.; Xie, N.; Cheng, X.; Liu, Y. Effect of Zeolite on Waste Based Alkali-Activated Inorganic Binder Efflorescence. *Constr. Build. Mater.* **2018**, *158*, 683–690. [CrossRef]
15. Miranda, T.; Leitão, D.; Oliveira, J.; Corrêa-Silva, M.; Araújo, N.; Coelho, J.; Fernández-Jiménez, A.; Cristelo, N. Application of Alkali-Activated Industrial Wastes for the Stabilisation of a Full-Scale (Sub)Base Layer. *J. Clean. Prod.* **2020**, *242*, 118427. [CrossRef]
16. Kang, S.-P.; Kwon, S.-J. Effects of Red Mud and Alkali-Activated Slag Cement on Efflorescence in Cement Mortar. *Constr. Build. Mater.* **2017**, *133*, 459–467. [CrossRef]
17. Karrech, A.; Dong, M.; Elchalakani, M.; Shahin, M.A. Sustainable Geopolymer Using Lithium Concentrate Residues. *Constr. Build. Mater.* **2019**, *228*, 116740. [CrossRef]
18. Baenla, J.; Bike Mbah, J.B.; Djon Li Ndjock, I.B.; Elimbi, A. Partial Replacement of Low Reactive Volcanic Ash by Cassava Peel Ash in the Synthesis of Volcanic Ash Based Geopolymer. *Constr. Build. Mater.* **2019**, *227*, 116689. [CrossRef]
19. Kryvenko, P.; Rudenko, I.; Konstantynovskyi, O. Design of Slag Cement, Activated by Na (K) Salts of Strong Acids, for Concrete Reinforced with Steel Fittings. *EEJET* **2020**, *6*, 26–40. [CrossRef]
20. Krivenko, P. Why Alkaline Activation–60 Years of the Theory and Practice of Alkali-Activated Materials. *J. Ceram. Sci. Technol.* **2017**, *8*, 323–334. [CrossRef]
21. Provis, J.L.; Van Deventer, J.S.J. (Eds). Alkali Activated Materials: State-of-the-Art Report, RILEM TC 224-AAM; RILEM State-of-the-Art Reports; Springer Netherlands: Dordrecht, The Netherlands, 2014; Volume 13. [CrossRef]
22. Palomo, A.; Maltseva, O.; Garcia-Lodeiro, I.; Fernández-Jiménez, A. Portland Versus Alkaline Cement: Continuity or Clean Break: "A Key Decision for Global Sustainability". *Front. Chem.* **2021**, *9*, 705475. [CrossRef] [PubMed]
23. MO + Khort, M.A. Formation of Structure and Properties of Alkaline Geocements. In *Tagungsbericht 14 Internationale Baustofftagung "Ibausil"*; Bauhaus Universitat Weimar: Weimar, Germany, 20–23 September 2000; Volume 1, pp. 1-0553–1-0560.
24. Krivenko, P.; MO + Khort, M.; Petropavlovskyi, O.; Vozniuk, G. Geocement Glues and Composite Materials: Practical Application. In Proceedings of the 2007—International Conference Alkali Activated Materials—Research, Production and Utilization, Praha, Czech Republic, 21–22 June 2007; pp. 397–412.
25. Pavlasova, S.; Skvara, F. High-Temperature Properties of Geopolymer Materials. In Proceedings of the 2007—International Conference Alkali Activated Materials—Research, Production and Utilization, Praha, Czech Republic, 21–22 June 2007; pp. 523–524.
26. MO + Khort, M.A.; Popel, G.N. The Studies on Structure-Formation Processes and Microstructure of an Artifical Stone on Alkaline Aluminosilicate Binders-Geocements. In *Alkaline Cements and Concretes*; Oranta Ltd.: Kyiv, Ukraine, 1999; pp. 291–302.

27. Davidovits, J. *Geopolymers of the First Generation: Siliface-Process, Geopolymer*; U.T.C. Université Technologique Compiègne: Compiègne, France, 1988; Volume 1, pp. 49–67.
28. Davidovits, J. Synthetic Inorganic Polymer of the Silicoaluminate Family and Process for the Preparation Thereof; Moulded Articles Containing This Polymer, and Process for Their Preparation. EP0026687A2, 8 April 1981.
29. MO + Khort, M.A. *Einfluß von Ausgangzustand und Erhärtungsbedingungen bei der Verfestingung von Geozementen*; Bauhaus Universitat Weimar: Weimar, Germany, 2003; Volume 1, pp. 137–143.
30. MO + Khort, M.; Tsibulya, Y. Experience of Application of Geocements for Manufacturing of Inorganic Basalt and Organic-Mineral Jute Compisites. In Proceedings of the 2007—International Conference Alkali Activated Materials—Research, Production and Utilization, Praha, Czech Republic, 21–22 June 2007; pp. 483–492.
31. MO + Khort, M.A.; Tsibulya, Y.L. Geocement Composites Based on Basalt Fabric and Alkaline Aluminosilicate Binder. In *Cement Combination for Durable Concrete*; Thomas Telford Publish: London, UK, 2005; pp. 287–292.
32. Gontchar, V.P. *The Studies on Thermo-Mechanical Characteristics of the Alkaline Aluminosilicate Binders Modified by a Silicon Carbide*; Oranta Ltd. Publish: Kyiv, Ukraine, 1999; pp. 303–312.
33. Baranovskii, A.V. The Use of Alkaline Aluminosilicate Adhesive in the Production of Wood-Based Materials. *News Acad. Constr. Ukr.* **1997**, *3*, 44–48.
34. Baranovskii, A.V. Wood-Based Materials from Alkaline Aluminosilicate Adhesive. Ph.D. Thesis, Kyiv National University of Construction and Architecture, Kyiv, Poland, 1997.
35. MO + Khort, M.A.; Dhir, R.K.; Hewlett, P.C.; Csetenyi, L.J. Slag Alkaline and Geocement Concretes on Basis of Organic Filler with High Durability. In *Innovations and Developments in Concrete Materials and Construction*; Thomas Telford Publishing: London, UK, 2002; pp. 597–605.
36. Voznyuk, G.V. *Environmentally Friendly Composites Based on Geocements for Foundry*; Oranta Ltd. Publish: Kyiv, Ukraine, 1999; pp. 237–247.
37. Torgal, F.P.; Labrincha, J.; Leonelli, C.; Palomo Sánchez, A.; Chindaprasirt, P. *HandbO + K of Alkali-Activated Cements, Mortars and Concretes*; Woodhead Publishing Series in Civil and Structural Engineering; Elsevier Woodhead Publishing: Cambridge, UK, 2015.
38. Krivenko, P.; Guzii, S.; Rudenko, I.; Konstantynovskyi, O. Intumescent Fireproof Coatings Based on Zeolite-Like Cement Matrices. *CE/Papers* **2023**, *6*, 923–929. [CrossRef]
39. Gorshkov, B.; Timashev, V.; Savel'ev, V. *Methods of Physical-Chemical Analysis of Binders*; High School: Moscow, Russia, 1984.
40. Barrer, B. *Hydrothermal Chemistry of Zeolites*; World: Moscow, Russia, 1985.
41. Krivenko, P.; MO + Khort, M. Process of Phisico-Chemical Structure Formation in Modified Geocements. In Proceedings of the 2007—International Conference Alkali Activated Materials—Research, Production and Utilization, Praha, Czech Republic, 21–22 June 2007; pp. 379–396.
42. Breck, D.W. *Zeolite Molecular Sieves*; Wiley: New York, NY, USA, 1973.
43. Scrivener, K.; Snellings, R.; Lothenbach, B. (Eds.) *A Practical Guide to Microstructural Analysis of Cementitious Materials*; First issued in paperback; CRC Press: Boca Raton, FL, USA; London, UK; New York, NY, USA, 2017.
44. Krivenko, P.; Rudenko, I.; Konstantynovskyi, O.; Vaičiukynienė, D. Feasibility of Incorporating SO_4^{2-}-Ions in Zeolite-like Matrices Based on Alkaline Aluminosilicate Binders. *Constr. Build. Mater.* **2023**, *391*, 131878. [CrossRef]

materials

MDPI

Article

An Expanded Wing Crack Model for Fracture and Mechanical Behavior of Sandstone Under Triaxial Compression

Esraa Alomari, Kam Ng * and Lokendra Khatri

Civil and Architectural Engineering and Construction Management, University of Wyoming, Laramie, WY 82071, USA; ealomari@uwyo.edu (E.A.); lokendra.khatri320@gmail.com (L.K.)
* Correspondence: kng1@uwyo.edu

Abstract: A new model is developed to predict the mechanical behavior of brittle sandstone under triaxial compression. The proposed model aims to determine the normalized critical crack length (L_{cr}), through which the failure strength (σ_f) of sandstone can be estimated based on fracture mechanics applied to secondary cracks emanating from pre-existing flaws, while considering the interaction of neighboring cracks. In this study, the wing crack model developed by Ashby and Hallam (1986) was adopted to account for the total stress intensity at the crack tip (K_I) as the summation of the stress intensity due to crack initiation and crack interaction. The proposed model is developed by first deriving the L_{cr} and then setting the crack length equal to the L_{cr}. Next, the total stress intensity is set equal to the rock fracture toughness in the original equation of K_I, resulting in an estimate of the σ_f. Finally, to evaluate the performance of the proposed model on predicting σ_f, theoretical results are compared with laboratory data obtained on sandstone formations collected from Wyoming and the published literature. Moreover, the σ_f predicted by our proposed model is compared with those predicted from other failure criteria from the literature. The comparison shows that the proposed model better predicts the rock failure strength under triaxial compression, based on the lowest RMSE and MAD values of 36.95 and 30.93, respectively.

Keywords: failure model; confining pressure; wing crack model; normalized critical crack length; type-I fracture toughness

1. Introduction

Brittle fracture in compression is a complex behavior that has attracted many past research developments on different failure criteria to describe fracture mechanisms [1]. Various research has proposed different approaches to model the brittle fracture of rocks under compression. Although some analytical models agree quite well with measured experimental data, the main contribution of most experimental studies is limited to uniaxial or biaxial compression tests conducted on pre-cracked artificial specimens [2]. Further research investigations are needed to develop reliable analytical models for describing the mechanical behavior of natural rock specimens under triaxial compression [3–5].

Modeling the brittle failure of rocks is one of the greatest challenges due to the heterogeneous nature of rocks containing pores, flaws, and other weaknesses [6–8]. Furthermore, such discontinuities play a significant role in the initiation and propagation of new cracks [9,10], which affect their strength behaviors. In a study by Lu and Yan of the concrete fracturing behavior (which represents brittle materials behavior such as rocks) fracture, they investigated the effect of the chevron notch inclination on the failure strength. The study indicated that the maximum failure load increased with increasing the inclination angle of the notch. Furthermore, the fracture propagation angle also increased with the notch inclination angle, and wing fractures were observed [11]. However, it is difficult to trace crack initiation and propagation during experiments due to the non-transparency nature of the rock.

Advanced measurement techniques such as acoustic emissions (AE) detection [12,13], computerized tomography (CT) scanning system [14], and scanning electron microscope (SEM) [15,16] have been applied to monitor the fracture process. However, the AE signals can be affected by various factors such as noise, attenuation, reflection, and sample geometry. CT is too expensive and time-consuming, and the SEM cannot capture the image of wet samples as they may be damaged by the vacuum required during operation.

A fracture mechanism due to pre-existing microcracks was proposed by Schulson et al. [17]. Unlike the pore-emanated cracking model for brittle failure of sedimentary rocks under triaxial compression [18], this fracture mechanism is concerned with wing cracks growing from an initial inclined flaw. In addition to wing crack initiation at the tip of preexisting flaws, they considered the effect of additional cracks emanating adjacent to wing cracks (splay cracks). The model states that the failure of rock mass is assumed to be initiated by the bending of microcolumns formed by splay and wing cracks, and the applied stress to trigger faulting under uniaxial loading is expressed in terms of the type-I fracture toughness (K_{IC}), coefficient of friction (μ), slenderness ratio of the microcolumns, and the angle between the direction of the principal stress and wing crack. However, this model underestimated the failure stress due to considering a single microcolumn for crack propagation and neglecting the frictional resistance gained by adjacent columns.

Paliwal and Ramesh [19] proposed a micro-mechanical damage model for brittle failure under uniaxial compression based on frictional sliding of pre-existing flaws. As a result of compressive loading, wing cracks are nucleated at the flaw tips, in addition to interactions among the growing cracks, which give an effective stress field around the crack. The model describes that the compressive strength increases with the increases in mean flaw size and the μ. However, the triaxial compression condition was not included in the study in addition to the several assumptions made in developing the model due to the lack of experimental data. Zhang et al. [20] have developed a modified sliding crack model to estimate the stress intensity factor (SIF) and the crack length (l) under the effect of external stresses. The SIF has been derived in the context of Linear Elastic Fracture Mechanics (LEFM) and weight function. The formula of SIF shows that the sliding crack is highly influenced by lateral stress. However, a functional relationship between the crack length and principal stresses (i.e., σ_1 and σ_3) was obtained without a quantitative estimation of the failure strength.

Hoek and Brown [21] have proposed an empirical failure criterion that provides practical means to estimate the rock triaxial strength using experimental data from triaxial compression tests. This failure criterion is adequate and easily applied for strength prediction. However, the criterion application requires a measured uniaxial compressive strength (UCS) of the rock to estimate the triaxial compressive strength. Another common failure criterion is the Mohr–Coulomb (MC) criterion. The MC criterion assumes that failure occurs at a particular combination of the maximum and minimum principal stresses. However, this criterion requires conducting several triaxial compression tests in the calculation of the shear strength parameters (i.e., friction angle and cohesion) before applying the failure model and estimating the failure strength.

Conventional triaxial compression tests can be employed to obtain the parameters of failure criteria to define a rock strength for relevant rock engineering projects. However, they are time-consuming and expensive for testing many samples, and therefore, not readily available at an early stage of a project. Congruent to the above-mentioned limitations, there is an interest in developing an analytical model that generally provides a good prediction of rock triaxial compressive strength based on easily measured or available rock properties.

This research is motivated by the fact that fracture of brittle materials is often induced by the growth and propagation of wing-like flaws based on a micromechanical damage criterion originally formulated by Ashby and Hallam [22]. The total stress intensity at the crack tip K_I was adopted from the original wing crack model, and an expanded analytical model was proposed for the normalized critical crack length (L_{cr}), which is substituted in the K_I equation to estimate the corresponding failure strength. The proposed model enables a systematical study of the effect of the μ, stress ratio (λ), and K_{IC} on the failure strength (σ_f) under a triaxial compression condition. This study also investigates the sensitivity of the critical crack length (i.e., the crack length at the unstable growth stage) to the initial damage level of the rock and confining pressure. The capability of the proposed model in estimating compressive strength is compared with those of other failure criteria that are commonly employed in rock engineering for an extensive database of triaxial compression tests conducted on sandstone samples collected from Wyoming and published data from the literature for different sandstone formations.

2. Materials and Methods

2.1. Wing Crack Model

Ashby and Hallam [22] have developed a wing crack model for the growth and interaction of cracks in brittle solids under compression. Consider a plate containing a pre-existing crack of length 2a subjected to major and minor principal stresses of σ_1 and σ_3, respectively, the crack is inclined with an angle of Ψ from the major principal stress σ_1 direction. The sign convention states that stresses are considered positive if tensile and negative if compressive. The shear stress tends to make the cracks slide but the frictional stress opposes the sliding (i.e., the coefficient of friction μ). The field is predominantly shear on the plane of the initial cracks. However, for the planes that lie at an angle from the crack tip, a normal stress appears that tends to cause type-I wing cracks. Therefore, the type-I fracture toughness is found by seeking the plane of the maximum tensile stress.

The wing crack model attributes the rock brittle failure to tensile cracks that emanate from the tips of pre-existing microcracks. The wing cracks initiate at an angle of 45° with respect to the pre-existing crack tip, where the tensile stress is maximum. The model states that the stress required for crack initiation depends on the initial crack half-length (a) and orientation, the μ, and the stress state. Once wing cracks are initiated, and as stress increases, further sliding of the main crack causes them to grow further parallel to the applied maximum stress [22]. The stress intensity (K_I) due to cracks initiation is given by the following equation:

$$K_I = \frac{\sigma_1\sqrt{\pi a}}{\sqrt{3}} \times \left\{ (1-\lambda)\left(1+\mu^2\right)^{1/2} - (1+\lambda)\,\mu \right\} \tag{1}$$

where a is the half-length of the pre-existing crack, λ is the stress ratio (σ_3/σ_1), and μ is defined as $tan(\varphi)$. With the increase in stress, these cracks begin to interact and extend further. Extension of wing cracks in an array of micro-cracks divides the solid into slender micro columns. This allows the resultant forces to be treated as shear forces, axial forces, and bending moments. Based on a comprehensive semi-circular bend test on concrete samples, Lu and Yan [12] found that the type-I crack mode is only guaranteed when the crack initiation angle (Ψ) is less than 30° or when the theoretical angle is over 70° in the triaxial case. The mixed type-I and type-II failure modes will be experienced when the Ψ angle falls between 45° and 70°, and the pure type-II mode is guaranteed when the Ψ angle exceeds 70.53°. Hence, it is important to note that the proposed wing crack model derived in Section 2.2 based on K_I for type-I mode is mostly applicable to rocks with Ψ angles less than 30°.

The governing equation for the stress intensity factor (K_I^I) due to microcracks interaction is presented in Equation (2). The term outside the curly brackets is the crack-opening stress intensity caused by the major principal stress σ_1, while the term inside represents the crack-closing stress intensity caused by the minor principal stress σ_3 based on λ

$$K_I^I = \frac{\sqrt{2D\times(L+\alpha)\times\pi a}}{\pi}\sigma_1 \times \sqrt{\left\{\left[1-\frac{8}{\pi}\,D\,\lambda\,(L+\alpha)^3\right]\left[1-\frac{2}{\pi}\,D\,\lambda\,(L+\alpha)^3\right]\right\}} \tag{2}$$

where $\alpha = cos(\Psi) = \frac{1}{\sqrt{2}}$ since the angle Ψ under the maximum tensile stress is 45°; therefore, $\alpha = 0.71$, and D measures the initial level of damage and equals $\pi a^2 N_A$, where N_A represents the number of initial cracks per unit area.

2.2. Derivation of the Proposed Model

Short cracks do not often interact, and the stress intensity at the crack tip is the same as that of an isolated crack given by Equation (1). However, as the cracks grow, they start to interact by mechanisms involving bending and buckling which increase the stress intensity at the crack tip [23]. The total stress intensity (K_I) at the crack tip given by Equation (3) below is derived by adding the stress contributions from external compression loading, given by Equation (1), and crack interaction, given by Equation (2):

$$K_I = \frac{\sigma_1\sqrt{\pi a}}{\sqrt{3}} \times \left\{(1-\lambda)\left(1+\mu^2\right)^{1/2} - (1+\lambda)\,\mu\right\} + \frac{\sqrt{2\,D_o \times (L+\alpha) \times \pi a}}{\pi}\sigma_1 \times \sqrt{\left\{\left[1-\frac{8}{\pi}D_o\,\lambda\,(L+\alpha)^3\right]\left[1-\frac{2}{\pi}D_o\,\lambda\,(L+\alpha)^3\right]\right\}} \quad (3)$$

The brittle deformation of rocks is predominantly governed by the growth and coalescence of cracks [24]. The higher the stress intensity at the crack tip, the easier the wing cracks grow. Increasing the major principal stress σ_1 will increase the stress intensity at the crack tip, which promotes crack propagation. However, confinement decreases the stress intensity at the crack tip and inhibits crack propagation [25]. The maximum growth length of the crack before failure is known as the critical crack length, which is reached due to crack interaction as crack growth is always stable under triaxial compression. The stress intensity (K_I^I) caused by cracks interaction, given by Equation (2), is differentiated with respect to crack length L as $\frac{dK_I^I}{dL}$, given by Equation (4):

$$\frac{dK}{dL} = \frac{0.029\,\left(\sigma_1\sqrt{\pi a}\right)^2 \times \left\{(L+\alpha) - 0.48\,\lambda\,(L+\alpha)^4 + 0.038\,\lambda^2 (L+\alpha)^7\right\}}{2k} \quad (4)$$

Setting Equation (4) to zero leads to the following:

$$0 = 0.2(L+0.71) - 0.65\lambda D(L+0.71)^4 + 0.33\lambda^2 D^2 (L+0.71)^7 \quad (5)$$

Solving Equation (5) using algebra, the $L = L_{cr}$ can be expressed in terms of D and λ by the following equation:

$$Lcr = \sqrt[3]{\frac{0.382}{\lambda D}} - 0.71 \quad (6)$$

The L_{cr} obtained from Equation (6) indicates the unstable growth regime of cracks emanating from flaw tips. Replacing L in Equation (3) with the Lcr determined from Equation (6) and setting $K_I = K_{IC}$, a new model describing the rock failure that accounts for the triaxial condition is proposed as given by Equation (7). This proposed model predicts the λ and therefore the failure strength σ_f.

$$\frac{k_{Ic}\lambda}{\sigma_3\sqrt{\pi a}} = \frac{1}{\sqrt{3}} \times \left\{(1-\lambda)(1+\mu^2)^{1/2} - (1+\lambda)\mu\right\} + \frac{\sqrt{2D \times \left(\left(\sqrt[3]{\frac{0.382}{\lambda D}} - 0.71\right) + \alpha\right)}}{\pi} \times \sqrt{\left\{\left[1-\frac{8}{\pi}D\lambda\left(\left(\sqrt[3]{\frac{0.382}{\lambda D}} - 0.71\right) + \alpha\right)^3\right]\left[1-\frac{2}{\pi}D\lambda\left(\left(\sqrt[3]{\frac{0.382}{\lambda D}} - 0.71\right) + \alpha\right)^3\right]\right\}} \quad (7)$$

Substituting the $\alpha = 0.71$ [21], Equation (7) is simplified to Equation (8):

$$\frac{k_{Ic}\lambda}{\sigma_3\sqrt{\pi a}}=\frac{1}{\sqrt{3}}\times\left\{(1-\lambda)\left(1+\mu^2\right)^{1/2}-(1+\lambda)\mu\right\}+\frac{\sqrt{2D\times\left(\sqrt[3]{\frac{0.382}{\lambda D}}\right)}}{\pi}\times\sqrt{\left\{\left[1-\frac{8}{\pi}D\lambda\times\frac{0.382}{\lambda D}\right]\left[1-\frac{2}{\pi}D\lambda\times\frac{0.382}{\lambda D}\right]\right\}} \quad (8)$$

Further simplifying Equation (8) results in the following:

$$\frac{k_{Ic}\lambda}{\sigma_3\sqrt{\pi a}}=\frac{1}{\sqrt{3}}\times\left\{(1-\lambda)\left(1+\mu^2\right)^{1/2}-(1+\lambda)\mu\right\}+\frac{\sqrt{2D\times\left(\sqrt[3]{\frac{0.382}{\lambda D}}\right)}}{\pi}\times\sqrt{\left\{1-3.19D\lambda\frac{0.382}{\lambda D}+1.632\lambda^2D^2\frac{0.382^2}{\lambda D}\right\}} \quad (9)$$

Rearranging Equation (9) leads to the following:

$$\frac{k_{Ic}\lambda}{\sigma_3\sqrt{\pi a}}=\frac{1}{\sqrt{3}}\left((1-\lambda)\left(1+\mu^2\right)^{1/2}-(1+\lambda)\mu\right)+\frac{0.927\sqrt{D}}{\pi}\sqrt[6]{\frac{0.382}{\lambda D}} \quad (10)$$

Solving the λ value by equating the two sides of Equation (10), the maximum stress σ_1 can be determined according to the corresponding confining pressure since $\lambda=\sigma_3/\sigma_1$.

2.3. Determination of the Model Parameters

For a practical application of the proposed model given by Equation (10), we will need the model parameters: a, K_{Ic}, D, and μ. The K_{Ic} is assumed to be 1.27 MPa$\sqrt{\text{m}}$, based on experimental results of type-I fracture toughness tests conducted on 34 sandstone samples collected from three wells drilled in Piceance Basin, Garfield County, Colorado using short-rod specimens [26]. The μ is calculated as $tan\varphi$, where φ is the internal friction angle of the rock, or the average μ can be generally assumed as 0.55, based on experimental measurements of friction angle for different sandstones [27].

For the determination of a, the basic form of crack initiation in confined rocks appears in the literature with the form given by Equation (11):

$$\sigma_1=c_1\,\sigma_3-\sigma_o \quad (11)$$

where c_1 and σ_o are material properties, and σ_1 and σ_3 are the axial stress and confining pressure, respectively. According to Nemat-Nasser and Horri [28] and Ashby and Hallam [22], cracks initiate when the stress intensity at the crack tip K_I reaches the K_{IC} of the rock. The stress (σ_1) at cracks initiation is detected experimentally by the start of Acoustic Emissions (AE), the first non-linearity of the stress–strain curve, or by the dilation of the sample [28] and can be calculated according to Equation (12):

$$\sigma_1=\frac{\left(1+\mu^2\right)^{1/2}+\mu}{\left(1+\mu^2\right)^{1/2}-\mu}\sigma_3-\left(\frac{\sqrt{3}}{\left(1+\mu^2\right)^{1/2}-\mu}\right)\left(\frac{K_{IC}}{\sqrt{\pi a}}\right) \quad (12)$$

By fitting the linear relationship between confining pressure and the axial crack initiation stress represented in Equation (12) with experimental data of crack initiation detected by the start of acoustic emissions, the parameter a can be determined [27]. Experimental data of acoustic emissions for crack initiation is plotted on the axis of σ_1 and σ_3 with σ_1 representing the Y-axis and σ_3 representing the X-axis to allow for comparison with the crack initiation criterion to give (a) as the intercept of the linear relationship. Results of crack initiation of triaxial compression experiments conducted on medium-grained Buntsandstone by Gowd and Rummel [27] showed that $2a$ is 2 mm.

For the determination of D, Ashby and Sammis [29] developed a microcrack damage evolution model in brittle solids based on the initiation and propagation of microfractures from preexisting cracks and described the evolution of damage with strain. The same experimental data are fitted in the damage model proposed by Ashby and Sammis [29] to produce a D of 0.15. This value is based on a specific sandstone formation and may not be representative of all sandstones; however, a parametric study is presented in Section 4 of this paper to show how sensitive the L_{cr} is to the D.

Substituting the constants of the proposed model (i.e., K_{Ic}, a, μ, and D), the λ can be calculated from Equation (10), through which the failure strength σ_f can be determined under different σ_3. By adopting the proposed model, no experimental tests are required to determine the model parameters in the prediction of the rock failure strength.

2.4. Validation of the Proposed Model

2.4.1. Triaxial Test Data on Sandstone from Wyoming

To evaluate the performance of the proposed model in predicting rock compressive strength, an extensive database of triaxial compression test results on different sandstone formations obtained from our experimental program and collected from published literature was developed. The failure of compressive strengths σ_f estimated using the proposed model are compared with the corresponding measured σ_f from triaxial experiments. Sandstone cores collected from Wyoming, USA were used for a series of triaxial compressive tests. The dry density of the rock specimens varies from 2.153 to 2.659 g/cm^3. Macroscale defects on specimens were not observed before testing [30]. ASTM D7012 (2014) standard [31] was adopted for the triaxial compression tests at the engineering laboratory at the University of Wyoming as shown in Figure 1a. Tested specimens after failure are shown in Figure 1b.

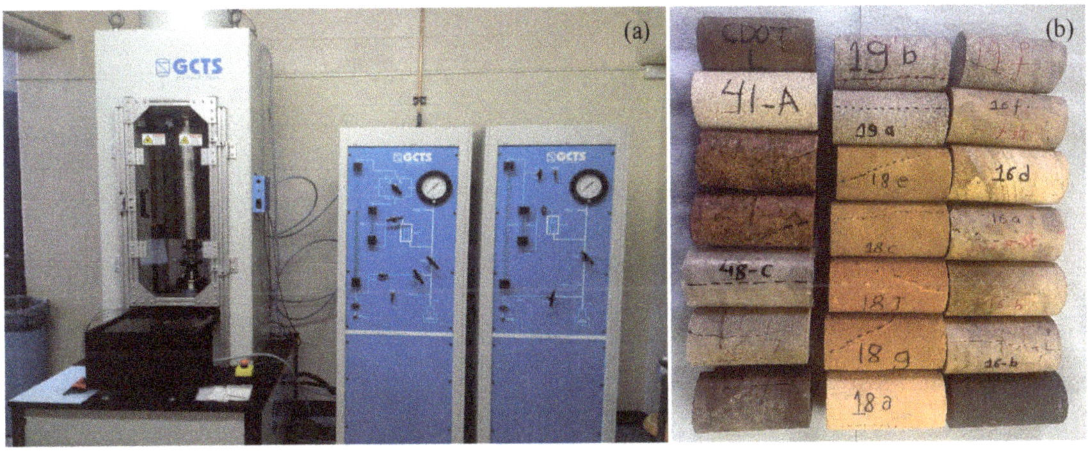

Figure 1. (**a**) Triaxial testing system at the University of Wyoming, and (**b**) Tested specimen.

Each triaxial test was conducted in two stages. In the first stage, a target confining pressure was increased at a rate of 5 MPa/min. In the second stage, deviatoric stress was applied until the specimen failed or reached a 5% axial strain limit. Before testing, Linear Variable Differential Transformers (LVDTs) were instrumented on each specimen to measure the axial and radial LVDT to measure the radial strain. Figure 2 shows an example of a stress–strain plot of Arikaree Sandstone showing the axial, volumetric, and radial strain curves.

The sandstone formation, geological age, core depth, physical properties, and strength properties of the tested rock samples from Wyoming are summarized in Table 1. The samples were tested under confinements ranging from 4 to 10 MPa since they were collected from shallow depths and have a compressive strength ranging from 14.00 to 143.53 MPa.

Table 1. Summary of triaxial compression test results of sandstones from Wyoming.

Sample ID	Formation	Geological Age	Depth m	n [1] %	σ_3 [2] MPa	σ_f [3] MPa	c [4] MPa	φ [5] Degree
16	Flathead	Cambrian	4.76	2.20	4	95.59	20.34	37
			5.85	3.50	8	120.12		
19	Aspen	Cretaceous	6.10	3.43	10	98.89	5.38	49
23	Lance	Cretaceous	Surface	12.18	4	36.50	11.00	8
31	Tensleep	Pennsylvanian	Surface	13.20	4	59.88	13.10	48
32	Arikaree	Lower Miocene	Surface	11.70	4	45.86	2.76	51
33	Arikaree	Lower Miocene	Surface	12.20	4	46.49	3.10	54
				11.10	10	111.03		
39	Hanna	Paleocene	44.15	13.10	4	33.19	2.48	46
			44.60	15.20	10	70.72		
41	Hanna	Paleocene	23.32	14.80	4	31.47	2.34	44
			24.09	16.80	6	42.96		
43	Wind River	Eocene	Surface	14.50	4	14.00	8.96	56
				14.00	10	21.11		
49	Bridger	Eocene	Surface	27.20	4	34.68	4.48	29
				24.20	10	42.44		
50	Fort Union	Paleocene	Surface	12.20	4	31.48	1.38	47
51	Fort Union	Paleocene	Surface	3.92	4	143.53	5.52	55
				5.92	10	53.79		
56	Casper	Permian	Surface	11.10	4	72.56	7.93	53
				10.20	10	132.23		

1 Porosity in percentage, 2 confining pressure in MPa, 3 failure strength in MPa, 4 cohesion in MPa, 5 internal friction angle in degree.

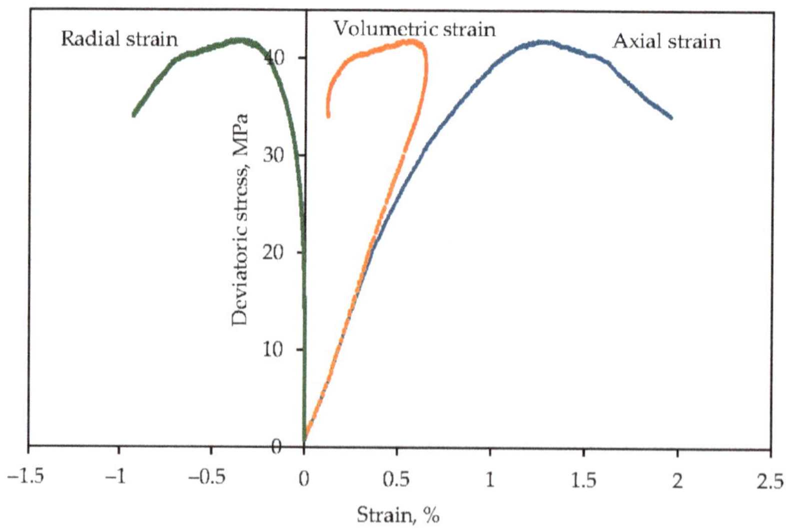

Figure 2. Axial, radial, and volumetric strain curves for Arikaree sandstone.

2.4.2. Triaxial Test Data on Sandstone from the Published Literature

Experimental data of different sandstones under triaxial compression were collected from the literature and used in this validation study. Nine different sandstone formations collected from the surface and various depths and have porosities ranging from 1.5 to 23% were included in this study to create a wide representative dataset in order to better evaluate the performance of the proposed model on predicting the failure strength σ_f. These sandstones have porosities ranging from 1.5 to 23% and were tested under confinements ranging from 3 to 60 MPa with compressive strengths ranging from 58 to 241.81 MPa.

Table 2 summarizes the formation, porosity, and strength properties of sandstones from the literature.

Table 2. Summary of triaxial compression test results of sandstones from the literature.

Formation (Location)	n [1] %	σ_3 [2] MPa	σ_f [3] MPa	c [4] MPa	φ [5] Degree	Reference
Vosges, (France)	22.00	10–60	83.00–178.00	21.50	35.00	[32]
Red sandstone, (China)	8.90	5–35	115.10–236.10	22.42	37.80	[2]
Hawkesbury, (Australia)	13.00	4–25	58.00–116.00	16.50	28.00	[33]
Gosford, (Australia)	16.00	10–30	109.90–172.40	22.60	31.50	[34]
Jurassic sandstone	-	3–7	84.00–169.00	2.96	63.89	[35]
Black sandstone, (NA) [6]	1.50	10–60	131.86–241.81	22.50	31.97	[36]
Red sandstone, (NA) [6]	2.00	10–60	93.40–131.49	13.00	26.25	
-	-	5–25	95.50–220.50	-	67.50	[37]
-	8.70	5–25	116.50–202.00	26.40	36.47	[38]
-	16.24	5–15	105.50–126.00	33.10	21.10	[39]

1 Porosity in percentage, 2 confining pressure in MPa, 3 compressive strength in MPa, 4 cohesion in MPa, 5 internal friction angle in degree, 6 not Available.

3. Discussion

Four failure criteria that have gained wide acceptance in rock engineering are selected for comparison with the proposed model in the prediction of the compressive strength of sandstone. The four criteria are the Hoek–Brown criterion, Griffith criterion, Renshaw and Schulson criterion, and Wiebols and Cook criterion. Hoek–Brown and Griffith's criteria are widely accepted criteria for rock engineering. Renshaw and Schulson's criterion was expanded from the wing crack model, which is similar to our proposed model, except that they take into consideration the slenderness ratio of the secondary crack rather than the density of preexisting cracks. The Wiebols and Cook criterion requires measured parameters, such as the internal friction angle and UCS for the determination of the failure strength. The main difference between our proposed model and the selected failure criteria is the derivation of the failure strength, as each criterion includes different parameters to estimate σ_1.

The performances of the proposed model and four failure criteria in predicting σ_f are evaluated based on the RMSE and MAD values determined from Equations (13) and (14), respectively, where $y_{(i)}$ is the measured failure strength, $\hat{y}_{(i)}$ is the predicted failure strength, and *num* is the number of data points. It is desirable to have a small RMSE and MAD for a reasonable candidate model. The RSME and MAD values are inserted in Figure 3 for comparison. In addition, their performances are evaluated based on the data scatteredness along the 1:1 lines of the plots comparing measured to predicted σ_1 in Figure 3. A total of 67 triaxial compression test data on sandstones were used to evaluate the performances of the proposed model and the other four failure criteria. Among the 67 data, 46 triaxial sandstone data were collected from literature and 21 triaxial data were determined for sandstones from Wyoming.

$$\mathrm{RMSE} = \sqrt{\frac{\Sigma\left(y_{(i)} - \hat{y}_{(i)}\right)^2}{\mathrm{num}}} \quad (13)$$

$$\mathrm{MAD} = \frac{\Sigma\left|y_{(i)} - \hat{y}_{(i)}\right|}{\mathrm{num}} \quad (14)$$

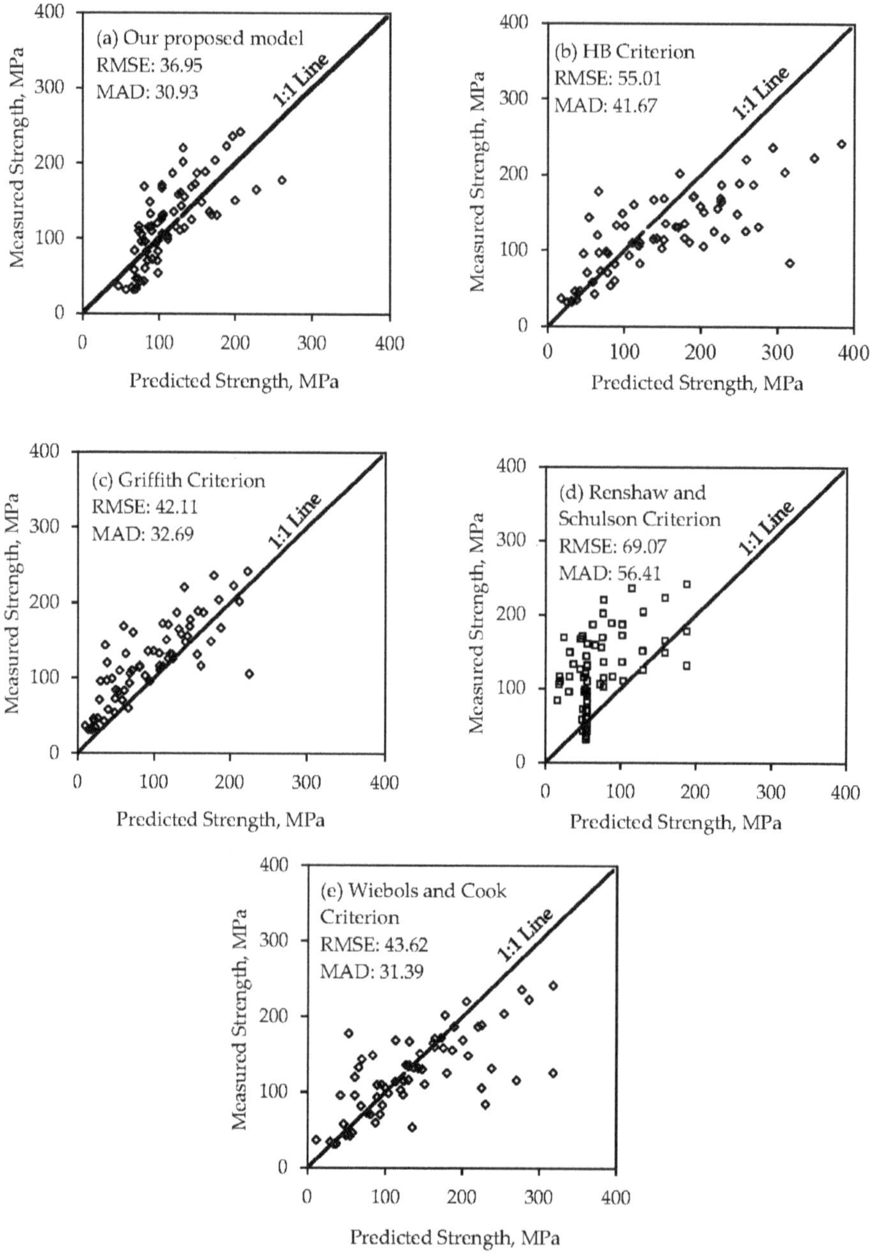

Figure 3. Measured strength versus predicted for (**a**) our proposed model, (**b**) HB criterion, (**c**) Griffith criterion, (**d**) Renshaw and Schulson criterion, and (**e**) Wiebols and Cook criterion.

3.1. Our Proposed Model

Based on the statistical results of predicted strength, our proposed model gives the best prediction of σ_f based on the lowest RMSE and MAD values of 36.95 and 30.93, respectively, with 52% of the data points above the 1:1 line and 48% of the data points under the 1:1 line, as shown in Figure 3a, indicating that the model slightly overestimates the compressive strength σ_f. This is attributed to the moderate to high range of confining pressures under which the rocks are tested (i.e., σ_3 ~20–60 MPa) that results in higher measured to predicted strength ratio and therefore, overestimated σ_f. The mean measured to the predicted ratio equals 1.06, indicating a good estimation of the σ_f.

Figure 4 shows the ratio of measured to predicted ratio versus the confining pressure for all sandstones from Wyoming and the literature. The measured-to-predicted ratio decreases with the increase in confining pressure, indicating that higher confinements result in an overestimation of the predicted σ_f using Equation (10) that indicates higher failure strength under a higher stress ratio λ. For Wyoming sandstones collected from surface and shallow depths and tested at lower confining pressures ranging from 4 to 10 MPa, the strength ratios are mostly lower than 1.0, indicating the overprediction of σ_f from the proposed model. Although these sandstones were tested under relatively low confinements, they have high internal friction angles (i.e., Average $\varphi = 45^{\circ}$). This explains the higher predicted σ_1 due to the proportional relationship between the failure strength and the friction coefficient according to Equation (10).

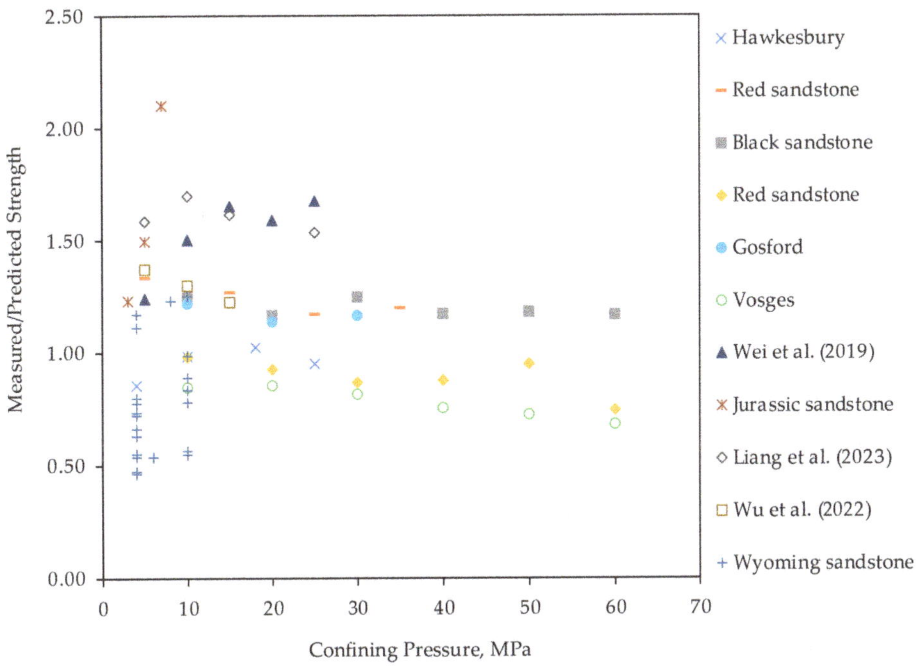

Figure 4. Measured to predicted strength ratio versus confining pressure [37–39].

3.2. Hoek Brown Criterion

The Hoek–Brown (HB) criterion is widely accepted for estimating the strength of jointed rock masses [21]. The Hoek and Brown criterion is developed empirically by fitting the results of triaxial tests conducted on intact rock samples and used to describe the nonlinear increase in the peak strength with confinement. The HB criterion for intact rock is expressed in terms of major and minor principal stresses (σ_1 and σ_3), respectively, by Equation (15):

$$\sigma_1 = \sigma_3 + \sigma_{ci}\sqrt{m_i\frac{\sigma_3}{\sigma_{ci}} + 1} \tag{15}$$

where σ_{ci} is the rock UCS, and m_i is a material constant for intact rock. Letting m_i = 17 for sandstone [40] and using test data from one uniaxial and triaxial compression tests, the failure strength σ_f can be predicted using this criterion. The HB criterion overpredicts the σ_f under high confinements (i.e., σ_3 > 40 MPa), as shown in Figure 3b and as seen in the case of Black and Red sandstones [41]. As shown in Figure 3b, the HB criterion overpredicts 76% of the σ_f data with having higher predicted strength than the measured, especially at higher confinements, according to Equation (15), which states that the failure strength is proportional to the confining pressure; therefore, higher confinement results in higher calculated strength.

Due to the adequacy of strength prediction and the convenient application of this criterion, the material constants are determined empirically and there is no fundamental relationship between them and the rock's physical properties [42]. Moreover, the HB criterion requires the estimation or measurement of the rock UCS, which is not required in the proposed model.

3.3. Griffith Criterion

Griffith theory is one of the most significant advances in rock fracture mechanisms that is physically plausible and developed to derive a failure criterion under compressive stresses [43]. The Griffith criterion has been applied to triaxial compression tests in the form of the following:

$$(\sigma_1 - \sigma_3)^2 - 8\,T_o\,(\sigma_1 + \sigma_3) = 0 \tag{16}$$

where T_o is the uniaxial tensile strength or σ_t. According to the original Griffith theory, the rock's compressive strength is eight times the tensile strength $\frac{\sigma_c}{\sigma_t} = 8$, where σ_c is the uniaxial compressive strength. Using test data from one uniaxial and triaxial compression tests, the failure strength σ_f can be predicted using this criterion, given by Equation (16). The Griffith criterion underpredicts the failure strength for 60 data points, which represents 90% of the dataset as shown in Figure 3c. The poor performance of the Griffith criterion in predicting σ_f is attributed to the limitation of the failure envelope curvature coefficient (n) to ½, which depends on the mechanical properties of the rock [44]. The application of this criterion requires the measurement of the UCS of the rock while the proposed model does not require the measurement of the UCS. However, Griffith's criterion yields the second-best estimate of σ_f with an RMSE value of 42.11.

3.4. Renshaw and Schulson Criterion

Renshaw and Schulson [7] proposed a failure criterion based on the brittle behavior of crystalline materials such as ice and rocks. The criterion is based on the concept of secondary cracks emanating from primary cracks creating a parallel set of microcolumns. Considering a primary crack of length $2c$ inclined at 45° to the major principal stress direction (σ_1), as shown in Figure 5, the length of secondary cracks is h with the spacing between them referred to as w. The slenderness ratio of the microcolumns is governed by the length and spacing of microcolumns and is equal to $\alpha = h/w$. Schulson et al. [17] have shown in experimental observations conducted on ice (which is used as a model material for rocks) that the slenderness ratio α ranges between 3.1 and 7.2; therefore, an average α value of 5.2 is adopted in this study.

The onset of failure is triggered by the bending-induced failure of these microcolumns. This bending and failure of microcolumns happen due to sliding along the primary crack, which is like the break of teeth in a comb under a sliding thumb. The frictional sliding along the primary crack will occur if the confinement ratio $\lambda = \frac{\sigma_3}{\sigma_1} < \frac{(1-\mu)}{(1+\mu)}$, since higher confinement suppresses the frictional sliding and leads to ductile failure. The failure strength σ_f using this criterion is predicted as follows:

$$\frac{\sigma_f\sqrt{c}}{K_{Ic}} = \frac{2}{\left\{\sqrt{\sqrt{1+\left(1-\mu\,\frac{(1+\lambda)}{(1-\lambda)}\right)^{2/3}}-1}\right\}\left\{\sqrt{1+3\,\mu^2\alpha^2(1-\lambda)^2}\right\}} \tag{17}$$

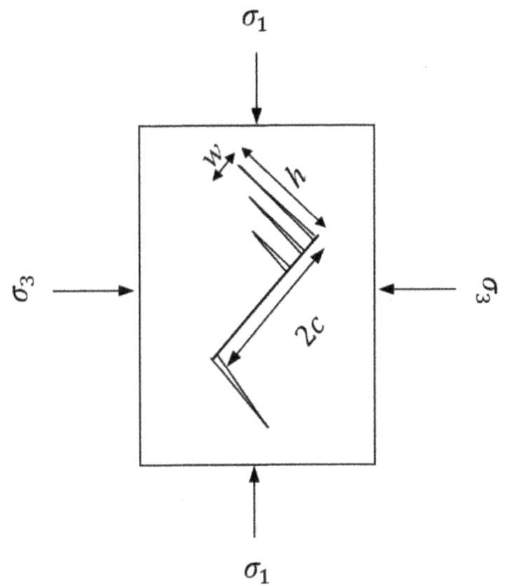

Figure 5. Microcracks emanating from pre-existing flaws.

Renshaw and Schulson [7] have adopted a value of 1 mm for $2c$ and a value of $1\ MPa\sqrt{m}$ for the K_{Ic} of sandstone. Substituting the value of $\mu = 0.5$ based on an experimental study conducted by Barnes et al. [45], where they reported that the μ for ice under -10 °C and sliding velocity of around 10^{-5} m/s and substituting $\alpha = 5.2$, the σ_f can be predicted for different confinements. The scatteredness of the results shown in Figure 3d is attributed to the approximation adopted for the criterion parameters, such as the slenderness ratio of the microcolumns, and the μ, since this criterion was developed originally for ice, which slightly differs in nature from rocks. The RMSE and MAD values using Renshaw and Schulson criteria were the highest among the other failure criteria in this study with 69.07 and 56.41 for RMSE and MAD, respectively.

3.5. Modified Wiebols and Cook Criterion

Zhou [46] proposed a failure criterion which is an extension of the circumscribed Drucker–Prager criterion with features similar to the criterion of Wiebols and Cook [47]. This criterion describes the sliding microcrack surfaces that cause failure when the stress condition meets the frictional criterion. The failure criterion proposed by Zhou predicts rock failure strength when

$$J_2^{1/2} = A + B\,J_1 + C\,J_1^2 \tag{18}$$

where J_1 is the mean effective confining pressure and equals $\frac{1}{3} \times (\sigma_1 + \sigma_2 + \sigma_3)$, and $J_2^{1/2} = \sqrt{\frac{1}{6}(\sigma_1 - \sigma_2)^2 + (\sigma_1 - \sigma_3)^2 + (\sigma_2 - \sigma_3)^2}$. The parameters A, B and C are determined by Equations (19), (20) and (21), respectively, such that Equation (18) is constrained by rock strengths under biaxial $(\sigma_1 = \sigma_2)$ and triaxial conditions $(\sigma_2 = \sigma_3)$.

$$A = \frac{C_o}{\sqrt{3}} - \frac{C_o}{3}\,B - \frac{{C_o}^2}{9}\,C \tag{19}$$

$$C = \frac{\sqrt{27}}{2\,C_1 + (q-1)\sigma_3 - C_o} \times \left(\frac{C_1 + (q-1)\,\sigma_3 - C_o}{2C_1 + 2(q+1)\,\sigma_3 - C_o} - \frac{q-1}{q+2}\right) \tag{20}$$

$$B = \frac{\sqrt{3}\,(q-1)}{q+2} - \frac{C}{3}\,(2\,C_o + (q+2)\sigma_3\Big) \tag{21}$$

where C_o is the UCS of the rock, $q = tan^2(\frac{\pi}{4} + \varphi/2)$, and $C_1 = (1 + 0.6\mu_o)C_o$.

The predicted strength using Wiebols and Cook criterion was determined for each rock under triaxial condition $(\sigma_2 = \sigma_3)$, by first calculating the constants *A*, *B* and *C* using Equations (19), (20) and (21), respectively. Knowing the measured UCS for each rock, the $J_2^{1/2}$ is calculated using Equation (18), through which σ_f can be estimated. The estimated strength using the Wiebols and Cook criterion was calculated based on the measured friction angle and UCS; however, our proposed model does not require UC tests for the determination of UCS and predicting the failure strength. Despite this, it still resulted in a lower RMSE and slightly higher MAD values of the predicted strength, which better validates our proposed model.

The modified Wiebols and Cook criterion tends to overpredict the failure strength as shown in Figure 3e. The fitting of this criterion is good for a low range of σ_3 (i.e., σ_3~4–10 MPa), while others present a poor fit due to the varying dependence of the intermediate stress σ_2 under different σ_3. However, this criterion is the third best fit of experimental data after our proposed model and Griffith criterion based on the RMSE and MAD values of 43.62 and 31.39, respectively.

4. Parametric Study

Under triaxial compression with moderate confining pressure, local tensile stresses develop leading to microcracks generation [48]. However, under higher confinement, stress variations are within a purely compressive stress field. Therefore, higher confining pressure suppresses the growth and interaction of cracks, as shown in Figure 6, where the L_{cr} decreases as the confining pressure increases. For example, when the confining pressure increases from 10 to 50 MPa, the normalized critical crack length decreases from 2.02 to 0.89, indicating a 56% decrease. The failure strength increases as the confining pressure increases due to the strengthening effect of confinement on the rock strength.

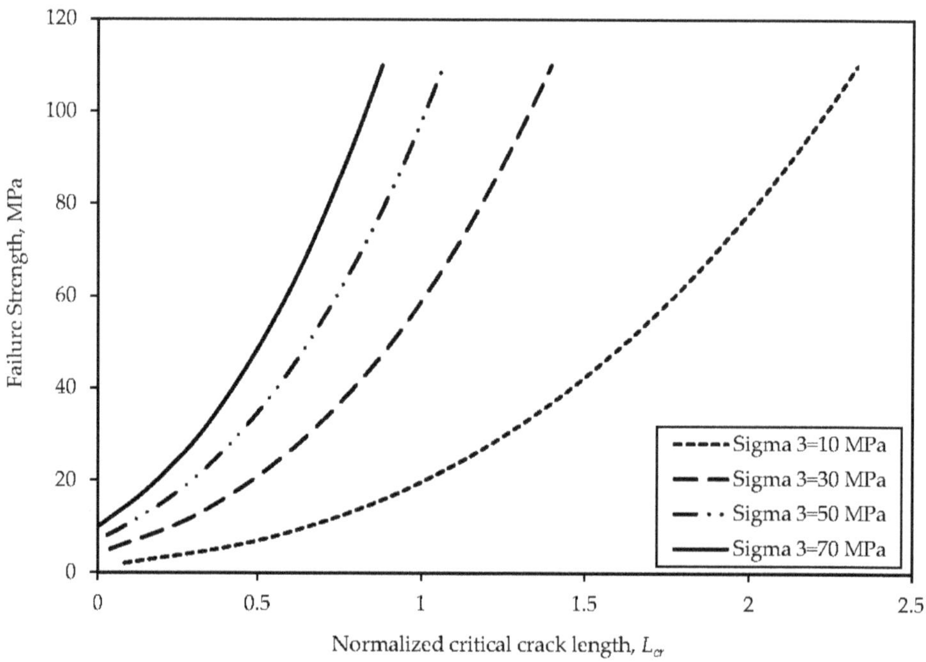

Figure 6. Relationship between the failure strength and normalized critical crack length.

Referring to Equation (6), two parameters that influence the L_{cr} are the λ and D. To determine the qualitative effect of λ, the initial constant D is assumed for a given rock type in this study. Figure 7 shows that the L_{cr} decreases as the λ increases and decreases with the increase in D at the same λ. In other words, the failure strength of the rock increases with the increase in the L_{cr}. In addition, increasing the λ leads to an increase in the failure strength. However, the increase in D decreases the failure strength.

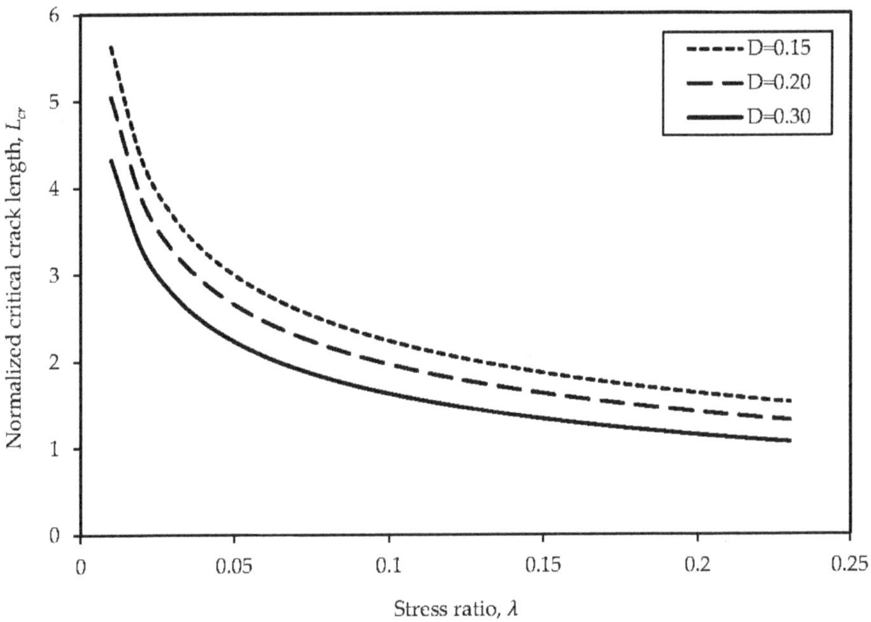

Figure 7. Relationship between the normalized critical crack length and stress ratio under constant damage levels.

For example, at a constant D of 0.15, the L_{cr} decreases from 2.23 to 1.63 when λ increases from 0.1 to 0.2, indicating a 27% decrease in the L_{cr}. On the other hand, at a constant λ of 0.1, the L_{cr} decreases from 1.63 to 1.14 when the initial D increases from 0.15 to 0.30, indicating a 30% decrease in the L_{cr}, and therefore, a decrease in the failure strength. This finding is consistent with a study by Zhang et al. [20] where they proved that the confining pressure has a significant effect on long-wing cracks based on polyaxial and true triaxial compression tests conducted on jointed rock masses. They indicated that it is hard to form long-wing cracks under high confinement because tensile fractures are suppressed under high confining pressure compression.

5. Conclusions

The present study investigated the mechanical and fracture behavior of sandstone under triaxial compression. A new analytical model was proposed to determine the normalized critical crack length (L_{cr}), through which the failure strength can be estimated based on fracture mechanics applied to wing cracks emanating from the tips of pre-existing cracks. Many earlier studies have focused on conducting expensive and time-consuming laboratory tests to capture the mechanical and fracture behavior of rocks, while other studies focused on the uniaxial compression condition only ignoring the effect of confinement on the mechanical behavior of rocks. Therefore, this study was motivated by the need to develop a reliable analytical model to describe the mechanical behavior of rocks under triaxial compression using easily measured parameters. However, it is important to note that the proposed wing crack model is mostly applicable to rocks with Ψ angles less than 30° that guarantees a tension failure mode. The main findings drawn from this study are as follows:

1. The proposed model describes that the L_{cr} decreases as the λ and D of the rock increase since higher confinements and higher levels of damage suppress the growth of cracks. The parametric study indicated that at a constant D of 0.15, the L_{cr} decreases from 2.23 to 1.63 when λ increases from 0.1 to 0.2, indicating a 27% decrease in the L_{cr}. On the other hand, at a constant λ of 0.1, the L_{cr} decreases from 1.63 to 1.14 when the D increases from 0.15 to 0.30, indicating a 30% decrease in the L_{cr}, and therefore, a decrease in the failure strength;
2. The proposed critical crack length is employed to predict the failure strength of sandstone. The total stress intensity is set equal to the rock K_{IC} and the normalized crack length is set equal to the L_{cr}; therefore, the corresponding strength represents the predicted rock failure strength. The proposed model describes that the failure strength increases as the λ and K_{IC} of the rock increases and decreases with the increase in D and the length of the pre-existing crack;
3. The proposed model assumes a value of 1 mm, 1.27 MPa$\sqrt{\text{m}}$, and 0.15 for the a, K_{IC}, and D for sandstone. The only unknown is the confining pressure. Therefore, no laboratory tests are required to determine the model parameters and predict the σ_f;
4. The proposed model was compared with widely accepted failure criteria to determine which parameters best describe the behavior of sandstone by minimizing the difference between the predicted and measured failure strength. The comparative study showed that our proposed model provides a better prediction of compressive strength based on the lowest RMSE and MAD values;
5. The HB and the modified Weibols and Cook criteria overpredict the failure strength, the Griffith criterion underpredicted the strength of 90% of the sandstones, and the Renshaw and Schulson criterion yields the greatest scatteredness of the strength prediction with the highest RMSE and MAD of 69.07 and 56.41, respectively.

Author Contributions: Conceptualization, K.N.; Validation, E.A.; Formal analysis, E.A.; Investigation, L.K.; Resources, L.K.; Data curation, L.K.; Writing—original draft, E.A.; Writing—review & editing, K.N. and L.K.; Supervision, K.N.; Project administration, K.N.; Funding acquisition, K.N. All authors have read and agreed to the published version of the manuscript.

Funding: The authors extend their gratitude to the Wyoming Department of Transportation for their valuable research support under Grant No. RS09220 and Mountain Plains Consortium.

Informed Consent Statement: Not applicable.

Data Availability Statement: The original contributions presented in this study are included in the article. Further inquiries can be directed to the corresponding author.

Conflicts of Interest: The authors declare no conflicts of interest.

List of Abbreviations

μ	Coefficient of friction
c	Cohesion
σ_3	Confining pressure
Ψ	Crack inclination angle
σ_f	Failure compressive strength
a	Half-length of the preexisting crack
l	Crack length

D	Initial level of damage
φ	Internal friction angle
σ_1	Major principal stress
m_i	Material constant for intact rock
J_1	Mean effective confining pressure
L_{cr}	Normalized critical crack length
N_A	Number of initial cracks per unit area
n	Porosity
K_I	Stress intensity factor due to crack initiation
K_I^I	Stress intensity factor due to crack interaction
λ	Stress ratio
σ_{ci}	Rock uniaxial compressive strength
K_{IC}	Type-I fracture toughness
T_o	Uniaxial tensile strength

References

1. Wang, E.Z.; Shrive, N.G. Brittle fracture in compression: Mechanisms, models and criteria. *Eng. Fract. Mech.* **1995**, *6*, 1107–1126. [CrossRef]
2. Yang, S.Q.; Jing, H.W. Evaluation on strength and deformation behavior of red sandstone under simple and complex loading paths. *Eng. Geol.* **2013**, *164*, 1–17. [CrossRef]
3. Kemeny, J.M.; Cook, N.G.W. *Crack Models for the Failure of Rocks in Compression*; Lawrence Berkeley National Laboratory: Berkeley, CA, USA, 1986.
4. Gratz, E.T.; Schulson, E.M. Brittle failure of columnar saline ice under triaxial compression. *J. Geophys. Res. Solid Earth* **1997**, *B3*, 5091–5107. [CrossRef]
5. Golshani, A.; Okui, Y.; Oda, M.; Takemura, T.A. micromechanical model for brittle failure of rock and its relation to crack growth observed in triaxial compression tests of granite. *Mech. Mater.* **2006**, *4*, 287–303. [CrossRef]
6. Nemat-Nasser, S.; Horii, H. Rock failure in compression. *Int. J. Eng. Sci.* **1984**, *22*, 999–1011. [CrossRef]
7. Renshaw, C.E.; Schulson, E.M. Universal behavior in compressive failure of brittle materials. *Nature* **2001**, *412*, 897–900. [CrossRef]
8. Wong, T.F.; Baud, P. The brittle-ductile transition in porous rock: A review. *J. Struct. Geol.* **2012**, *44*, 25–53. [CrossRef]
9. Park, C.H.; Bobet, A. Crack coalescence in specimens with open and closed flaws: A comparison. *Int. J. Rock Mech. Min. Sci.* **2009**, *5*, 819–829. [CrossRef]
10. Janeiro, R.P.; Einstein, H.H. Experimental study of the cracking behavior of specimens containing inclusions (under uniaxial compression). *Int. J. Fract.* **2010**, *164*, 83–102. [CrossRef]
11. Lockner, D.A.; Byerlee, J.D.; Kuksenko, V.; Ponomarev, A.; Sidorin, A. Chapter 1 Observations of quasistatic fault growth from acoustic emissions. *Int. Geophys.* **1992**, *51*, 3–31.
12. Lu, X.; Yan, G. A Quasi-2D exploration of mixed-mode fracture propagation in concrete semi-circular chevron-notched disks. *Buildings* **2023**, *13*, 2633. [CrossRef]
13. Cox, S.J.D.; Meredith, P.G. Microcrack formation and material softening in rock measured by monitoring acoustic emissions. *Int. J. Rock Mech. Min. Sci. Geomech. Abstr.* **1993**, *30*, 11–24. [CrossRef]
14. Kawakata, H.; Cho, A.; Kiyama, T.; Yanagidani, T.; Kusunose, K.; Shimada, M. Three-dimensional observations of faulting process in Westerly granite under uniaxial and triaxial conditions by X-ray CT scan. *Tectonophysics* **1999**, *3*, 293–305. [CrossRef]
15. Tapponnier, P.; Brace, W.F. Development of stress-induced microcracks in Westerly granite. *Int. J. Rock Mech. Min. Sci. Geomech. Abstr.* **1976**, *13*, 103–112. [CrossRef]
16. Wu, X.Y.; Baud, P.; Wong, T.F. Micromechanics of compressive failure and spatial evolution of anisotropic damage in Darley Dale sandstone. *Int. J. Rock Mech. Min. Sci.* **2000**, *37*, 143–160. [CrossRef]
17. Schulson, E.M.; Iliescu, D.; Renshaw, C.E. On the initiation of shear faults during brittle compressive failure: A new mechanism. *J. Geophys. Res. Solid* **1999**, *104*, 695–705. [CrossRef]
18. Alomari, E.; Khatri, L.; Ng, K.W. Expanded pore-emanated cracking model for brittle failure of sedimental rocks under triaxial compression. *J. Rock Mech. Geotech. Eng.* **2024**, *in press*.
19. Paliwal, B.; Ramesh, K.T. An interacting micro-crack damage model for failure of brittle materials under compression. *J. Mech. Phys. Solids* **2008**, *56*, 896–923. [CrossRef]
20. Zhang, Y.; Cui, S.; Yu, Z.; Cheng, J. Fracture characteristics of sliding crack in brittle rock: Analysis based on an improved equivalent crack model. *Front. Earth Sci.* **2022**, *10*, 893549. [CrossRef]
21. Hoek, E.; Brown, E.T. Empirical strength criterion for rock masses. *J. Geotech. Eng. Div.* **1980**, *106*, 1013–1035. [CrossRef]
22. Ashby, M.F.; Hallam, S.D. The failure of brittle solids containing small cracks under compressive stress states. *Acta Metall.* **1986**, *34*, 497–510. [CrossRef]
23. Sammis, C.G.; Ashby, M.F. The failure of brittle porous solids under compressive stress states. *Acta Metall.* **1986**, *34*, 511–526. [CrossRef]
24. Paterson, M.S.; Wong, T.F. *Experimental Rock Deformation: The Brittle Field*; Springer: Berlin/Heidelberg, Germany, 2005; p. 347.

25. Zhu, S.; Zheng, J.; Zhu, Z.; Zhu, Q.; Zhou, L. Experiments on three-dimensional flaw dynamic evolution of transparent rock-like material under osmotic pressure. *Tunn. Undergr. Space Technol.* **2022**, *128*, 104624. [CrossRef]
26. Senseny, P.E.; Pfeifle, T.W. Fracture toughness of sandstones and shales. In Proceedings of the 25th US Rock Mechanics/Geomechanics Symposium, Evanston, IL, USA, 25–27 June 1984; p. ARMA-84-0390.
27. Gowd, T.N.; Rummel, F. Effect of confining pressure on the fracture behavior of a porous rock. *Int. J. Rock Mech. Min. Sci. Geomech. Abstr.* **1980**, *17*, 225–229. [CrossRef]
28. Nemat-Nasser, S.; Horii, H. Compression-induced nonplanar crack extension with application to splitting, exfoliation, and rock burst. *J. Geophys. Res. Solid Earth* **1982**, *87*, 6805–6821. [CrossRef]
29. Ashby, M.F.; Sammis, C.G. The damage mechanics of brittle solids in compression. *Pure Appl. Geophys.* **1990**, *133*, 489–521. [CrossRef]
30. Alomari, E.M.; Kam, W.N.; Lokendra, K.; Shaun, S.W. Effect of Physical Properties on Mechanical Behaviors of Sandstone under Uniaxial and Triaxial Compressions. *Materials* **2023**, *16*, 4867. [CrossRef]
31. *ASTM D7012*; Standard Test Methods for Compressive Strength and Elastic Moduli of Intact Rock Core Specimens under Varying States of Stress and Temperature. American Society for Testing and Materials: West Conshohocken, PA, USA, 2014.
32. Bésuelle, P.; Desrues, J.; Raynaud, S. Experimental characterization of the localization phenomenon inside a Vosges sandstone in a triaxial cell. *Int. J. Rock Mech. Min. Sci.* **2000**, *37*, 1223–1237. [CrossRef]
33. Wasantha, P.L.P.; Ranjith, P.G. Water-weakening behavior of Hawkesbury sandstone in brittle regime. *Eng. Geol.* **2014**, *178*, 91–101. [CrossRef]
34. Roshan, H.; Masoumi, H.; Regenauer-Lieb, K. Frictional behavior of sandstone: A sample-size dependent triaxial investigation. *J. Struct. Geol.* **2017**, *94*, 154–165. [CrossRef]
35. Cong, L.; Hu, X. Triaxial rheological property of sandstone under low confining pressure. *Eng. Geol.* **2017**, *231*, 45–55. [CrossRef]
36. Zhou, Z.; Cai, X.; Ma, D.; Chen, L.; Wang, S.; Tan, L. Dynamic tensile properties of sandstone subjected to wetting and drying cycles. *Constr. Build. Mater.* **2018**, *182*, 215–232. [CrossRef]
37. Wei, S.; Yang, Y.; Su, C.; Cardosh, S.R.; Wang, H. Experimental study of the effect of high temperature on the mechanical properties of coarse sandstone. *Appl. Sci.* **2019**, *9*, 2424. [CrossRef]
38. Liang, X.; Tang, S.; Tang, C.A.; Hu, L.; Chen, F. Influence of water on the mechanical properties and failure behaviors of sandstone under triaxial compression. *Rock Mech. Rock Eng.* **2023**, *56*, 1131–1162. [CrossRef]
39. Wu, L.Y.; Wang, Z.; Ma, D.; Zhang, J.W.; Wu, G.; Wen, S.; Zha, M.; Wu, L. A continuous damage statistical constitutive model for sandstone and mudstone based on triaxial compression tests. *Rock Mech. Rock Eng.* **2022**, *55*, 4963–4978. [CrossRef]
40. Hoek, E. Estimating Mohr-Coulomb friction and cohesion values from the Hoek-Brown failure criterion. *Int. J. Rock Mech. Min. Sci. Geomech. Abstr.* **1990**, *27*, 227–229. [CrossRef]
41. Zhang, L.; Zhu, H. Three-dimensional Hoek-Brown strength criterion for rocks. *J. Geotech. Geoenvironmental Eng.* **2007**, *133*, 1128–1135. [CrossRef]
42. Pariseau, W.G. Fitting failure criteria to laboratory strength tests. *Int. J. Rock Mech. Min. Sci.* **2007**, *44*, 637–646. [CrossRef]
43. Griffith, A.A. Theory of rupture. In Proceedings of the First International Congress for Applied Mechanics, Delft, The Netherlands, 22–26 April 1924; pp. 55–63.
44. Yu, H.; Ng, k.; Grana, D.; Alvarado, V.; Kaszuba, J.; Campbell, E. A generalized power-law criterion for rocks based on Mohr failure theory. *Int. J. Rock Mech. Min. Sci.* **2020**, *128*, 104274. [CrossRef]
45. Barnes, P.; David, T.; Walker, J.C.F. The friction and creep of polycrystalline ice. *Proc. R. Soc. Lond. A Math. Phys. Sci.* **1971**, *324*, 127–155.
46. Zhou, S. A program to model the initial shape and extent of borehole breakout. *Comput. Geo-Sci.* **1994**, *20*, 1143–1160. [CrossRef]
47. Wiebols, G.A.; Cook, N.G.W. An energy criterion for the strength of rock in polyaxial compression. *Int. J. Rock Mech. Min. Sci. Geomech. Abstr.* **1968**, *5*, 6. [CrossRef]
48. Janach, W.; Guex, L.H. In-plane propagation of shear microcracks in brittle rocks under triaxial compression. *J. Geophys. Res. Solid Earth* **1980**, *85*, 2543–2553. [CrossRef]

materials

MDPI

Article

Evaluating the Accuracy of Bonded Block Models for Prediction of Rockmass Analog Mechanical Behavior

Isabella West [1,2], Gabriel Walton [1,*] and Sankhaneel Sinha [1,3]

1 Department of Geology and Geological Engineering, Colorado School of Mines, Golden, CO 80401, USA; igwest@mines.edu (I.W.); sankhaneelsinha@mines.edu (S.S.)
2 WSP USA Inc., Lakewood, CO 80226, USA
3 Equilibrium Mining, Kolkata 700042, India
* Correspondence: gwalton@mines.edu

Abstract: Large-scale rock formations, referred to as "rockmasses", consist of intact rock separated by pre-existing discontinuities (i.e., joints). The mechanical behavior of rockmasses is difficult to directly test in the laboratory due to the required specimen scale. Instead, Synthetic Rockmass Modeling (SRM) is often used to simulate field-scale rockmass behavior. SRM requires a calibrated discrete element model (DEM) of intact rock combined with a Discrete Fracture Network (DFN). While the SRM concept has been informally determined to provide reasonable results based on practitioner experience, detailed and peer-reviewed validation is lacking. The goal of this study was to evaluate the predictive capabilities of the SRM method. Previously available data on intact and rockmass analog laboratory specimens of Blanco Mera granite containing DFNs with two joint sets were used as a basis for the SRM created in this study. Specifically, the intact DEM was a Bonded Block Model (BBM), generated to match the grain structure and composition of Blanco Mera granite and the model's input parameters were calibrated so that the behavior of the BBM matched that of the intact laboratory specimens. The predictive capabilities of the model were evaluated by recreating the DFN from the jointed laboratory specimens within the intact BBM and comparing the behavior of the jointed models back to the jointed laboratory specimens, which has not been previously studied in the literature. The BBM was found capable of approximately predicting the behavior of rockmass analog specimens containing a pre-existing DFN without further calibration, which shows potential for the use of SRM in both industry and academia. Specifically, the BBM predicted the strength, dilatancy, and microfracturing behavior of the jointed laboratory specimens.

Keywords: rockmass behavior; numerical modeling; mechanical properties; triaxial testing; bonded block models; UDEC Voronoi; artificially jointed specimens; forward prediction; prediction model; synthetic rockmass model

1. Introduction

Excavation of rock is necessary for the construction of many engineering structures like deep foundations, road cuts, and surface and underground mines. At scales relevant to these engineering structures, rock is not a homogeneous, intact material; it contains pre-existing structural discontinuities like joints, shear zones, etc. These structural discontinuities reduce the strength of the material compared to purely intact rock and also change other aspects of its mechanical behavior [1]. Rockmasses can be characterized mechanically by attributes such as strength, stiffness, and brittleness. Such material attributes can be estimated through laboratory testing on representative rockmass specimens.

However, since field-scale rock formations (i.e., rockmasses) are composed of sections of intact rock separated by a network of pre-existing discontinuities (e.g., joints, faults), direct experimental research on rockmass mechanical behavior is limited due to the large sampling scale necessary to adequately represent most rockmasses. Although large-scale laboratory apparatus exist, they typically allow for specimens up to only 1 m in diameter [2–5]. Additionally, such large-scale testing is expensive and difficult to perform.

Due to these scale limitations, two alterative techniques for researching rockmass behavior have been used in the literature: (1) testing of laboratory-scale rockmass analog specimens (pre-existing joints sawed/cured into small-scale intact rock/material core), where the influence of pre-existing joints on mechanical behavior can be directly evaluated in the laboratory [6–11] and (2) Synthetic Rockmass Modeling (SRM), which uses numerical models to simulate rockmasses at the field-scale [12–14].

SRM avoids the practical limitations associated with laboratory testing as rockmasses with any joint geometry and properties can be easily simulated using numerical models. The concept of SRM involves the use of a calibrated discrete element model of intact rock, typically calibrated based on laboratory data (at the decimeter scale) [15]. A discrete fracture network (DFN) of pre-existing joints is incorporated within the calibrated intact material and used to predict the behavior of the rockmass at large scales and/or with the presence of pre-existing joint sets [16]. SRM is based on the premise that if a model can be calibrated so that the micromechanical damage mechanisms (i.e., inelastic yield of blocks and the formation of cracks along block boundaries) occurring at a small-scale in real rock, in addition to matching emergent material attributes (i.e., strength, stiffness, dilatancy), then the model will act in a realistic manner under other conditions. Although numerous SRMs have been developed in the literature that have applied the SRM approach (as well as rock mechanics modeling studies more generally), they have either focused on back-analysis/calibration [14,17–19] or practical forward modeling in field-scale case studies [12,13,20–22]. Such studies have performed limited and/or qualitative comparison to field data thus do not provide a rigorous evaluation of model predictive performance. Typically, the DFNs used in literature are complex, modeled after in situ rockmasses (and often after case studies) [20,21,23]. However, direct and controlled validation of these large-scale and complex SRMs is not possible, as such rockmasses cannot be tested in the laboratory. Accordingly, the goal of this study is to evaluate the predictive capabilities of SRM using laboratory-scale rockmass analog specimens. Such an evaluation has not, to the best of our knowledge, been documented in the literature.

The laboratory data from [24], a previously published study using rockmass analog laboratory specimens of Blanco Mera granite containing smooth, saw-cut joints, have been used as the basis for the SRM developed in this study. Bonded Block Modeling (BBM) was the modeling method chosen for SRM development, as it has been demonstrated to be an effective approach for modeling laboratory triaxial compression tests on intact rock [25–30].

Numerical models—both BBMs and otherwise—have been previously created and calibrated to the Blanco Mera granite laboratory data from Ref. [24] [17–19,31–33]. However, each of these studies calibrated their models to both intact and jointed specimen data simultaneously, which does not ensure that the model is predictive under conditions different from which the model was calibrated for [34] (see Section 3.5 for further discussion on this topic). This method of using a single data set to both calibrate and evaluate the model performance is common practice in other SRM scenarios as well, both at the laboratory-scale [35,36] and the field-scale [14,17–19]. In contrast, the goal of the current study was to create a BBM of intact Blanco Mera granite calibrated to intact laboratory data of [24] and evaluate its predictive capabilities to match laboratory behavior of the jointed rockmass analog specimens.

In this study, a BBM was generated to match the intact rock grain structure of Blanco Mera granite and the model input parameters for two different models (with elastic or inelastic blocks) were calibrated so that the behavior of the intact model matched that of the intact laboratory data. Next, pre-existing smooth joints were added to the BBM in the same locations and orientations as the joints in the jointed laboratory specimens from [24] and no further calibration was performed. The mechanical behavior of the jointed models was then compared to the behavior of the jointed laboratory specimens to evaluate the ability of the two different models (elastic and inelastic blocks) to predict rockmass analog mechanical behavior outside the conditions for which the initial model was calibrated.

The remainder of this paper is organized into six main sections: (1) background information related to the Bonded Block Modeling (BBM) method applied in this study; (2) relevant information on previous laboratory testing and numerical modeling of Blanco Mera granite; (3) an overview of the methods performed to generate and calibrate the BBM; (4) the results of the BBM calibration; (5) the evaluation of the predictive capabilities of the BBM for jointed specimen cases; and (6) discussion of the results and their future implications. Ultimately, the comparison between model predictions and laboratory data for the jointed specimen cases presents a direct evaluation of the predictive capabilities of the type of BBMs created in this study, which has not been previously documented in the literature.

We note that the development of the model geometry used in this study and details of the calibration of the intact Blanco Mera granite BBM with elastic blocks presented in this paper were previously presented by [37]. Critical aspects of this previous work are summarized in this paper. The calibrated intact BBM with inelastic zones and the jointed model results are fully unique to this study.

2. Background on Bonded Block Modeling (BBM)

A common discrete element modeling (DEM) approach used to represent intact rock in SRMs is Bonded Block Modeling (BBM) [29,38–40]. BBMs have been demonstrated to be effective for modeling laboratory triaxial compression tests on intact rock [25–28]. Due to its ability to be used in both field-scale and laboratory-scale applications, BBM was the chosen method used to develop the models in this study.

In BBMs, material is broken into a set of distinct, polygonal elements that are initially bonded to and interact with each other via contacts. BBMs are effective at modeling the mechanical behavior of intact rock because the polygonal blocks can be generated with specified shapes and sizes that can reasonably approximate the grain structure of low-porosity rocks [41,42]. At the laboratory-scale, blocks can be assigned as certain minerals to match the composition of the real rock and therefore the interaction of the blocks with adjacent ones is analogous to the interaction between mineral grains within real rock [25]. Contacts of blocks within BBMs can break, simulating the loss of strength that occurs at grain-to-grain contacts in rock as damage accumulates. Therefore, BBMs simulate material failure through the formation and propagation of grain-scale fractures along the block boundaries [28].

The blocks in a BBM can be modeled as elastic or inelastic. Elastic blocks have been most commonly used in the literature [25–27,30,42–45], but BBMs with inelastic blocks have been found to more accurately replicate the behavior of intact rock, particularly in the post-peak regime and under higher confinements [28,46]. This is because the use of inelastic blocks allows the model to replicate the increased intragranular deformation and damage that occurs under these conditions. As both block types can be reasonably used in different scenarios (and the simpler elastic blocks are preferable in cases where they are applicable), two BBMs were created for this study: one using elastic block and the other using inelastic blocks.

As discussed previously, BBMs simulate material failure through the formation and propagation of grain-scale fractures along the block boundaries (i.e., "contacts"). Both the elastic and inelastic block BBMs in this study use an elastoplastic constitutive model to define the deformation of the contacts (i.e., the boundary between two adjacent blocks). The strength of the contacts is defined using the Mohr–Coulomb failure criterion and both peak and residual strength properties are assigned to the contacts, where the strength of the contacts instantaneously decrease to residual levels after their peak strength has been reached (the "Coulomb Slip Model" per [47]). Please refer to [47] for a complete explanation of the mechanisms involved in the Coulomb Slip Model and other constitutive models available in UDEC.

Deformation also occurs within blocks in a BBM. The difference between elastic and inelastic blocks is the constitutive models used to define the deformation of the blocks, which are governed by different sets of input parameters. Elastic blocks use an elastic, isotropic model where the blocks can deform elastically but have infinite strength. This constitutive model requires only density and elastic moduli as inputs. On the other hand, inelastic blocks use an elastoplastic constitutive model, which requires density, elastic moduli, strength, and post-yield dilation input parameters. Specifically, the Mohr–Coulomb criterion was used to define the peak and residual strengths of the inelastic blocks (the "Strain-Softening Mohr–Coulomb Model" per [47]). Per this elastoplastic constitutive model, inelastic blocks will first deform elastically, defined by their elastic moduli inputs. With further loading, blocks begin to experience inelastic (i.e., permanent) deformation once the stress within the block becomes equivalent to its strength, as defined by the Mohr–Coulomb strength criterion inputs. The block strength will decay to the residual level after the plastic (i.e., permanent) shear strain in the block becomes equivalent to the critical shear strain input parameter. Therefore, for BBMs with inelastic blocks, the overall deformation of the model is controlled by the formation of grain-scale fractures along contacts as well as the inelastic (i.e., permanent) deformation of blocks.

This study uses calibrated BBMs to simulate intact rock, and then joints were added to these models to test the predictive capabilities of the model. Only two-dimensional models were considered for this research due to simulation run-time limitations. Although three-dimensional BBMs have been developed in the literature, large block sizes are used in order to shorten the simulation run-times. Therefore, such BBMs do not approximate the grain structure of the rock being was modeled [40,48–50].

3. Blanco Mera Granite: Previous Laboratory Testing and Numerical Modeling

Ref. [24] presented the first study to test laboratory-scale rockmass analog specimens containing more than one degree of jointing with more than one joint set. Additionally, their study was relatively unique in that the analog specimens prepared were composed of real rock. This is in contrast to most other laboratory-scale rockmass analog studies using non-rock materials like plaster and gypsum [8,9,51,52], which tend to exhibit less brittle damage mechanisms than real rock [51,53].

In [24], compression tests were performed on specimens of Blanco Mera granite containing smooth joints with two different geometries: (1) one sub-vertical joint with two sub-horizontal joints and (2) two sub-vertical joints with three sub-horizontal joints. Trends were analyzed regarding the specimens' elastic moduli, peak and residual strengths, Mohr–Coulomb and Hoek–Brown strength envelopes, and Geological Strength Index (GSI) values. The data from these experiments were later analyzed by [54] for trends in pre-peak damage thresholds and post-peak dilatancy. The relevant data from these two studies [24,54] are summarized in the following subsections.

3.1. Petrology

Blanco Mera is a coarse-grained granite from Spain [11] containing predominantly quartz, plagioclase, alkali feldspar, and mica mineral grains. Petrographic analysis was performed on the rock by [55]. Refer to Table A1 in Appendix A for a summary of the constituent mineral percentages in Blanco Mera granite and their grain sizes.

3.2. Mechanical Behavior of Intact Specimens

Blanco Mera granite has been thoroughly tested in the laboratory [11,24,54,55]. The data from [24,54] were used as the basis for model calibration in this study, as they contain the most complete data set. Refer to Table A2 in Appendix A for a summary of the mechanical properties of intact Blanco Mera granite.

3.3. Jointed Specimens

Two specimen types representing two different pre-existing fracture network densities were tested by [24]: the 1 + 2 series, which contains 1 sub-vertical joint and 2 sub-horizontal joints (see Figure 1a) and the 2 + 3 series, which contains 2 sub-vertical joints and 3 sub-horizontal joints (see Figure 1b). All jointed were sawed into the granite at standardized locations. The sub-vertical joints are oriented 78 degrees from the horizontal and the sub-horizontal joints are oriented 23 degrees from the horizontal, as indicated in Figure 1. The joints were found to have a friction angle of approximately 30 degrees per tilt testing on the specimens. The confinement levels used for the triaxial tests on jointed specimens were 0.5, 1, 2, 4, 6, 10, and 12 MPa [24].

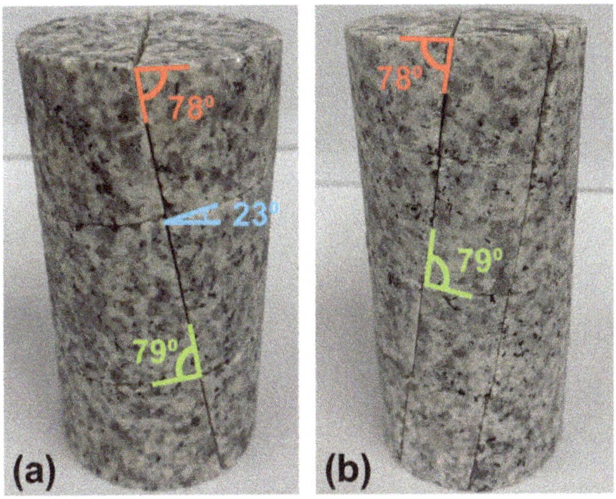

Figure 1. Images of the two jointed specimen types tested in the laboratory: (**a**) 1 sub-vertical joint and 2 sub-horizontal joints (1 + 2) and (**b**) 2 sub-vertical joints and 3 sub-horizontal joints (2 + 3). The dip of the sub-vertical joints is labeled in red, sub-horizontal joints is labeled in blue, and the angles between the two joint sets are labeled in green. Note that all joints are fully persistent. Photos courtesy: Dr. Leandro Alejano, University of Vigo.

Figure 1 shows specimens with the joints oriented such that the angle between the sub-vertical and sub-horizontal joints is approximately 79 degrees. An alternative geometry of the 2 + 3 jointed specimens were also tested such that the angle between the two joint sets was 55 degrees (i.e., both the sub-vertical and sub-horizontal joint sets dipped in the same direction, rather than opposite directions as in Figure 1b). The authors of [24] found no notable difference in mechanical behavior between the two types of 2 + 3 jointed specimens.

3.4. Mechanical Behavior of Jointed Specimens

The results of the triaxial tests performed on the jointed specimens demonstrated that the presence of pre-existing joints notably influenced the mechanical behavior of the rock (see Figure 2). The results of the previous studies on Blanco Mera granite found an increase in jointing resulted in the following changes to mechanical properties, under all confinement levels:

- Decrease in Young's modulus;
- Increase in Poisson's ratio;
- Decrease in peak strength;
- Decrease in CI and CD values;
- No change in residual strength;
- Increase in apparent ductility;

- Decrease in Peak Dilation Angle;
- No change in Post-Peak Dilatancy at large strains.

More discussion on these trends can be found in [24,54].

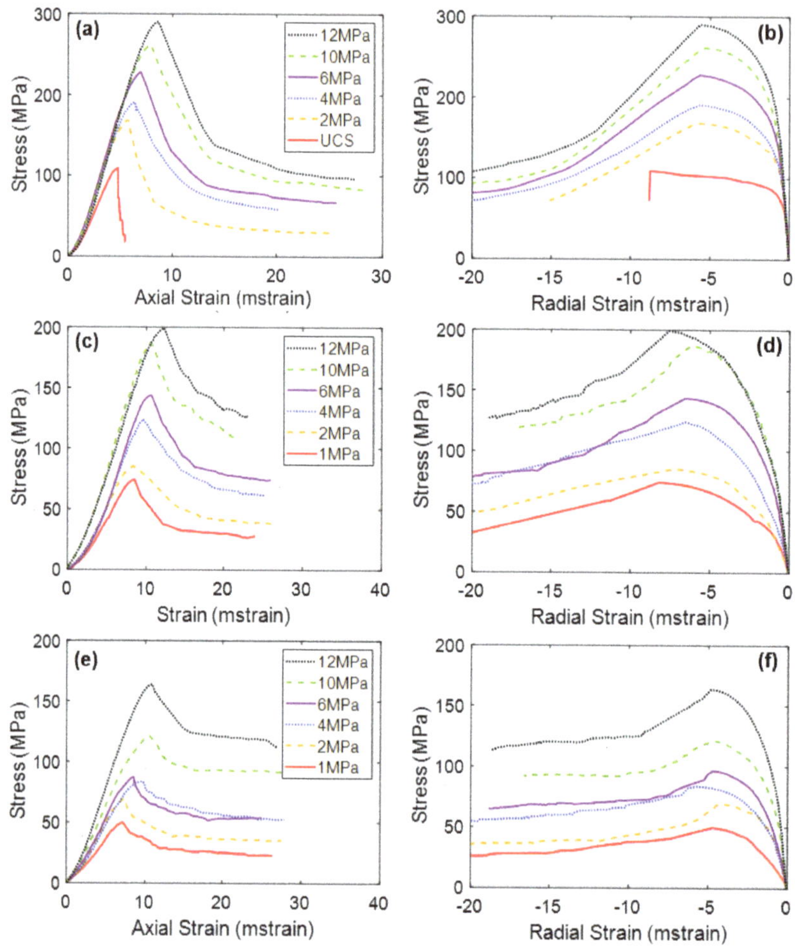

Figure 2. Average stress–axial strain curves (**a**,**c**,**e**) and average stress–radial strain curves (**b**,**d**,**f**) of Blanco Mera granite under various confinement levels, as tested in the laboratory [24]. Graphs (**a**,**b**) show intact specimen behavior, (**c**,**d**) show 1 + 2 specimen behavior, and (**e**,**f**) show 2 + 3 specimen behavior.

3.5. Previously Calibrated Numerical Models of Jointed Blanco Mera Granite

Multiple numerical models have been developed to simulate the mechanical behavior of the jointed rockmass analog laboratory specimens tested by [24]. However, none of these studies have replicated all the mechanical attributes observed in the laboratory (see Table 1).

Each of these prior studies have one common aspect of their calibration process: a single calibration was performed to identify parameters that could replicate both the intact and jointed specimen behavior. Ref. [34] has noted the issue with using this sort of model for prediction purposes, where the model is calibrated to all known data without verification of the failure mode. For such models, its predictive capabilities outside the condition of calibration (i.e., for different rockmass conditions) are uncertain [56].

Table 1. Previous studies that aimed to model the rockmass analog laboratory specimens with smooth pre-existing joints from [24]. E = Young's Modulus and v = Poisson's Ratio. CI = Crack Initiation Stress and CD = Crack Damage Stress [57–59]. BPM = Bonded Particle Model per [60,61].

		----------Calibrated Material Attributes----------						
Study	**Model Type**	**E**	***v***	**CI/CD**	**Peak Strength**	**Dilatancy**	**Residual Strength**	**Calibrated to Jointed Data?**
[17]	LS-SRM (BPM)	x			x		x	x
[18]	BBM	x *			x			x
[19]	BPM	x			x		x **	x
[31]	BBM	x *			x			x
[32]	BPM	x *	x		x			x
[33]	<u>Discrete</u>: based on zero thickness interface elements	x			x		x	x
	<u>Continuum</u>: based on a multilaminate model	x			x	x	x	x

* Calibrated to secant Young's Modulus rather than tangent Young's Modulus. ** Did not match residual strength under higher confinement levels.

With all this in mind, this study has aimed to develop Bonded Block Models of the Blanco Mera granite specimens tested by [24] that accurately match the grain structure of the real rock, and is calibrated to match material attributes (i.e., stiffness and strength) as well as the micromechanical failure mechanism, as characterized by CI, CD, and post-peak stress–strain behavior. This study has also used forward simulation to predict the behavior of the jointed rockmass analog specimens. This process entails first back-calibrating an intact BBM (e.g., with no pre-existing joints) to laboratory data, including the micromechanical failure processes. Next, pre-existing joints are added to the model and no further calibration is performed. The degree of agreement (or disagreement) between the simulated predictions for the jointed specimen mechanical behavior and the observed behavior is representative of the predictive capabilities of the model.

4. Methods: Bonded Block Model (BBM) Development

A Bonded Block Model (BBM) with heterogeneous block and contact properties (by mineral type) was the chosen model type used in this study due to its ability to replicate the micromechanical behavior of real rock [28]. Details of BBM generation and calibration are included in the subsequent subsections.

4.1. Voronoi Generation

A three-dimensional (3D) Bonded Block Model (BBM) of intact Blanco Mera granite was generated using the software Neper Polycrystal v3.5 [62]. The blocks were stochastically generated using input parameters of mean diameter, standard deviation of diameter, average sphericity, and standard deviation derived from petrographic analysis of real Blanco Mera granite specimens [63]. Diameter inputs in Neper Polycrystal were determined using a statistical method that incorporated estimated standard deviations of grain size diameters for each constituent mineral in Blanco Mera granite (refer to [37] for more details on this process). A variety of sphericity values and their standard deviations were tested to obtain an approximate visual match with the actual shapes of grains within Blanco

Mera granite, where the grains of the model are relatively round like the real grains of Blanco Mera granite, but many of the smaller grains are more rectangular in shape, which is typical for micas. Figure 3 shows the BBM generated in Neper (trimmed to scale) beside a photograph of real Blanco Mera granite for comparison.

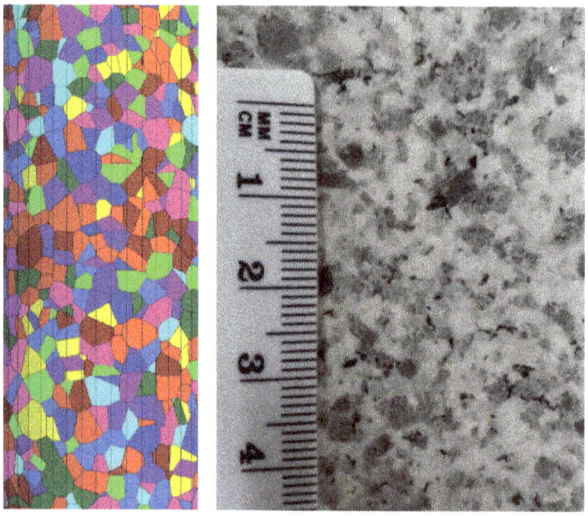

Figure 3. Neper-generated geometry (**left**) vs. real Blanco Mera granite (**right**) with approximately the same scale. The color-coding of the grains within the Neper model is arbitrary. Images from: [37].

Two-dimensional (2D) cross sections of the 3D Neper model were generated using 3DEC. AutoCAD was used to prepare these cross sections to be built in UDEC v6, the software used to simulate compression tests.

4.2. Block Assignment to Mineral Groups

After the 2D Voronoi structure was built in UDEC, each block was assigned to a mineral group. A statistical method was used to assign blocks to mineral groups to achieve similar grain size distributions for each individual mineral in the model as in the real rock (see [37] for more detail). A comparison of the sizes and proportions of blocks representing different mineral groups within the BBM against those of the real Blanco Mera granite is presented in Table 2.

Table 2. The final BBM's average grain diameters (i.e., equivalent spherical diameter computed based on the area of each block) and proportions for each mineral group compared to the real rock. Table from: [37].

Mineral Group	Rock Proportion	BBM Proportion	Rock Average Grain Diameter (mm)	BBM Average Block Diameter (mm)
Plagioclase	37%	37.03%	3.5	3.50
Quartz	20%	20.18%	3.5	3.47
Alkali Feldspar	27%	27.12%	3.0	2.65
Mica	16%	15.67%	1.9	1.91

4.3. BBM with Pre-Existing Joints

In addition to the generation and calibration of an intact BBM of Blanco Mera granite, this study generated BBMs in UDEC with pre-existing smooth joints similar to the rockmass analog specimens tested in the laboratory (see Section 3.3). The exact locations of the joints corresponding to the 1 + 2 (Figure 4a) and the 2 + 3 (Figure 4b) joint patterns were determined by visually approximating their locations on the laboratory specimens (Figure 1). As discussed in Section 3.3, some jointed laboratory specimens were also tested with an opposite joint orientation. Although [24] found no difference in the strengths of these two specimen geometries, for completeness, the present study has also generated BBMs with the opposite joint geometry for mechanical behavior analysis (see Figure 4c,d).

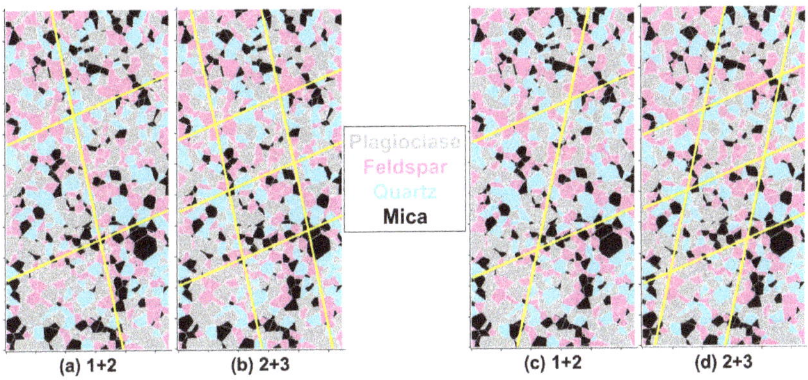

Figure 4. Bonded Block Models (BBMs) of Blanco Mera granite with the two joint patterns: (**a**,**c**) 1 + 2 and (**b**,**d**) 2 + 3. (**a**,**b**) are the main joint orientation geometry and (**c**,**d**) are the opposite joint orientation geometry. Joints are smooth, like in the corresponding laboratory specimens (see Figure 1).

Once each block was assigned to one of the four mineral groups (plagioclase, quartz, alkali feldspar, or mica), input parameters were assigned to each block, as well as each corresponding contact, as depicted in Figure 5.

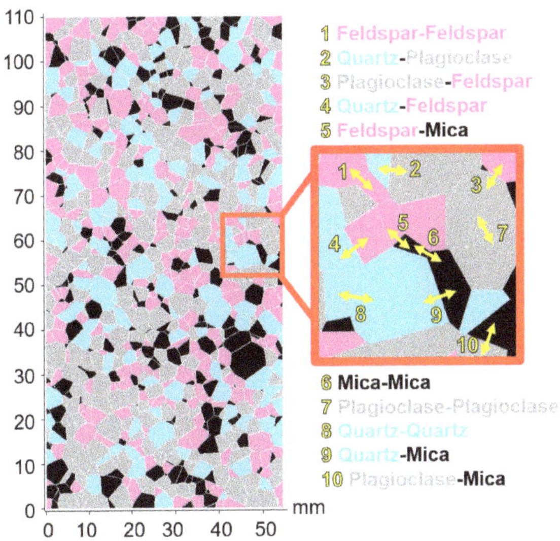

Figure 5. The BBM of Blanco Mera granite used in this study. It contains four mineral (block) types: alkali feldspar (pink), quartz (blue), plagioclase (gray), and mica (black). Ten unique contact types exist, labeled in yellow. Figure from: [64].

4.4. Compression Test Simulation

Two-dimensional UCS and triaxial test simulations were performed using the intact and jointed BBMs. All blocks were zoned with a dense mesh. Even the smallest blocks

(0.005 mm diameter) were meshed with at least three distinct zones. A boundary condition restricting movement in the vertical direction was applied to the bottom of the model, which represents the base platen used in real laboratory. A constant velocity boundary condition of 0.01 m/s was applied to the top of the model in the vertical direction to simulate quasi-static loading (note that this value is consistent with previous numerical studies and cannot be directly compared to loading velocities used in physical laboratory tests [31,65,66]). A uniformly distributed horizontal load was applied to the lateral boundaries to simulate confining pressure. These boundary conditions represent the principal stress directions (the first principal stress is the axial loading, and the third principal stress is the lateral confining pressure). No boundary conditions were applied to the horizontal velocity stress at the top and bottom of the model or the vertical velocity or stress at the sides of the model, consistent with the actual conditions associated with compression testing.

During simulations, stress, axial strain, lateral strain, and the number of contacts failing in tension and shear were tracked to monitor macroscopic properties of the stress–strain behavior of the BBM. In the case of the inelastic block BBM, the number of zones yielding in tension and shear were also monitored. Stress was determined by calculating the average internal stress value within each zone of the BBM. Strains were calculated by computing the average displacement of the edge points of the BBM (in both the axial and lateral directions) from their original positions and then normalizing to the original corresponding dimension of the specimen.

4.5. *Evaluation of Model Outputs*

The macroscopic mechanical behavior of each BBM from the beginning of loading through peak strength was characterized using the following six material attributes for all confining stresses evaluated: Young's modulus, Poisson's ratio, Crack Initiation (CI) and Crack Damage (CD) parameters, peak strength, and peak dilation angle. The definition of each attribute, including references with further details on specific methods for estimation of values, are summarized in Table 3. The attributes were determined for the BBM using the stress and strain data tracked during each simulation. In addition to these properties, the stress–strain behavior post-failure was qualitatively assessed by visual assessment of stress–strain plots. Evaluation of these attributes for each model and the comparison with the corresponding laboratory data is part of the BBM calibration process (described in Section 4.7).

4.6. *Hoek–Brown Fits*

For the strength parameters discussed in Section 4.5, a model was fit to the data as a function of confinement using the Hoek–Brown failure envelope (Equation (1)).

$$\sigma_1 = \sigma_3 + \sigma_{ci} \times \left(m \times \frac{\sigma_3}{\sigma_{ci}} + s\right)^a \quad (1)$$

where σ_1 is the maximum effective principal stress at a given strength threshold (either peak strength, CI, or CD), σ_3 is the minimum effective principal stress (confinement), σ_{ci} is the value of the given strength threshold (peak strength, CI, or CD) of the intact rock under unconfined conditions, m is the Hoek–Brown material parameter, and s and a are rockmass characteristic parameters.

For the intact specimen data, s was set equal to 1 and a was set equal to 0.5 [67]. σ_{ci} and m were fit to stress data (σ_1) as a function of confinement (σ_3) using a least-squares fitting method. For the jointed models (1 + 2 and 2 + 3), σ_{ci} was set equal to that of the intact rock and s, m, and a were allowed to independently vary. All four Hoek–Brown fit parameters are reported along with the results in Sections 5 and 6.

Table 3. Summary of the material attributes used to quantify the mechanical behavior of the specimens in this study. The definition, method of determination, and references are included for each attribute.

Material Attribute	Method of Determination	References
Young's Modulus	The slope of the axial stress–axial strain curve within the linear elastic portions of the axial strain data	[68]
Poisson's Ratio	The slope of the linear elastic portion of the lateral strain–axial strain curve	[69]
Crack Initiation (CI) Threshold	The point of non-linearity in the lateral strain–axial strain or lateral strain–axial stress plot	[49,57,70,71]
Crack Damage (CD) Threshold	The point of non-linearity in the axial stress–axial strain plot	[49,57,70]
Peak Strength	The largest stress achieved	-
Peak Dilation Angle	Maximum value of the dilation angle, where dilation angle varies as a function of plastic shear strain and confinement	[54,72,73]
Brittleness	Slope of post-peak strength decrease; high rate = an increased slope indicates more brittle behavior	[74,75]

4.7. Model Calibration to Intact Specimen Laboratory Data

Both the elastic block and inelastic block BBMs were calibrated by iteratively adjusting their input parameters. To ensure that the model's behavior matched laboratory data under the full range of confinement levels in which data are available, the behavior of the BBM under unconfined and confined conditions with 12 MPa of confinement were analyzed for each input parameter set used during the calibration process. Once the BBM was calibrated to match the laboratory data of the lower and upper bounds of confinements, the BBM was run under intermediate confinement levels to evaluate the match to all available laboratory data under each confinement level.

Although most of the model input parameters cannot be determined directly from laboratory testing or field measurements, a reasonable range of values for each input parameter can be constrained based on the results of previous studies using BBMs. Starting input parameters were selected from the literature, and efforts were made to avoid making large changes to any individual input parameter.

Each time a model was run with a new set of input parameters, the mechanical attributes described in Section 4.5 were determined and compared to laboratory data [24,54]. Once the emergent behavior matched that of the intact specimen laboratory data for Blanco Mera granite, the calibration was deemed complete. The procedures followed for calibration of both the elastic and inelastic block BBMs are documented in greater detail in the following subsections.

The elastic block BBM was calibrated first, as it requires all the same inputs as inelastic block BBMs other than block yield and dilation parameters. Therefore, the elastic block and contact input parameters calibrated during elastic block BBM calibration were used as the starting point for the inelastic block BBM calibration (more on this in Section 4.7.2).

4.7.1. Elastic Block BBM Calibration

For the elastic block BBM calibration, Young's Modulus, Poisson's Ratio, CI, CD, and peak strength were matched to laboratory data. No attempt was made to reproduce post-peak attributes, as post-peak behavior cannot be matched using elastic blocks. Elastic block BBMs neglect the intra-granular fracturing process, which is an important damage mechanism in the post-peak region of stress–strain curves, as well as for stress–strain curves under higher confinement levels [28].

Since the Blanco Mera BBM is heterogeneous, meaning that there is more than one type of block (corresponding to the four mineral groups) and more than one type of contact (see Figure 5), each type of block and contact can be assigned different values as input parameters. Therefore, an input parameter can be varied in the BBM by changing each type's value by the same amount so that the average value changes, or by changing each type's values systematically so that the spread of values changes (i.e., change in heterogeneity). Refer to [64] for an in depth parametric study on BBM input parameter heterogeneity.

A summary of the elastic block BBM calibration process for Blanco Mera granite in sequential order is as follows:

1. **Poisson's Ratio**: Macroscopic Poisson's Ratio is controlled by the block elastic moduli and ratio of contact shear stiffness (jks) to contact normal stiffness (jkn). Using starting block moduli values from other BBMs containing grains of the same mineral types, a Poisson's Ratio of 0.17 (see Table A2) could not be attained without modifying the jks to jkn ratio to be outside the range reported by previous studies using elastic block BBMs. Therefore, the jks to jkn ratio was set to 0.65, which is towards the upper bound of what has been used in previous studies. Instead, the block moduli were lowered until the correct Poisson's Ratio was achieved.
2. **Young's Modulus**: jks and jkn values were increased proportionally to maintain the same ratio until the BBM replicated the macroscopic Young's Modulus.
3. **Crack Initiation (CI)**: CI corresponds to the onset of tensile fracturing, so this property is predominantly controlled by the tensile strength of block contacts. Increasing the contact tensile strength increases the value of CI under all confinements. Additionally, the overall heterogeneity of the BBM (in terms of both blocks and contacts) can be increased to decrease the value of CI [64].
4. **Unconfined Peak Strength**: UCS is primarily controlled by contact peak cohesion of contacts where increasing peak cohesion increases the UCS of the BBM. Ref. [64] also found that increasing the heterogeneity of contact peak cohesion decreased the BBM's UCS.
5. **Crack Damage (CD)**: Since CD corresponds to the onset of shear fracturing, increasing the shear strength of the contacts increases the CD value of the BBM. Therefore, CD is controlled by contact peak cohesion and peak friction angle, where contact peak cohesion has a greater influence on the unconfined CD and peak friction angle has a greater influence on the confined CD. Additionally, both the heterogeneity of contact peak cohesion and the overall heterogeneity of the BBM (in terms of both block and contact properties) can be increased to decrease the value of CD [64].
6. **Confined Peak Strength**: Peak and residual contact friction angles predominately control the confined strength of the BBM. It should be noted that increasing the peak friction angle increases the confined peak strength but will also increase the CD of the BBM. On the other hand, increasing contact residual friction angle increases the confined peak strength of the BBM, but does not notably affect the model's CD.

The calibrated block mineral group and contact input parameters are included in Tables 4 and 5, respectively.

Table 4. Mineral group densities and elastic parameters assigned to blocks within the BBM. Table from: [37].

Mineral Group	Property	Value	Unit
Plagioclase	Density	2630	kg/m^3
	Young's Modulus (E)	55	GPa
	Poisson's Ratio (ν)	0.18	-
	Bulk Modulus (K)	28.65	GPa
	Shear Modulus (G)	23.31	GPa

Table 4. *Cont.*

Mineral Group	Property	Value	Unit
Quartz	Density	2650	kg/m^3
	Young's Modulus (E)	66	GPa
	Poisson's Ratio (ν)	0.06	-
	Bulk Modulus (K)	25.00	GPa
	Shear Modulus (G)	31.13	GPa
Alkali Feldspar	Density	2650	kg/m^3
	Young's Modulus (E)	42	GPa
	Poisson's Ratio (ν)	0.15	-
	Bulk Modulus (K)	20.00	GPa
	Shear Modulus (G)	18.26	GPa
Mica	Density	3050	kg/m^3
	Young's Modulus (E)	20	GPa
	Poisson's Ratio (ν)	0.18	-
	Bulk Modulus (K)	10.42	GPa
	Shear Modulus (G)	8.48	GPa

Table 5. Contact parameters of the calibrated elastic block BBM. Parameter abbreviations are: jkn = joint normal stiffness, jks = joint shear stiffness, c = cohesion, ϕ = friction angle, jt = tensile strength, c_{res} = residual cohesion, jt_{res} = residual tensile strength, Ψ = dilation angle. Table from: [37].

Property	jkn (GPa/m)	jks (GPa/m)	c (MPa)	Φ (^{0})	jt (MPa)	c_{res} (MPa)	Φ_{res} (^{0})	jt_{res} (MPa)	Ψ (^{0})
Mica-Mica	1.7×10^6	1.1×10^6	38.3	60.0	23.6	0.0	10.0	0.0	10.0
Quartz-Quartz	3.6×10^6	2.4×10^6	62.2	65.0	32.6	0.0	10.0	0.0	10.0
Plagioclase-Plagioclase	3.2×10^6	2.1×10^6	53.6	64.5	34.5	0.0	10.0	0.0	10.0
Feldspar-Feldspar	3.0×10^6	1.9×10^6	52.6	64.5	32.6	0.0	10.0	0.0	10.0
Quartz-Plagioclase	3.0×10^6	1.9×10^6	43.0	62.0	19.6	0.0	10.0	0.0	10.0
Quartz-Mica	3.0×10^6	1.9×10^6	38.3	55.0	35.0	0.0	10.0	0.0	10.0
Plagioclase-Mica	3.0×10^6	1.9×10^6	25.8	55.0	20.9	0.0	10.0	0.0	10.0
Plagioclase-Feldspar	2.7×10^6	1.8×10^6	51.7	62.0	29.8	0.0	10.0	0.0	10.0
Feldspar-Mica	3.0×10^6	1.9×10^6	28.7	55.0	10.6	0.0	10.0	0.0	10.0
Feldspar-Quartz	3.5×10^6	2.3×10^6	36.4	62.0	26.3	0.0	10.0	0.0	10.0

4.7.2. Inelastic Block BBM Calibration

Inelastic block BBMs require all the same inputs as elastic block BBMs, but also require block strength properties. The starting inelastic block parameters were taken from a previously calibrated BBM of Creighton granite by [28]. The same set of inelastic parameter values was assigned to all block types (i.e., mineral types), despite each type having unique elastic moduli values, as it has previously been determined that heterogeneous inelastic block parameters are not necessary to match all mechanical attributes at the laboratory-scale [28]. Additionally, assigning each block type different values of inelastic block input parameters would significantly increase the number of inputs in the model, making the calibration process more difficult and increasing the potential for model non-uniqueness.

The iterative process of inelastic block BBM calibration by trial-and-error is identical to that of the elastic block BBM calibration but involves more input parameters. Unlike elastic block BBMs, inelastic block BBMs are able to match the post-peak behavior of rock, where the inelastic block input parameters influence the post-peak behavior. In fact, for the range of input parameter values tested during this calibration process, the inelastic block parameters were not found to have a notable effect on pre-peak mechanical attributes of the BBM (Young's Modulus, Poisson's Ratio, CI, or CD). Rather, these input parameters only affected the peak strength, and post-peak behavior. Additionally, the elastic block and contact input parameters controlled the pre-peak mechanical attributes in the same way as summarized in Section 4.7.1. Prior to final model calibration, Ref. [76] documented the results of a sensitivity analysis evaluating the influences of inelastic block input parameter on the peak strength and post-peak behavior of the Blanco Mera inelastic block BBM, which are summarized in Table A3 in Appendix A.

During the inelastic block BBM calibration process, emphasis was placed on the inelastic block input parameters rather than the contact input parameters. However, contact input parameters were also varied in certain cases where the inelastic block input parameters did not induce a large enough change on the model behavior. In particular, contact residual friction angle and dilation angle were found to have an effect on the post-peak behavior of the BBM, so these input parameters were frequently changed during the calibration process in order to match the BBM's post-peak brittleness to laboratory data.

The final values calibrated for the inelastic block BBM of Blanco Mera granite are listed in Table 4 (block elastic moduli), Table 6 (block inelastic parameters), and Table 7 (contact parameters).

Table 6. Inelastic input parameters of the inelastic block BBM that match laboratory data from [24,54]. eps = critical plastic shear strain parameter, which dictates the shear strain at which zones within the blocks decay from peak values to residual.

Inelastic Input Parameter	Value	Units
Peak Cohesion	100	MPa
Peak Friction Angle	57	Degrees
Peak Tensile Strength	40	MPa
Peak Dilation Angle	10	Degrees
Residual Cohesion	50	MPa
Residual Friction Angle	40	Degrees
Residual Tensile Strength	0.5	MPa
Residual Dilation Angle	5.0	Degrees
eps	0.2	-

Table 7. Contact parameters of the inelastic block BBM. Parameter abbreviations are: jkn = joint normal stiffness, jks = joint shear stiffness, c = cohesion, ϕ = friction angle, jt = tensile strength, c_{res} = residual cohesion, jt_{res} = residual tensile strength, Ψ = dilation angle.

Property	jkn (GPa/m)	jks (GPa/m)	c (MPa)	Φ (^{0})	jt (MPa)	c_{res} (MPa)	Φ_{res} (^{0})	jt_{res} (MPa)	Ψ (^{0})
Mica-Mica	1.7×10^6	1.1×10^6	35.0	60.0	23.6	0.0	3.0	0.0	7.5
Quartz-Quartz	3.6×10^6	2.4×10^6	33.1	65.0	32.6	0.0	3.0	0.0	7.5
Plagioclase-Plagioclase	3.2×10^6	2.1×10^6	34.9	64.5	34.5	0.0	3.0	0.0	7.5
Feldspar-Feldspar	3.0×10^6	1.9×10^6	38.7	64.5	32.6	0.0	3.0	0.0	7.5
Quartz-Plagioclase	3.0×10^6	1.9×10^6	37.4	62.0	19.6	0.0	3.0	0.0	7.5
Quartz-Mica	3.0×10^6	1.9×10^6	36.4	55.0	35.0	0.0	3.0	0.0	7.5
Plagioclase-Mica	3.0×10^6	1.9×10^6	25.8	55.0	20.9	0.0	3.0	0.0	7.5
Plagioclase-Feldspar	2.7×10^6	1.8×10^6	41.7	62.0	29.8	0.0	3.0	0.0	7.5
Feldspar-Mica	3.0×10^6	1.9×10^6	28.4	55.0	10.6	0.0	3.0	0.0	7.5
Feldspar-Quartz	3.5×10^6	2.3×10^6	35.5	62.0	26.3	0.0	3.0	0.0	7.5

4.8. Jointed Model Input Parameters

After the intact BBM was calibrated for both the elastic and inelastic block cases, smooth pre-existing joints were added to the BBM in the same orientations as those in the laboratory tests (see Section 4.3). As discussed previously, the input values for the pre-existing joints were not calibrated but rather estimated from similar models in the literature or taken directly from laboratory data. The pre-existing joints were modeled using an elastic-plastic constitutive model with Coulomb slip. With this constitutive model, the strength of the joints is governed by the peak friction angle and does not decay to a residual level with slip [47]. The cohesion and tensile strength of a pre-existing smooth joint with no infilling are zero. Therefore, the required input parameters of the pre-existing joints are: jkn, jks, cohesion, friction angle, and tensile strength. The values used for these inputs are listed in Table 8.

Table 8. Input parameters of the pre-existing joints added to the BBM after the calibration process. jkn = joint normal stiffness, jks = joint shear stiffness.

Input Parameter	Value	Units
jkn	750	GPa/m
jks	75	GPa/m
Cohesion	0	MPa
Friction Angle	30	Degrees
Tensile Strength	0	MPa

The jks and jkn values were taken from a previous study that modeled intact and jointed Blanco Mera granite [19], as these values were within the middle of the range of values used in all of the previous studies that modeled jointed Blanco Mera granite and had comparable modeling methods including other BBMs [18,31] or Bonded Particle Models [19]. A small sensitivity analysis was performed on the persistent joint jkn and jks values for one of the 1 + 2 jointed BBMs. Stiffness values were varied from the values listed in Table 8 to analyze the effect that joint stiffness values had on the model strength. Increasing jkn from 500 to 1000 GPa/m and jks from 50 to 100 GPa/m resulted in an increase in model strength by 10% under low confinement (1 MPa) and an increase in model strength by 2% under high confinement (12 MPa). Although very large changes in stiffness value magnitude (on the order of multiple magnitudes) are expected to notably affect the model's strength, such values are outside the range of those used by previous studies and are not reasonable values to use for such joints.

A friction angle of 30 degrees was determined based on the basic friction angle of joints as determined in the laboratory [24]. Note that no calibration was performed for the jointed BBMs; all joint parameter values were estimated ahead of time based on values documented in the literature, as might be the case in practical scenarios where a model is being used to make forward predictions.

5. Intact BBM Calibration Results and Discussion

5.1. Mechanical Attributes Derived from Stress–Strain Curves

Figure 6 presents the stress–strain curves of the two intact BBMs (elastic and inelastic blocks) under various confinement levels. Note on Figure 6a,c that all simulations exhibited the same Young's modulus regardless of the confining pressure. This trend is not consistent with laboratory experiments (see Figure 2), which show an increasing Young's modulus with increasing confining pressure [24,54,55]. However, due to the inherent nature of Voronoi tessellations, BBMs do not contain pre-existing discontinuities that are responsible for the artificial reduction in the stiffness of the low-confinement laboratory specimens.

In addition to Young's modulus and Poisson's ratio, peak strength, CI, CD, and peak dilation angle in both model types were matched to laboratory data, the results of which are shown in Figure 7. The Hoek–Brown best fit parameters corresponding to the curves plotted on Figure 7 are included in Table A4 in Appendix A.

Note that the calibration target for the strength of the unconfined BBMs (both with elastic and inelastic blocks) was at the upper end of the range of laboratory data. Many data points of unconfined strength are believed to be artificially low due to the presence of small pre-existing fractures in the specimens, which are not modeled in the BBM. These pre-existing fractures do not notably affect the strength of specimens tested under confinement because the confining pressure closes these fractures prior to loading.

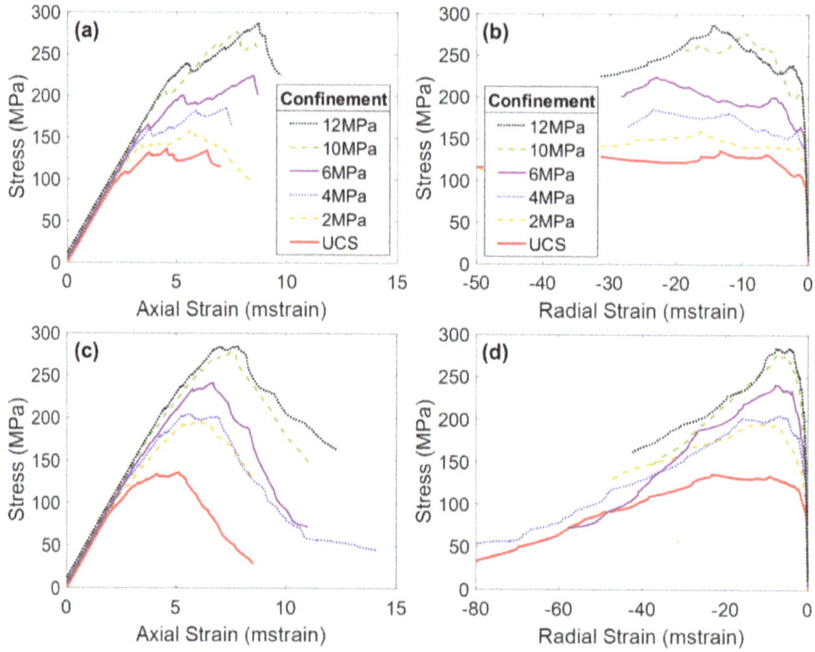

Figure 6. Axial stress–axial strain (**a**,**c**) and axial stress–radial strain (**b**,**d**) curves of the intact elastic block BBM (**a**,**b**) and the inelastic block BBM (**c**,**d**). Strain is plotted in milli-strain (mstrain).

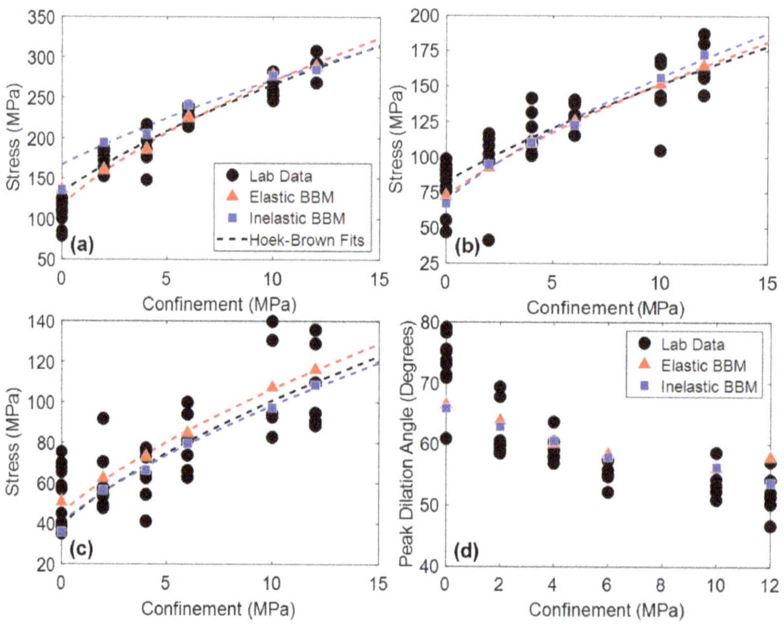

Figure 7. (**a**) Peak strength, (**b**) CD, (**c**) CI, and (**d**) peak dilation angle of laboratory data (black circles), the elastic block BBM (red triangles), and the inelastic block BBM (blue squares) of Blanco Mera granite. Hoek–Brown fits are included where applicable. Note that peak dilation angle data were not reported in [54] but were manually interpreted from the strain data provided.

Considering the material attributes discussed so far (Young's modulus, Poisson's ratio, CI, CD, peak strength, and peak dilation angle), both the elastic and inelastic block BBMs are virtually identical. One exception is the peak strength envelope of the inelastic block BBM. Although both the elastic and inelastic block BBMs were calibrated to the same values of peak strength under no confinement and under 12 MPa of confinement, the intermediate confinement simulations resulted in different peak strength values (see Figure 7a). This is expected, as the elastic and inelastic block BBMs behave differently at the grain-scale (e.g., inelastic blocks incur permanent deformation whereas elastic blocks do not). The peak strength of the inelastic block BBM is larger than that of the elastic block BBM under confinement levels of 2 MPa to 6 MPa. At larger confinements (10+ MPa), the peak strengths of the two BBMs are virtually the same.

The main difference between the stress–strain results of the two BBMs occurs between CD and peak strength, and into the post-peak. The elastic block BBM (Figure 6a) exhibits significant strain hardening, where the peak strength is not reached until large values of axial strain. Large amounts of strain hardening are commonly associated with elastic block BBMs [25,28,66]. This amount of strain hardening does not exist for the inelastic block BBM (see Figure 6c). As intragranular damage (zone yield) becomes more prevalent in the model after CD, the strength of the model is reduced. The internal deformation of blocks helps to decrease friction that is mobilized in the BBM as large sections of intact material break off from one another along failed contacts, which decreases the extent of strain hardening. The behavior of the inelastic block BBM is akin to that of the real Blanco Mera granite, which is brittle in behavior and does not exhibit so much strain hardening (see Figure 2). The post-peak brittleness of the inelastic block BBM (see Figure 6c) is realistic, where the unconfined simulation was relatively brittle (i.e., had a sharp drop in strength following the peak strength) and the sharpness of the drop in strength generally decreased with increasing confinement. On the other hand, the elastic block BBM did not exhibit brittle post-peak stress–strain behavior due to the previously described excessive strain hardening phenomenon.

5.2. Damage Mechanisms

As discussed in Section 4, the behavior of a BBM must replicate the micromechanical failure of real rock to be fully calibrated, as well as to be predictive. Brittle rock under low confinement tends to form tensile fractures, which dilates the specimen. Such rocks fail by axial splitting [77]. As the rock is placed under higher confinement levels, the formation of tensile fractures is suppressed, shear fractures become more prevalent, and less dilation occurs [78]. Such rocks fail by shear banding, which is a less brittle failure mechanism [75]. This switch from tensile fracture dominated failure to shear fracture dominated failure is captured by both the elastic and inelastic block BBMs, as indicated in Figure 8 where each model is shown at its peak strength.

Axial splitting occurs in both the elastic block and inelastic block BBMs under zero confinement conditions (indicated by the yellow arrows in the top left panel of Figure 8). There is large dilatancy caused by separation along the axially oriented fractures. On the other hand, under 12 MPa of confinement, less separation occurs along the fractures, which results in less dilation.

Although a macroscopic failure plane under higher confinements is not obvious in Figure 8, significant en echelon shearing has occurred (indicated by the yellow arrows in the top right panel of Figure 8) within both the elastic and inelastic block BBM, characterized by both shear fracture formation and, in the case of the inelastic block BBM, shear zone yield of the blocks. In the models under zero confinement, notably more tensile fractures (and tensile zone yield) exist compared to shear. In the models under 12 MPa of confinement, the number of tensile and shear fractures (and zone yield, in the case of the inelastic block BBM) are nearly equivalent. As such, the models are also behaving like real rock, where tensile fracture formation and propagation is suppressed by confinement and shear failure dominates.

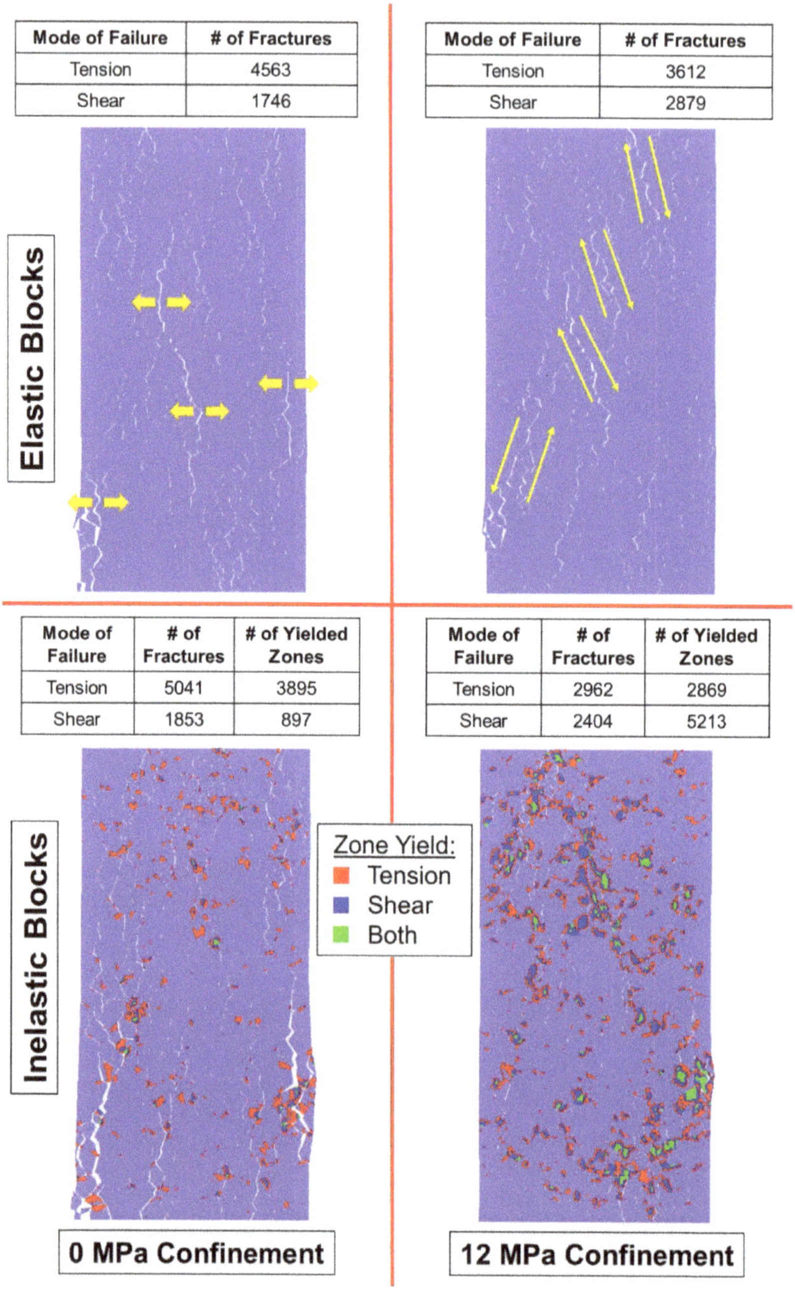

Mode of Failure	# of Fractures
Tension	4563
Shear	1746

Mode of Failure	# of Fractures
Tension	3612
Shear	2879

Mode of Failure	# of Fractures	# of Yielded Zones
Tension	5041	3895
Shear	1853	897

Mode of Failure	# of Fractures	# of Yielded Zones
Tension	2962	2869
Shear	2404	5213

Figure 8. Intact BBMs at failure (peak strength) showing the location of macroscopic fractures with visible separation. At (**top left**) is the elastic block BBM under unconfined conditions, (**top right**) is the elastic block BBM under 12 MPa of confinement, (**bottom left**) is the inelastic block BBM under unconfined conditions, and the (**bottom right**) is the inelastic block BBM under 12 MPa of confinement. Inelastic models indicate the location of zones that have yielded. Yellow arrows are included to highlight the tensile fracturing (**top left**) and shear en echelon fracturing (**top right**).

Ref. [28]'s study on elastic and inelastic block BBMs previously found that inelastic block yield was more important for the replication of brittle rock behavior under higher confinement levels than under lower confinement levels. When comparing the elastic and inelastic block BBMs' failure modes under 0 MPa of confinement (left panels on Figure 8), there is not a notable difference in the pervasive failure pathways between the two model types. However, when comparing the elastic and inelastic block BBMs' failure modes under 12 MPa of confinement (right panels on Figure 8), the inelastic block BBM exhibited notably less fracture separation and macroscopic coalescence. Instead, the inelastic block yield contributed to the model's pervasive failure pathways.

6. Jointed BBM Results and Discussion

6.1. Mechanical Attributes Derived from Stress–Strain Curves

The stress–strain results of the jointed BBMs are presented in Figures 9 and 10 for the 1 + 2 and 2 + 3 models, respectively. The most representative set of stress–strain curves of each of the two jointed models (Figure 4) are included for simplicity. Both joint orientations were found to produce similar macroscopic stress–strain responses in the BBMs have been included as Figures A1 and A2 in Appendix A, for the 1 + 2 and 2 + 3 jointed models, respectively.

When comparing the elastic regions of the stress–strain curves of the intact BBM (Figure 6) to the jointed BBMs (Figures 9 and 10), it is noted that the slope of this linear region of the stress–axial strain curves (i.e., Young's modulus) decreases as a function of increasing jointing. This same trend was noted for the laboratory specimens as well (see Figure 2). The pre-existing joints in the jointed BBMs decrease the stiffness of the overall model. This effect becomes less extreme with increased confinement, which increases the normal stress on the joints and inhibits slip, which thus increases the model's overall stiffness. The BBMs were able to match this decrease in Young's modulus, as well as the slight increase in Poisson's ratio as a function of increased jointing (see Figure 11). Note that the results of both joint geometries (see Figure 4) are included in Figure 11.

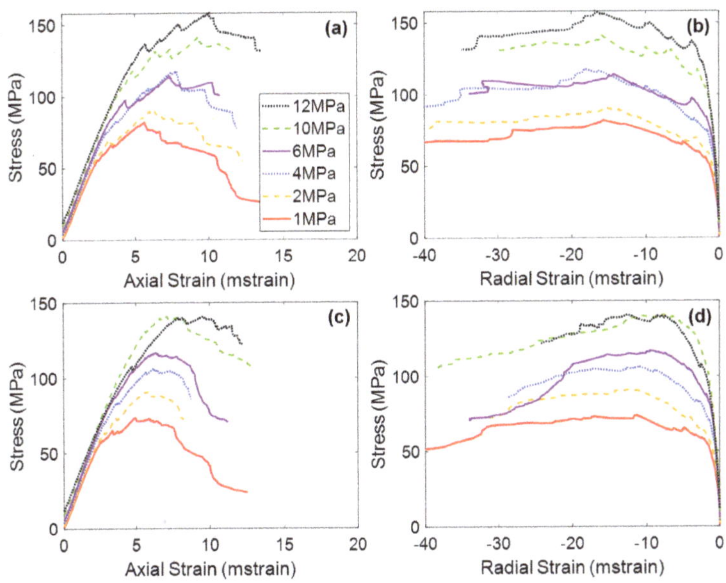

Figure 9. Axial stress–axial strain (**a**,**c**) and axial stress–radial strain (**b**,**d**) curves of the 1 + 2 elastic block BBM (**a**,**b**) and the inelastic block BBM (**c**,**d**). All curves correspond to the 1 + 2 jointed model geometry shown in Figure 4a. Strain is plotted in milli-strain (mstrain).

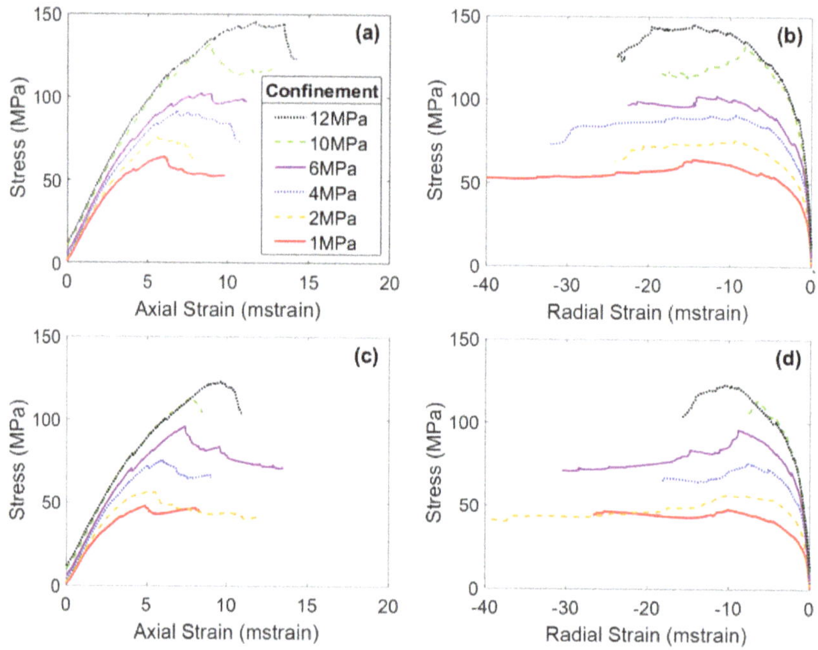

Figure 10. Axial stress–axial strain (**a**,**c**) and axial stress–radial strain (**b**,**d**) curves of the 2 + 3 elastic block BBM (**a**,**b**) and the inelastic block BBM (**c**,**d**). The curves on panels a and b correspond to the 2 + 3 jointed model shown in Figure 4b. The curves on panels c and d correspond to the 2 + 3 jointed model shown in Figure 4d. Strain is plotted in milli-strain (mstrain).

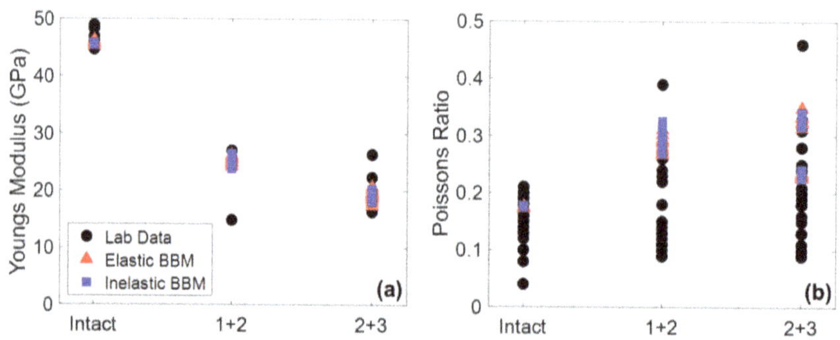

Figure 11. (**a**) Young's modulus and (**b**) Poisson's ratio of the elastic (red triangle) and inelastic (blue square) BBMs compared to laboratory data (black circles). The results intact, 1 + 2, and 2 + 3 series are included as well as both joint geometries (see Figure 4). Note that the laboratory Young's modulus data only includes data under 12 MPa of confinement.

Both the elastic and inelastic block BBMs were able to predict the Young's modulus and Poisson's ratio of Blanco Mera granite equally well. The models with joint geometries shown in Figure 4c,d were found to have higher values of Poisson's ratio than their geometric counterparts shown in Figure 4a,b. Although this difference in Poisson's ratio was small for the 1 + 2 models, this difference is more noticeable for the 2 + 3 models, where the two groups in data points on Figure 11b correspond to the two different geometry types. However, the Poisson's ratio values of both model joint geometries fall within the range of values of the laboratory specimens.

Peak strength, CI, CD, and peak dilation angle values of models of both joint geometries are shown in Figures 12 and 13 for the 1 + 2 and 2 + 3 joint cases, respectively. The Hoek–Brown best fit parameters corresponding to the curves plotted on Figures 12 and 13 are included in Tables A5 and A6 in Appendix A, respectively. No trend was observed regarding the effect of the jointed model geometry (see Figure 4) on the mechanical attributes plotted in Figures 12 and 13. As such, data points of both model geometries are included in Figures 12 and 13 and the Hoek–Brown best fit lines were created using all of the data.

Per Figures 12 and 13, both the elastic and inelastic block BBMs were able to reasonably predict the behavior of the jointed laboratory specimens in terms of strength, cracking thresholds, and dilatancy. Although the exact values of each material attribute were not able to be predicted, the BBMs exhibited similar trends in values as a function of confinement (the Hoek–Brown fits). One exception is the 1 + 2 jointed specimens under higher confinement levels (10 MPa of confinement and higher). Both BBMs underpredicted the trend (i.e., the Hoek–Brown fit) of peak strength and CD by as much as 20% in the case of the elastic block BBM and 27% in the case of the inelastic block BBM.

Regarding the post-peak behavior, the inelastic block BBM was able to better replicate the increase in apparent post-peak ductility as a function of increased jointing (see Figures 6c, 9c and 10c for BBM curves and Figure 2a,c,e for the corresponding laboratory curves). Due to the excessive strain hardening of the elastic block BBM, this increase in apparent ductility as a function of increased jointing is not clear (see Figures 6a, 9a and 10a). As discussed in Section 5, inelastic blocks are required to match the post-peak behavior of brittle rock. Per Figure 7, inelastic blocks were also required to match the post-peak behavior of the jointed laboratory specimens. The stress–strain curves of the inelastic block BBM exhibit less strain hardening, which better matches laboratory data (see Figure 2).

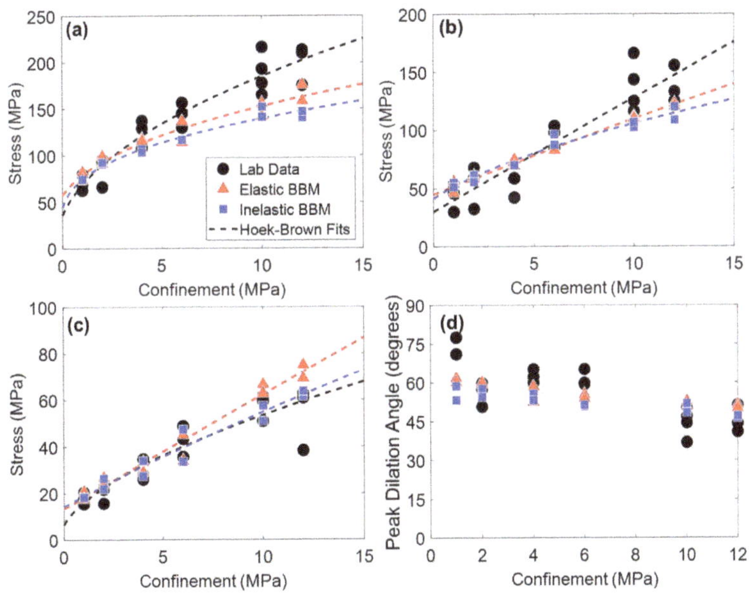

Figure 12. (**a**) Peak strength, (**b**) CD, (**c**) CI, and (**d**) peak dilation angle of laboratory data (black circles), the elastic block BBMs (red triangles), and the inelastic block BBMs (blue squares) of Blanco Mera granite containing the 1 + 2 joint pattern (see Figure 4). Hoek–Brown fits are included where applicable. The two BBM data points per confinement level correspond to the two joint orientations per Figure 4. Note that peak dilation angle data were not reported in [54] but were manually interpreted from the strain data provided.

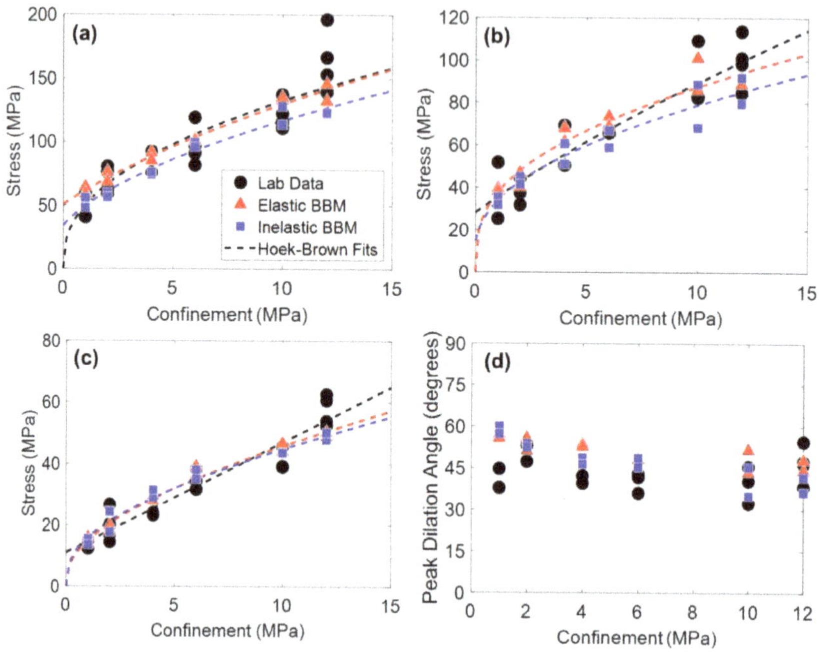

Figure 13. (**a**) Peak strength, (**b**) CD, (**c**) CI, and (**d**) peak dilation angle of laboratory data (black circles), the elastic block BBM (red triangles), and the inelastic block BBM (blue squares) of Blanco Mera granite containing the 2 + 3 joint pattern (see Figure 4). Hoek–Brown fits are included where applicable. The two BBM data points per confinement level correspond to the two joint orientations per Figure 4. Note that peak dilation angle data were not reported in [54] but were manually interpreted from the strain data provided.

Although the inelastic block BBM better predicts the post-peak behavior of Blanco Mera granite, elastic block BBMs require notably less run-time, which makes the elastic block BBM the most favorable for predictive purposes where accurately representing post-peak behavior is not necessary.

6.2. Damage Mechanisms

The presence of the pre-existing joints influenced the failure mechanisms of the laboratory specimens, where the rock failed via a mixed mode of intact brittle material failure and slip along the joints [24]. The intact brittle material fails via the same fracturing processes described in Section 5: material fails via axial splitting induced by the formation of tensile fracturing when the confinement level is low, and material fails via shear banding induced by the formation of shear fractures when the confinement level is high. However, the overall jointed specimens fail by both intact material failure as well as slip along the joints. The BBMs were able to replicate both the intact material failure, as well as slip along the pre-existing joints, as shown in Figures 14 and 15 for the 1 + 2 and 2 + 3 jointed models, respectively.

Figure 14 shows the 1 + 2 jointed models failing largely through intact material failure, where the low confinement models exhibited axially oriented tensile fracturing and the high confinement models exhibited the same en echelon shear fracturing as noted in the intact BBMs (Figure 8 in Section 5). Also like the intact BBMs, the inelastic 1 + 2 jointed BBMs (bottom panels in Figure 14) exhibited zone yield in lieu of some of the contact fracturing seen in the elastic BBMs. The pervasive failure pathway in the inelastic models is composed of both fractures and zone yield.

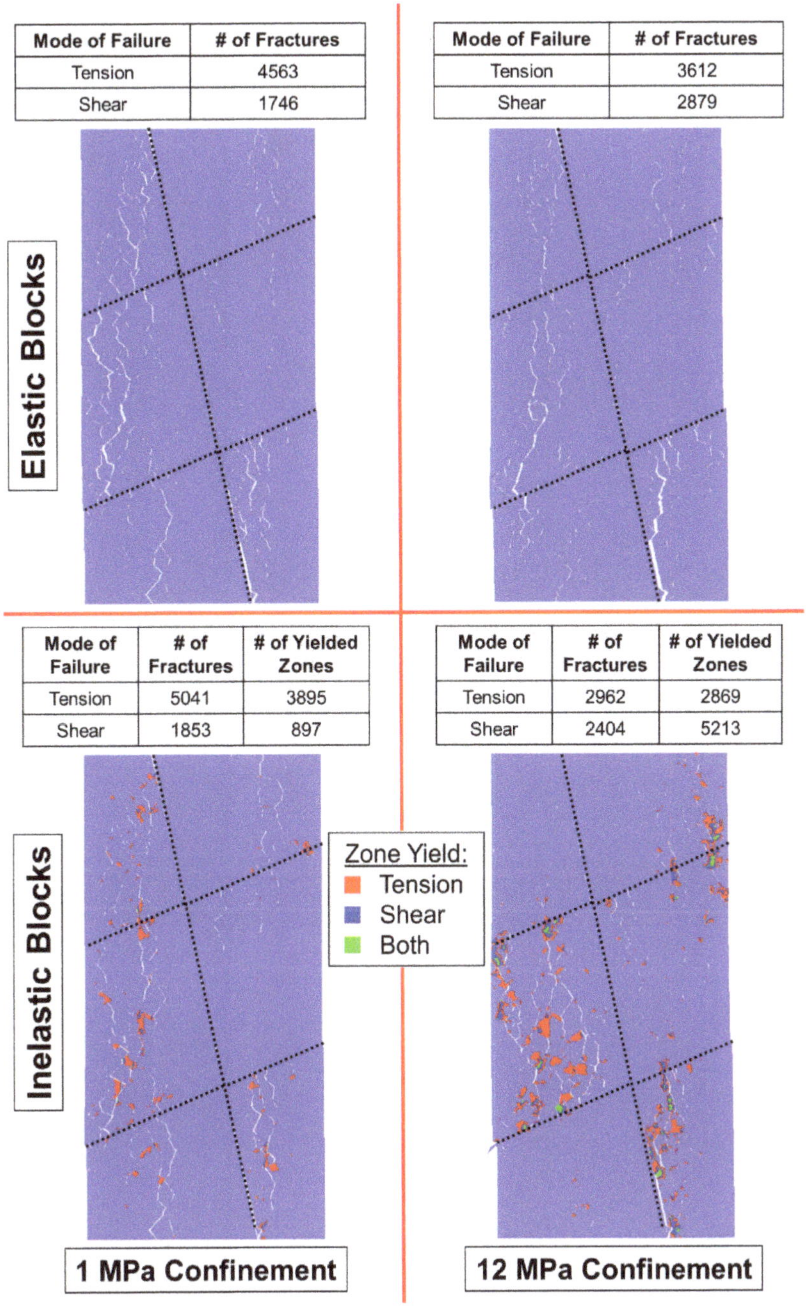

Figure 14. **The** 1 + 2 jointed BBMs at failure (peak strength) showing the location of macroscopic fractures with visible separation: (**top left**) is the elastic block BBM under unconfined conditions, (**top right**) is the elastic block BBM under 12 MPa of confinement, (**bottom left**) is the inelastic block BBM under unconfined conditions, and the (**bottom right**) is the inelastic block BBM under 12 MPa of confinement. Inelastic models indicate the location of zones that have yielded.

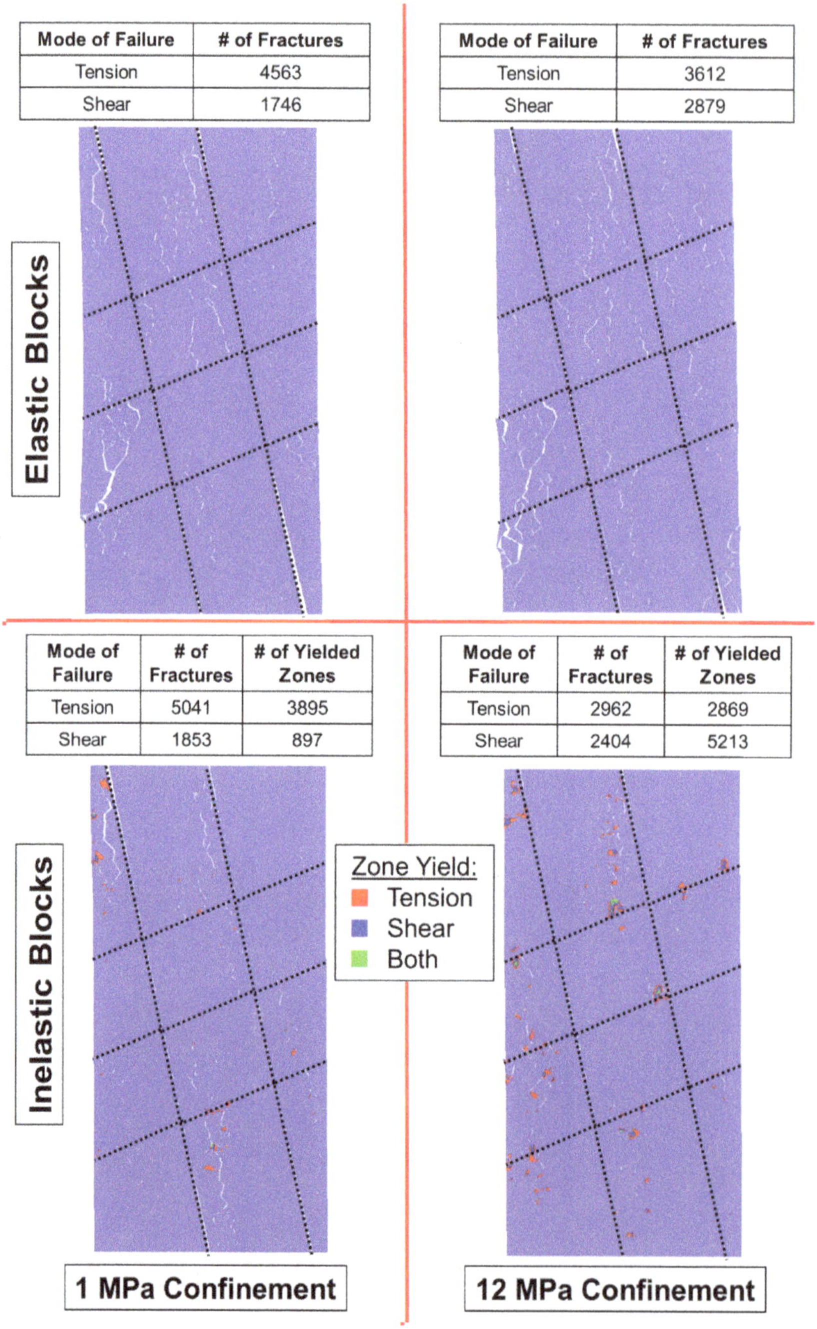

Figure 15. **The** 2 + 3 jointed BBMs at failure (peak strength) showing the location of macroscopic fractures with visible separation: (**top left**) is the elastic block BBM under unconfined conditions, (**top right**) is the elastic block BBM under 12 MPa of confinement, (**bottom left**) is the inelastic block BBM under unconfined conditions, and the (**bottom right**) is the inelastic block BBM under 12 MPa of confinement. Inelastic models indicate the location of zones that have yielded.

Different trends are observed in Figure 15, which shows the 2 + 3 jointed models failing less through the formation of axial fracturing and more en echelon shearing than the intact and 1 + 2 models (Figures 8 and 14, respectively), even under the low confinement level. This decrease in brittle failure as a function of increased jointing has been noted previously by [12] in the context of SRM and can also be observed from the Blanco Mera granite laboratory results from [24]. The 2 + 3 jointed models also do not exhibit large, pervasive failure pathways (formed by either fractures in the case of the elastic block models or a mixture of fractures and zone yield in the case of the inelastic block models). This indicates that the dominant mode of failure in these models is slip along the persistent joints. Therefore, slip along the persistent joints becomes a more important component of specimen failure as the degree of jointing increases.

7. Discussion and Implications for Future Work

The results of this study have quantitatively evaluated the concept of Synthetic Rockmass Modeling (SRM), where a DFN can be added to a calibrated discrete model of intact rock to simulate rockmass behavior. Ultimately, we found that a well-calibrated BBM can be used to predict the behavior of rock under jointing conditions that are different from those for which the BBM was calibrated (i.e., intact rock). This BBM of Blanco Mera granite was calibrated to intact laboratory data and its predictive capabilities were verified by adding smooth, persistent joints to the model and comparing the behavior of the jointed models to respective jointed laboratory specimen data. The BBM was able to replicate the behavior of the rock under these new conditions (i.e., with smooth, persistent joints) because the intact model was calibrated to reproduce the micromechanical failure mechanisms of the real rock. As such, we expect that joints of other geometries (i.e., density, orientation) and mechanical properties (i.e., persistence, roughness, infilling) could also be added to the BBM and the model would still approximate the behavior of a real specimen with such a joint network.

The intact BBM developed in this study is relatively unique, as it is uncommon for models of intact rock to be calibrated to a large range of mechanical attributes (elastic moduli, CI, CD, peak strength, dilatancy, and brittleness) over a wide range of confining stresses [13,25–29]. However, this rigorous calibration is required to ensure that the micromechanical failure mechanisms of the model approximate those of the real rock [25–29,44,50]. Many of the previous SRMs developed after the jointed Blanco Mera granite specimens (see Table 1) were unable to capture some of the failure mechanisms seen in the jointed Blanco Mera granite laboratory specimens [17–19,31,33], in contrast to the more extensively calibrated models developed in this study. For example, in [18,31], the models were not calibrated to match CI and CD thresholds. The results show that, in fact, CD was nearly coincident with peak strength, and we interpret this to have led to a greater confinement dependency of strength than was exhibited by the actual specimens; because of this, peak strength was overestimated under higher confinement levels (6 MPa and above).

Another example is the discrete model developed by [33], which failed predominantly through displacement along the pre-existing joints, rather than through intact material damage, as was observed in the jointed specimens of Blano Mera granite [24]. The strength of the intact model was calibrated to be too high compared to the strength of the pre-existing joints, which resulted in incorrect failure mechanisms of the jointed models.

As noted in Section 3.5, it is not possible to verify the predictive capability of these previously developed models of jointed Blanco Mera granite because their intact and jointed models were calibrated simultaneously. Validation is only possible when the models can be calibrated to an intact data set and then validated using a second data set corresponding to a different condition (i.e., the jointed specimen data). Looking beyond studies of Blanco Mera granite, it is common practice in rock mechanics research to calibrate numerical models to a complete laboratory data set without a separate validation [14,36,79–82]. Such an approach does not provide any evaluation of the model using data that were not considered as part of the calibration.

There are cases (particularly field-scale scenarios) where a second data set is not always available to verify a model's predictive capability. In such cases, when utilizing an SRM approach, it is recommended that the model for intact rock be calibrated to a comprehensive set of mechanical attributes, which we interpret to be critical to ensuring the predictive capabilities demonstrated in this study.

Note that this study has only considered one relatively brittle rock with negligible porosity (Blanco Mera granite). The findings of this study will not necessarily apply to more ductile rocks or rocks with non-negligible porosity. Additionally, the numerical models created in this study are two-dimensional, which neglects out-of-plane stress and strains. Further advances in computational capabilities and parallelization of modeling software will make it increasingly feasible to conduct similar studies in three dimensions in the future.

8. Conclusions

The idea that a mechanistically accurate numerical model will behave realistically in other scenarios proved to be true for the BBMs created in this study. The addition of pre-existing smooth joints to the intact BBM (with no further calibration) resulted in model behavior that predicted the behavior of the jointed laboratory specimens after which they were modeled without further calibration. Although the BBMs were not able to predict exact values of material attributes, with the exception of a small (i.e., 27% or less) underprediction of peak strength and CD in the case of the higher confinement level values for the 1 + 2 jointing case, the predicted mechanical attribute values were typically within 10% of the model trend derived from the laboratory data.

Two types of BBMs were used in this study: one with blocks modeled via an elastic constitutive model and one with blocks modeled via an inelastic constitutive model. Both BBMs were able to reasonably predict pre-peak attributes and peak strength of the jointed models. However, the pre-peak hardening and post-peak stress–strain behavior were only able to be replicated using the inelastic block BBM. Although the dominant failure mechanism of rock is the formation of fractures along grain boundaries, as significant damage accumulates in rock (particularly after peak strength) intragranular damage becomes increasingly common. This intragranular damage is represented in BBMs by block yield, which is unique to inelastic blocks.

Although the inelastic block BBM was better able to predict the pre-peak hardening and post-peak mechanical behavior of jointed Blanco Mera granite, the elastic and inelastic block BBMs were found to equally predict pre-peak attributes and peak strength of the jointed specimens. Elastic block BBMs are computationally less expensive than inelastic block BBMs, so this is the preferred model type for prediction purposes in cases where accurately simulating post-peak behavior is not critical.

Overall, this study has verified the concept of SRM, where the BBMs were able to replicate the behavior of the rock under new conditions (i.e., with smooth, persistent joints). This finding suggests that joints of other geometries and mechanical properties can also be added to the BBM and the model will approximately predict the mechanical behavior of such a specimen. Since creating rockmass analog laboratory specimens with heavily dense, impersistent, rough, and infilled joints is difficult (if not impossible), the use of numerical models to simulate varying rockmass conditions is a powerful tool for future research into rockmass mechanical behavior.

Author Contributions: Conceptualization, I.W. and G.W.; methodology, I.W. and G.W.; software, I.W. and S.S.; validation, I.W. and G.W.; formal analysis, I.W.; investigation, I.W.; resources, G.W.; data curation, I.W.; writing—original draft preparation, I.W.; writing—review and editing, G.W. and S.S.; visualization, I.W.; supervision, G.W. and S.S.; funding acquisition, G.W. All authors have read and agreed to the published version of the manuscript.

Funding: The research conducted for this study was partially funded by the National Institute of Occupational Safety and Health (NIOSH), Grant Number 200-2016-90154.

Institutional Review Board Statement: Not applicable.

Informed Consent Statement: Not applicable.

Data Availability Statement: For access to the data used in this study, please email the corresponding author, G. Walton at gwalton@mines.edu.

Acknowledgments: The authors would like to thank the Itasca Consulting Group for providing educational licenses to UDEC, which were used to run some of the models for this study, as well as Leandro Alejano and Manuel Gonzalez-Fernandez from the University of Vigo, who provided part of the data and materials used for this study.

Conflicts of Interest: Author I.W. was employed by WSP USA Inc. and author S.S. was employed by Equilibrium Mining while part of this study was conducted. All authors declare that the research was conducted in the absence of any commercial or financial relationships that could be construed as a potential conflict of interest.

Appendix A

Table A1. Available data on average grain diameters or ranges of diameters for each mineral present in Blanco Mera granite [63] and their proportion in the rock [55]. Note: chlorite exists as twinning within biotite. Table from: [37].

Mineral	Grain Diameter Data	Proportion Percentage
Quartz	1–6 mm	20%
Plagioclase	<6 mm	35%
Alkaline Feldspar	~3.0 mm	27%
Muscovite	~2.5 mm	7%
Biotite w/Chlorite	~1.5 mm	9%
Sericite	<0.2 mm	1%
Accessory Minerals	<0.1 mm	1%

Table A2. Mechanical properties of Blanco Mera granite from [24]. Density and tensile strength (σ_T) were reported in [55] and the Crack Initiation (CI) and Crack Damage (CD) fit parameters as well as Walton–Diederichs (WD) Dilation Parameters [83] were reported in [54]. Note: this value of Young's Modulus is an averaged value for rock specimens tested at confinements of 12 MPa, as values for Young's Modulus at lower confinements were interpreted to be artificially reduced by pre-existing flaws within the specimens, induced during specimen extraction [74,84].

	Property	Average Value
Density	Density (g/cm^3)	2.6
Elastic Properties	E (GPa)	47
	ν	0.17
Tensile Strength	σ_T (MPa)	6.12
Hoek–Brown Strength Parameters	σ_{ci} (MPa)	123.4
	m_i	41.7
	GSI (peak)	100
	GSI (residual)	41.7
Mohr–Coulomb Strength Parameters	UCS (MPa)	132.9
	φ_{peak} (degrees)	59.5
	c_{peak} (MPa)	18.1
	Unconfined Residual Strength (MPa)	15.8
	$\varphi_{residual}$ (degrees)	50.0
	$c_{residual}$ (MPa)	2.9

Table A2. *Cont.*

	Property	Average Value
Hoek–Brown Crack Damage (CD) Parameters	σ_{ci} (MPa)	77.1
	m_i	18.6
	GSI	100
Hoek–Brown Crack Initiation (CI) Parameters	σ_{ci} (MPa)	36.7
	m_i	19.6
	GSI	100
Walton–Diederichs Dilation Model Parameters	α	0–0.1
	γ_m	1–4
	β_0	1.0
	β'	0.1
	$\gamma_1{}'$	−12.2
	$\gamma_2{}'$	51.6

Table A3. Summary of [76] which investigated the effects of inelastic block input parameters on emergent BBM behavior (an increase or decrease in value per the up or down arrows, respectively). Input parameters not listed did not have notable or clear changes in values.

	Effect on Material Attribute (As a Function of Increasing the Block Input Value)			
Inelastic Block Property	**Unconfined Strength**	**Inelastic Block Property**	**Unconfined Strength**	**Inelastic Block Property**
Peak Cohesion	↑	↑	-	-
Peak Friction Angle	-	↑	-	-
Peak Tensile Strength	↓	↓	-	-
Critical Plastic Shear Strain	-	↑	-	-
Residual Cohesion	-	-	↓	↓
Residual Tensile Strength	-	-	↑	-

Table A4. Hoek–Brown fit parameters for intact Peak Strength, Crack Damage (CD), and Crack Initiation (CI) parameters for fits shown in Figure 7.

	LAB		
	Strength	**CD**	**CI**
σ_{ci}	132.2	82.3	38.9
s	1	1	1
m_i	36.5	16.1	17.3
a	0.5	0.5	0.5
	ELASTIC BBM		
	Strength	**CD**	**CI**
σ_{ci}	116.4	72.2	45.0
s	1	1	1
m_i	46.9	20.8	16.2
a	0.5	0.5	0.5
	INELASTIC BBM		
	Strength	**CD**	**CI**
σ_{ci}	167.3	69.3	40.3
s	1	1	1
m_i	24.4	24.1	15.4
a	0.5	0.5	0.5

Table A5. Hoek–Brown fit parameters for 1 + 2 Peak Strength, Crack Damage (CD), and Crack Initiation (CI) parameters for fits shown in Figure 12.

	LAB		
	Strength	**CD**	**CI**
σ_{ci}	132.2	82.3	38.9
s	1	1	1
m_i	36.5	16.1	17.3
a	0.5	0.5	0.5
	ELASTIC BBM		
	Strength	**CD**	**CI**
σ_{ci}	116.4	72.2	45.0
s	0.1	0.6	0.3
m_i	20.8	6.5	3.9
a	0.3	0.8	1.0
	INELASTIC BBM		
	Strength	**CD**	**CI**
σ_{ci}	167.3	69.3	40.3
s	0.0	0.3	0.3
m_i	6.2	12.5	3.5
a	0.3	0.4	0.8

Table A6. Hoek–Brown fit parameters for 2 + 3 Peak Strength, Crack Damage (CD), and Crack Initiation (CI) parameters for fits shown in Figure 13.

	LAB		
	Strength	**CD**	**CI**
σ_{ci}	132.2	82.3	38.9
s	0.0	0.2	0.3
m_i	10.9	6.1	2.6
a	0.4	0.7	1.0
	ELASTIC BBM		
	Strength	**CD**	**CI**
σ_{ci}	116.4	72.2	45.0
s	0.0	0.0	0.0
m_i	10.9	9.0	2.6
a	0.4	0.3	0.4
	INELASTIC BBM		
	Strength	**CD**	**CI**
σ_{ci}	167.3	69.3	40.3
s	0.0	0.0	0.0
m_i	5.5	6.8	2.7
a	0.4	0.3	0.4

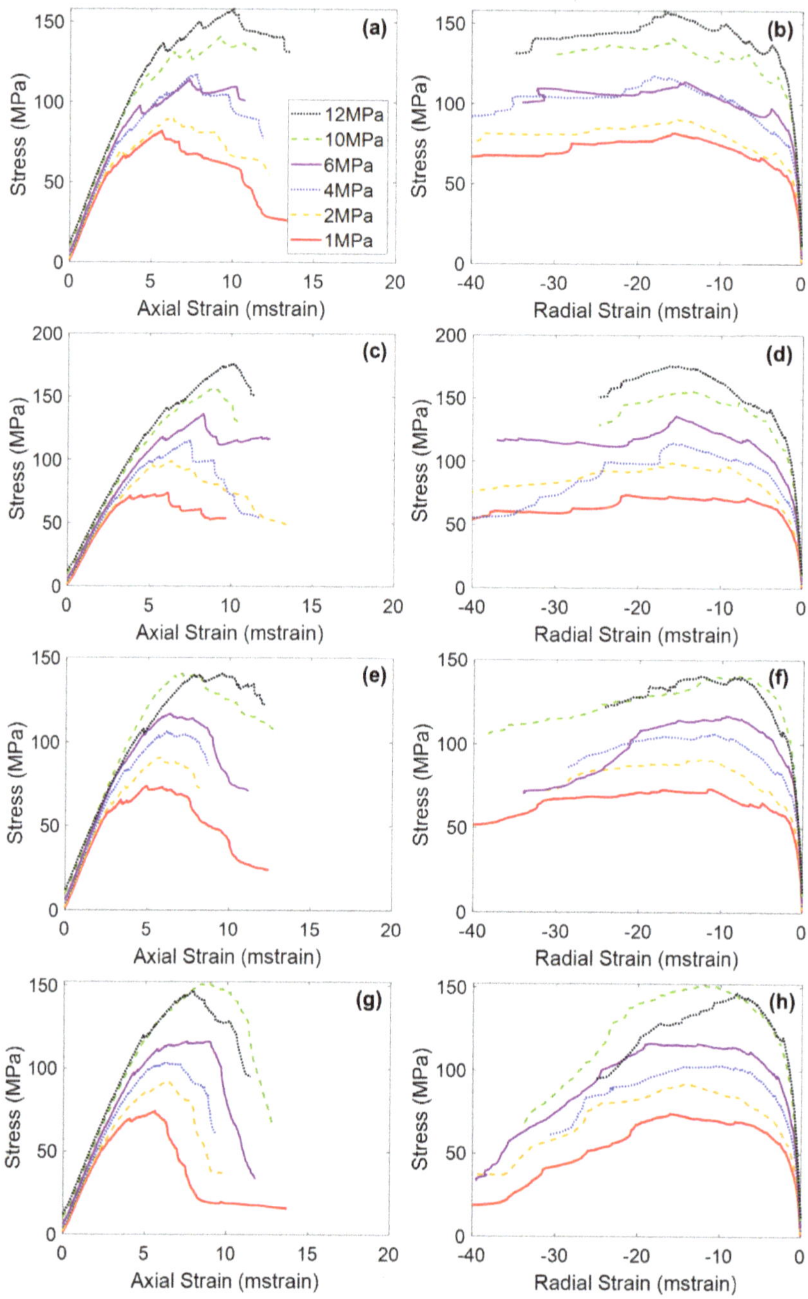

Figure A1. Axial stress–axial strain (**a**,**c**,**e**,**g**) and axial stress-radial strain (**b**,**d**,**f**,**h**) curves of the 1 + 2 elastic block BBM (**a**–**d**) and the inelastic block BBM (**e**–**h**). Strain is plotted in milli-strain (mstrain): (**a**–**d**) correspond to the model shown in Figure 4a (**e**–**h**) correspond to the model shown in Figure 4c.

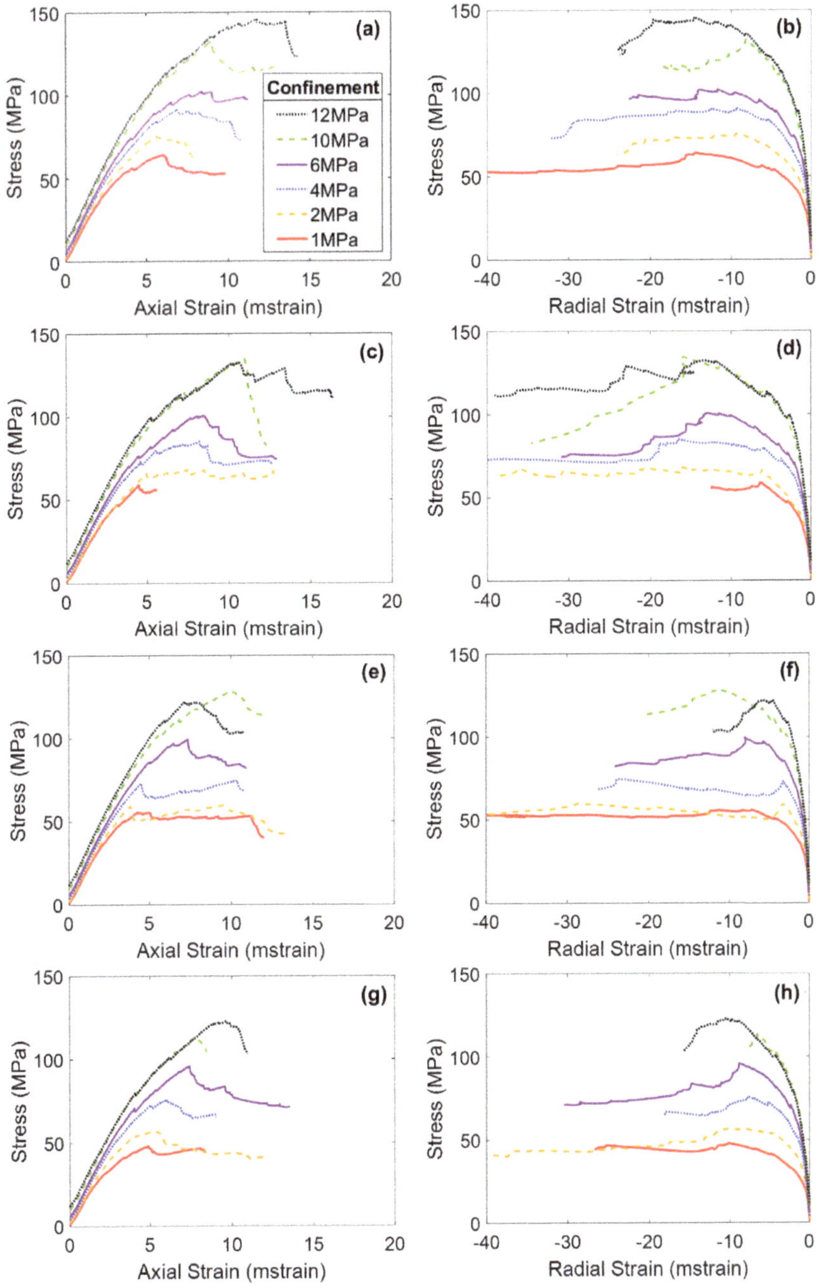

Figure A2. Axial stress–axial strain (**a**,**c**,**e**,**g**) and axial stress-radial strain (**b**,**d**,**f**,**h**) curves of the 2 + 3 elastic block BBM (**a**–**d**) and the inelastic block BBM (**e**–**h**). Strain is plotted in milli-strain (mstrain): (**a**–**d**) correspond to the model shown in Figure 4b (**e**–**h**) correspond to the model shown in Figure 4d.

References

1. Hoek, E.; Brown, E.T. Practical estimates of rock mass strength. *Int. J. Rock Mech. Min. Sci.* **1997**, *34*, 1165–1186. [CrossRef]
2. Natau, O.P.; Frohlich, B.O.; Muschler, T.O. Recent Developments of the Large-Scale Triaxial Test. In Proceedings of the ISRM International Symposium, Melbourne, Australia, 10–15 April 1983.
3. Muschler, T.O.; Natau, O.P. Further developments for the determination of the stress-strain behaviour of jointed rock mass by large scale tests. In Proceedings of the ISRM International Symposium, Aachen, Germany, 16–20 September 1991.
4. Singh, M.M.; Huck, P.J. Large Scale Triaxial Tests on Rock. In Proceedings of the 14th US Symposium on Rock Mechanics, University Park, PA, USA, 11–14 June 1972.
5. Vergara, M.R.; Kudella, P.; Triantafyllidis, T. Large Scale Tests on Jointed and Bedded Rocks Under Multi-Stage Triaxial Compression and Direct Shear. *Rock Mech. Rock Eng.* **2015**, *48*, 75–92. [CrossRef]
6. Duan, G.; Li, J.; Zhang, J.; Assefa, E.; Sun, X. Mechanical Properties and Failure Modes of Rock Specimens with Specific Joint Geometries in Triaxial Unloading Compressive Test. *Adv. Mater. Sci. Eng.* **2019**, *2019*, 1340934. [CrossRef]
7. Kulatilake, P.H.S.W.; He, W.; Um, J.; Wang, H. A Physical Model Study of Jointed Rock Mass Strength Under Uniaxial Compressive Loading. *Int. J. Rock Mech. Min. Sci.* **1997**, *34*, 3–4. [CrossRef]
8. Prudencio, M.; Van Sint Jan, M. Strength and failure modes of rock mass models with non-persistent joints. *Int. J. Rock Mech. Min. Sci.* **2007**, *44*, 890–902. [CrossRef]
9. Ramamurthy, T.; Arora, V.K. Strength predictions for jointed rocks in confined and unconfined states. *Int. J. Rock Mech. Min. Sci.* **1994**, *31*, 9–22. [CrossRef]
10. Shaunik, D.; Singh, M. Strength behaviour of a model rock intersected by non-persistent joint. *J. Rock Mech. Geotech. Eng.* **2019**, *11*, 1243–1255. [CrossRef]
11. Arzua, J.; Alejano, L.R.; Walton, G. Strength and dilation of jointed granite specimens in servo-controlled triaxial tests. *Int. J. Rock Mech. Min. Sci.* **2014**, *69*, 93–104. [CrossRef]
12. Mas Ivars, D.; Pierce, M.E.; Darcel, C. The synthetic rock mass approach for jointed rock mass modelling. *Int. J. Rock Mech. Min. Sci.* **2011**, *48*, 219–244. [CrossRef]
13. Wang, X.; Cai, M. A DFN—DEM Multi-scale Modeling Approach for Simulating Tunnel Excavation Response in Jointed Rock Masses. *Rock Mech. Rock Eng.* **2020**, *53*, 1053–1077. [CrossRef]
14. Sainsbury, B.A.; Pierce, M.; Mas Ivars, D. Analysis of Caving Behaviour Using a Synthetic Rock Mass—Ubiquitous Joint Rock Mass Modelling Technique. In Proceedings of the First Southern Hemisphere International Rock Mechanics Symposium, Perth, Australia, 16–19 September 2008.
15. Elmo, D.; Moffitt, K.; Carvalho, J. Synthetic rock mass modelling: Experience gained and lessons learned. In Proceedings of the 50th US Rock Mechanics/Geomechanics Symposium, Houston, TX, USA, 26–29 June 2016.
16. Pierce, M.E.; Cundall, P.A.; Potyondy, D.O.; Mas Ivars, D. A synthetic rock mass model for jointed rock. In Proceedings of the 1st Canada-US Rock Mechanics/Geomechanics Symposium—Rock Mechanics: Meeting Society's Challenges and Demands, Vancouver, CO, Canada, 27–31 May 2007.
17. Bastola, S.; Cai, M. Investigation of mechanical properties of jointed granite under compression using lattice-spring-based synthetic rock mass modeling approach. *Int. J. Rock Mech. Min. Sci.* **2020**, *126*, 104191. [CrossRef]
18. Shen, J.; Shu, Z.; Cai, M.; Du, S. A shear strength model for anisotropic blocky rock masses with persistent joints. *Int. J. Rock Mech. Min. Sci.* **2020**, *134*, 104430. [CrossRef]
19. Castro-Filgueira, U.; Alejano, L.R.; Ivars, D.M. Particle flow code simulation of intact and fissured granitic rock samples. *J. Rock Mech. Geotech. Eng.* **2020**, *12*, 960–974. [CrossRef]
20. Suner, M.C.; Tulu, I.B. Examining the Effect of Natural Fractures on Stone Mine Pillar Strength Through Synthetic Rock Mass Approach. *Min. Met. Explor.* **2022**, *39*, 1863–1871. [CrossRef]
21. Vallejos, J.A.; Brzovic, A.; Lopez, C.; Bouzeran, L.; Mas Ivars, D. Application of the synthetic rock mass approach to characterize rock mass behavior at the El Teniente Mine, Chile. In Proceedings of the 3rd International FLAC/DEM Symposium, Hangzhou, China, 24 October 2013.
22. Pierce, M.E.; Mas Ivars, D.; Sainsbury, D. Use of Synthetic Rock Masses (SRM) to Investigate Jointed Rock Mass Strength and Deformation Behavior. In Proceedings of the International Conference on Rock Joints and Jointed Rock Masses, Tucson, AZ, USA, 7–8 January 2009.
23. Esmaieli, K.; Hadjigeorgiou, J.; Grenon, M. Estimating geometrical and mechanical REV based on synthetic rock mass models at Brunswick Mine. *Int. J. Rock Mech. Min. Sci.* **2010**, *47*, 915–926. [CrossRef]
24. Alejano, L.R.; Arzua, J.; Bozorgzadeh, N.; Harrison, J.P. Triaxial strength and deformability of intact and increasingly jointed granite samples. *Int. J. Rock Mech. Min. Sci.* **2017**, *95*, 87–103. [CrossRef]
25. Farahmand, K.; Diederichs, M.S. A calibrated synthetic rock mass (SRM) model for simulating crack growth in granitic rock considering grain scale heterogeneity of polycrystalline rock. In Proceedings of the 49th US Rock Mechanics/Geomechanics Symposium, San Francisco, CA, USA, 29 June–1 July 2015.
26. Li, X.F.; Li, H.B.; Zhao, J. The role of transgranular capability in grain-based modelling of crystalline rocks. *Comput. Geotech.* **2019**, *110*, 161–183. [CrossRef]
27. Park, J.W.; Park, C.; Song, J.W.; Park, E.S.; Song, J.J. Polygonal grain-based distinct element modeling for mechanical behavior of brittle rock. *Int. J. Numer. Anal. Methods Geomech.* **2017**, *41*, 880–898. [CrossRef]

28. Sinha, S.; Walton, G. A study on Bonded Block Model (BBM) complexity for simulation of laboratory-scale stress-strain be-havior in granitic rocks. *Comput. Geotech.* **2020**, *118*, 103363. [CrossRef]
29. Wang, X.; Cai, M. A comprehensive parametric study of grain-based models for rock failure process simulation. *Int. J. Rock Mech. Min. Sci.* **2019**, *115*, 60–76. [CrossRef]
30. Chen, W.; Konietzky, H. Simulation of heterogeneity, creep, damage and lifetime for loaded brittle rocks. *Tectonophysics* **2014**, *633*, 164–175. [CrossRef]
31. Huang, F.; Shen, J.; Cai, M.; Xu, C. An Empirical UCS Model for Anisotropic Blocky Rock Masses. *Rock Mech. Rock Eng.* **2019**, *52*, 3119–3131. [CrossRef]
32. Gao, G.; Meguid, M.A. Microscale Characterization of Fracture Growth in Increasingly Jointed Rock Samples. *Rock Mech. Rock Eng.* **2022**, *55*, 6033–6061. [CrossRef]
33. Gonzalez-Molano, N.A.; Alvarellos, J.; Lakshmikantha, M.R.; Arzua, J.; Alejano, L.R. Numerical and experimental characterization of mechanical behaviour of an artificially jointed rock. In Proceedings of the ISRM International Symposium, Trondheim, Norway, 14–19 June 2020.
34. Stacey, T.R.; Wesseloo, J. Design and Prediction in Rock Engineering: The Importance of Mechanisms of Failure, with Focus on High Stress, Brittle Rock Conditions. *Rock Mech. Rock Eng.* **2022**, *55*, 1517–1535. [CrossRef]
35. Zhang, X.P.; Ji, P.Q.; Peng, J.; Wu, S.C.; Zhang, Q. A grain-based model considering pre-existing cracks for modelling mechanical properties of crystalline rock. *Comput. Geotech.* **2020**, *127*, 103776. [CrossRef]
36. Fan, X.; Kulatilake, P.H.; Chen, X. Mechanical behavior of rock-like jointed blocks with multi-non-persistent joints under uniaxial loading: A particle mechanics approach. *Eng. Geol.* **2015**, *190*, 17–32. [CrossRef]
37. West, I.G.; Walton, G.; Sinha, S. Simulating the Behavior of Compressively Loaded Blanco Mera Granite Using Bonded Block Models. In Proceedings of the 54th US Rock Mechanics/Geomechanics Symposium, Golden, CO, USA, 28 June–1 July 2020.
38. Turichshev, A.; Hadjigeorgiou, J. Development of Synthetic Rock Mass Bonded Block Models to Simulate the Behaviour of Intact Veined Rock. *Geotech. Geol. Eng.* **2017**, *35*, 313–335. [CrossRef]
39. Garza-Cruz, T.; Pierce, M.E.; Board, M. Effect of Shear Stresses on Pillar Stability: A Back Analysis of the Troy Mine Experience to Predict Pillar Performance at Montanore Mine. *Rock Mech. Rock Eng.* **2019**, *52*, 4979–4996. [CrossRef]
40. Garza-Cruz, T.; Pierce, M.E.; Kaiser, P.K. Use of 3DEC to study spalling and deformation associated with tunnelling at depth. In Proceedings of the Seventh International Conference on Deep and High Stress Mining, Sudbury, ON, Canada, 16–18 September 2014.
41. Contreras Inga, C.E.; Walton, G.; Holley, E. Statistical assessment of the effects of grain-structure representation and micro-properties on the behavior of bonded block models for brittle rock damage prediction. *Sustainability* **2021**, *13*, 7889. [CrossRef]
42. Lan, H.; Martin, C.D.; Hu, B. Effect of heterogeneity of brittle rock on micromechanical extensile behavior during compression loading. *J. Geophys. Res.* **2010**, *115*. [CrossRef]
43. Li, J.; Konietzky, H.; Fruhwirt, T. Voronoi-Based DEM Simulation Approach for Sandstone Considering Grain Structure and Pore Size. *Rock Mech. Rock Eng.* **2017**, *50*, 2749–2761. [CrossRef]
44. Nicksiar, M.; Martin, C.D. Factors affecting crack initiation in low porosity crystalline rocks. *Rock Mech. Rock Eng.* **2014**, *47*, 1165–1181. [CrossRef]
45. Chen, W.; Konietzky, H.; Tan, X.; Fruhwirt, T. Pre-failure damage analysis for brittle rocks under triaxial compression. *Comput. Geotech.* **2016**, *74*, 45–55. [CrossRef]
46. Sinha, S.; Shirole, D.; Walton, G. Investigation of the Micromechanical Damage Process in a Granitic Rock Using an Inelastic Bonded Block Model (BBM). *J. Geophys. Res. Solid Earth* **2020**, *125*, e2019JB018844. [CrossRef]
47. Itasca Consulting Group Inc. *Universal Distinct Element Code Constitutive Models*, 6th ed.; Itasca Consulting Group Inc.: Minneapolis, MN, USA, 2014.
48. Garza-Cruz, T.; Pierce, M.E. A 3DEC model for heavily veined massive rock masses. In Proceedings of the 48th US Rock Mechanics/Geomechanics Symposium, Minneapolis, MN, USA, 1–4 June 2014; Volume 3, pp. 2074–2082.
49. Ghazvinian, E.; Diederichs, M.S.; Quey, R. 3D random Voronoi grain-based models for simulation of brittle rock damage and fabric-guided micro-fracturing. *J. Rock Mech. Geotech. Eng.* **2014**, *6*, 506–521. [CrossRef]
50. Wang, X.; Cai, M. Modeling of brittle rock failure considering inter- and intra-grain contact failures. *Comput. Geotech.* **2018**, *101*, 224–244. [CrossRef]
51. Xia, Y.; Meng, Q.; Zhang, C. Application of 3D Printing Technology in the Mechanical Testing of Complex Structural Rock Masses. *Geofluids* **2021**, *2021*, 7278131. [CrossRef]
52. Yang, Z.Y.; Chen, J.M.; Huang, T.H. Effect of joint sets on the strength and deformation of rock mass models. *Int. J. Rock Mech. Min. Sci.* **1998**, *35*, 75–84. [CrossRef]
53. Fereshtenejad, S.; Song, J.J. Applicability of powder-based 3D printing technology in shear behavior analysis of rock mass containing non-persistent joints. *J. Struct. Geol.* **2021**, *143*, 104251. [CrossRef]
54. Walton, G.; Alejano, L.R.; Arzua, J.; Markley, T. Crack Damage Parameters and Dilatancy of Artificially Jointed Granite Samples under Triaxial Compression. *Rock Mech. Rock Eng.* **2018**, *51*, 1637–1656. [CrossRef]
55. Arzua, J.; Alejano, L.R. Dilation in granite during servo-controlled triaxial strength tests. *Int. J. Rock Mech. Min. Sci.* **2013**, *61*, 43–56. [CrossRef]
56. Lambe, T.W. Predictions in soil engineering. *Geotechnique* **1973**, *23*, 151–202. [CrossRef]

57. Diederichs, M.S.; Martin, C.D. Measurement of spalling parameters from laboratory testing. In Proceedings of the European Rock Mechanics EUROCK, Lausanne, Switzerland, 15–18 June 2010.
58. Ghazvinian, E.; Perras, M.A.; Diederichs, M.S.; Labrie, D. Formalized approaches to defining damage thresholds in brittle rock: Granite and limestone. In Proceedings of the 46th US Rock Mechanics/Geomechanics Symposium, Chicago, IL, USA, 24–27 June 2012.
59. Martin, C.D.; Chandler, N.A. The progressive fracture of Lac du Bonnet granite. *Int. J. Rock Mech. Min. Sci.* **1994**, *31*, 643–659. [CrossRef]
60. Potyondy, D.O. A flat-jointed bonded-particle material for hard rock. In Proceedings of the 46th US Rock Mechanics/Geomechanics Symposium, Chicago, IL, USA, 24–27 June 2012.
61. Potyondy, D.O.; Cundall, P.A. A bonded-particle model for rock. *Int. J. Rock Mech. Min. Sci.* **2004**, *41*, 1329–1364. [CrossRef]
62. Quey, R.; Dawson, P.R.; Barbe, F. Large-scale 3D random polycrystals for the finite element method: Generation, meshing and remeshing. *Comput. Methods Appl. Mech. Eng.* **2011**, *200*, 1729–1745. [CrossRef]
63. University of Vigo, Natural Resources and Environmental Engineering Department, Vigo, Spain. 2011, *Unpublished Petrology Report*.
64. West, I.; Walton, G. Quantitative Evaluation of the Effects of Input Parameter Heterogeneity on Model Behavior for Bonded Block Models of Laboratory Rock Specimens. *Rock Mech. Rock Eng.* **2023**, *56*, 7129–7146. [CrossRef]
65. Kazerani, T.; Zhao, J. Micromechanical parameters in bonded particle method for modelling of brittle material failure. *Int. J. Numer. Anal. Methods Geomech.* **2010**, *34*, 1877–1895. [CrossRef]
66. Stavrou, A.; Murphy, W. Quantifying the effects of scale and heterogeneity on the confined strength of micro-defected rocks. *Int. J. Rock Mech. Min. Sci.* **2018**, *102*, 131–143. [CrossRef]
67. Hoek, E.; Brown, E.T. Empirical strength criterion for rock masses. *J. Geotech. Eng.* **1980**, *106*, 1013–1035. [CrossRef]
68. Makowski, P.; Ostrowski, Å. The Methodology for the Young Modulus Derivation for Rocks and Its Value. *Procedia Eng.* **2017**, *191*, 134–141. [CrossRef]
69. Gercek, H. Poisson's ratio values for rocks. *Int. J. Rock Mech. Min. Sci.* **2007**, *44*, 1–13. [CrossRef]
70. Clark, M.D.; Day, J.J.; Diederichs, M.S. Assessing the geomechanical behaviours of skarn-related hydrothermal veins in intact laboratory tests. In Proceedings of the 53rd US Rock Mechanics/Geomechanics Symposium, New York, NY, USA, 23–26 June 2019.
71. Ghazvinian, E. Modelling and Testing Strategies for Brittle Fracture Simulation in Crystalline Rock Samples. Ph.D. Thesis, Queen's University, Kingston, ON, Canada, 2010.
72. Zhao, X.G.; Cai, M. A mobilized dilation angle model for rocks. *Int. J. Rock Mech. Min. Sci.* **2010**, *47*, 368–384. [CrossRef]
73. Alejano, L.R.; Alonso, E. Considerations of the dilatancy angle in rocks and rock masses. *Int. J. Rock Mech. Min. Sci.* **2005**, *42*, 481–507. [CrossRef]
74. Walton, G.; Arzua, J.; Alejano, L.R.; Diederichs, M.S. A laboratory-testing-based study on the strength, deformability, and dilatancy of carbonate rocks at low confinement. *Rock Mech. Rock Eng.* **2015**, *48*, 941–958. [CrossRef]
75. Walton, G.; Hedayat, A.; Kim, E.; Labrie, D. Post-yield Strength and Dilatancy Evolution Across the Brittle-Ductile Transition in Indiana Limestone. *Rock Mech. Rock Eng.* **2017**, *50*, 1691–1710. [CrossRef]
76. West, I.; Walton, G. Evaluating the Influence of Parameter Inputs on Macroscopic Behavior of Bonded Block Models with Inelastic Blocks. In Proceedings of the 56th US Rock Mechanics/Geomechanics Symposium, Santa Fe, NM, USA, 26–29 June 2022.
77. Diederichs, M.S. Instability of Hard Rockmasses. Ph.D. Thesis, University of Waterloo, Waterloo, ON, Canada, 1999.
78. Amitrano, D. Rupture by damage accumulation in rocks. *Int. J. Fract.* **2006**, *139*, 369–381. [CrossRef]
79. Gao, F.Q.; Stead, D.O. The application of a modified Voronoi logic to brittle fracture modelling at the laboratory and field scale. *Int. J. Rock Mech. Min. Sci.* **2014**, *68*, 1–14. [CrossRef]
80. Chong, W.L.; Haque, A.; Gamage, R.P.; Shahinuzzaman, A. Modelling of intact and jointed mudstone samples under uniaxial and triaxial compression. *Arab. J. Geosci.* **2013**, *6*, 1639–1646. [CrossRef]
81. Yang, J.P.; Chen, W.Z.; Yang, D.S.; Yuan, J.Q. Numerical determination of strength and deformability of fractured rock mass by FEM modeling. *Comput. Geotech.* **2015**, *64*, 20–31.
82. Guo, S.; Qi, S.; Zhan, Z.; Zheng, B. Plastic-strain-dependent strength model to simulate the cracking process of brittle rocks with an existing non-persistent joint. *Eng. Geol.* **2017**, *231*, 114–125. [CrossRef]
83. Walton, G.; Diederichs, M.S. A New Model for the Dilation of Brittle Rocks Based on Laboratory Compression Test Data with Separate Treatment of Dilatancy Mobilization and Decay. *Geotech. Geol. Eng.* **2015**, *33*, 661–679. [CrossRef]
84. Vazaios, I.; Farahmand, K.; Vlachopoulos, N.; Diederichs, M.S. Effects of confinement on rock mass modulus: A synthetic rock mass modelling (SRM) study. *J. Rock Mech. Geotech. Eng.* **2018**, *10*, 436–456. [CrossRef]

materials

MDPI

Article

New Minerals from Inclusions in Corundum Xenocrysts from Mt. Carmel, Israel: Magnéliite, Ziroite, Sassite, Mizraite-(Ce) and Yeite

Chi Ma [1,*], Fernando Cámara [2], Luca Bindi [3], Vered Toledo [4] and William L. Griffin [5,*]

1 Division of Geological and Planetary Sciences, California Institute of Technology, Pasadena, CA 91125, USA
2 Dipartimento di Scienze della Terra "A. Desio", Università degli Studi di Milano, Via Mangiagalli 34, I-20133 Milan, Italy; fernando.camara@unimi.it
3 Dipartimento di Scienze della Terra, Università degli Studi di Firenze, Via La Pira 4, I-50121 Florence, Italy; luca.bindi@unifi.it
4 Shefa Gems (A.T.M.) Ltd., Netanya 4210602, Israel; veredshefa@gmail.com
5 ARC Centre of Excellence for Core to Crust Fluid Systems and GEMOC, Earth and Environmental Sciences, Macquarie University, Macquarie Park, NSW 2109, Australia
* Correspondence: chima@caltech.edu (C.M.); bill.griffin@mq.edu.au (W.L.G.)

Abstract: Our nanomineralogical investigation of melt inclusions in corundum xenocrysts from the Mt. Carmel area, Israel has revealed seven IMA-approved new minerals since 2021. We report here four new oxide minerals and one new alloy mineral. Magnéliite ($Ti^{3+}{}_2Ti^{4+}{}_2O_7$; IMA 2021-111) occurs as subhedral crystals, ~4 μm in size, with alabandite, zirconolite, Ti,Al,Zr-oxide, and hibonite in corundum Grain 767-1. Magnéliite has an empirical formula $(Ti^{3+}{}_{1.66}Al_{0.13}Ti^{4+}{}_{0.15}Mg_{0.10}Ca_{0.01}Sc_{0.01})_{\Sigma 2.06}(Ti^{4+}{}_{1.93}Zr_{0.08})_{\Sigma 2.01}O_7$ and the triclinic $P\bar{1}$ Ti_4O_7-type structure with the cell parameters: a = 5.60(1) Å, b = 7.13(1) Å, c = 12.47(1) Å, α = 95.1(1)°, β = 95.2(1)°, γ = 108.7(1)°, V = 466(2) Å^3, Z = 4. Ziroite (ZrO_2; IMA 2022-013) occurs as irregular crystals, ~1–4 μm in size, with baddeleyite, hibonite, and Ti,Al,Zr-oxide in corundum Grain 479-1a. Ziroite has an empirical formula $(Zr_{0.72}Ti^{4+}{}_{0.26}Mg_{0.02}Al_{0.02}Hf_{0.01})_{\Sigma 1.03}O_2$ and the tetragonal $P4_2/nmc$ zirconia(HT)-type structure with the cell parameters: a = 3.60(1) Å, c = 5.18(1) Å, V = 67.1(3) Å^3, Z = 2. Sassite ($Ti^{3+}{}_2Ti^{4+}O_5$; IMA 2022-014) occurs as subhedral-euhedral crystals, ~4–16 μm in size, with Ti,Al,Zr-oxide, mullite, osbornite, baddeleyite, alabandite, and glass in corundum Grain 1125C1. Sassite has an empirical formula $(Ti^{3+}{}_{1.35}Al_{0.49}Ti^{4+}{}_{0.08}Mg_{0.07})_{\Sigma 1.99}(Ti^{4+}{}_{0.93}Zr_{0.06}Si_{0.01})_{\Sigma 1.00}O_5$ and the orthorhombic $Cmcm$ pseudobrookite-type structure with the cell parameters: a = 3.80(1) Å, b = 9.85(1) Å, c = 9.99(1) Å, V = 374(1) Å^3, Z = 4. Mizraite-(Ce) ($Ce(Al_{11}Mg)O_{19}$; IMA 2022-027) occurs as euhedral crystals, <1–14 μm in size, with Ce-silicate, Ti-sulfide, Ti,Al,Zr-oxide, ziroite, and thorianite in corundum Grain 198-8. Mizraite-(Ce) has an empirical formula $(Ce_{0.76}Ca_{0.10}La_{0.07}Nd_{0.01})_{\Sigma 0.94}(Al_{10.43}Mg_{0.84}Ti^{3+}{}_{0.60}Si_{0.09}Zr_{0.04})_{\Sigma 12.00}O_{19}$ and the hexagonal $P6_3/mmc$ magnetoplumbite-type structure with the cell parameters: a = 5.61(1) Å, c = 22.29(1) Å, V = 608(2) Å^3, Z = 2. Yeite (TiSi; IMA 2022-079) occurs as irregular-subhedral crystals, 1.2–3.5 μm in size, along with wenjiite (Ti_5Si_3) and zhiqinite ($TiSi_2$) in Ti-Si alloy inclusions in corundum Grain 198c. Yeite has an empirical formula $(Ti_{0.995}Mn_{0.003}V_{0.001}Cr_{0.001})(Si_{0.996}P_{0.004})$ and the orthorhombic $Pnma$ FeB-type structure with the cell parameters: a = 6.55(1) Å, b = 3.64(1) Å, c = 4.99(1) Å, V = 119.0(4) Å^3, Z = 4. The five minerals are high-temperature oxide or alloy phases, formed in melt pockets in corundum xenocrysts derived from the upper mantle beneath Mt. Carmel.

Keywords: magnéliite; $Ti^{3+}{}_2Ti^{4+}{}_2O_7$; ziroite; ZrO_2; sassite; $Ti^{3+}{}_2Ti^{4+}O_5$; mizraite-(Ce); $Ce(Al_{11}Mg)O_{19}$; yeite; TiSi; new minerals; corundum; Mt. Carmel; Israel

1. Introduction

Worldwide, super-reduced mineral assemblages are commonly associated with explosive volcanic events such as kimberlites, alkali basalts, and tholeiitic basalts, as well as in ophiolites linked to deep subduction along continental plate margins [1–3]. The origins of these assemblages have sparked debate, with some attributing them to human activities [4]. However, the extensively documented xenoliths and xenocrysts discovered in small Cretaceous volcanoes and Plio-Pleistocene gem placer deposits at Mt. Carmel, Israel, play a crucial role in this discussion. The geological context, along with thorough geochemical analysis and precise geochronological data, effectively refute any plausible notion of human interference [3,5,6]. Many super-reduced minerals are identified as inclusions within xenoliths composed of corundum aggregates. The relationships between these different phases within melt inclusions have been crucial in interpreting the genesis of super-reduced magma-fluid systems.

A nanomineralogical investigation of melt inclusions found in corundum xenocrysts coming from volcanic centers and associated alluvial deposits in the Mt. Carmel area, Israel, has allowed us to discover seven IMA-approved new minerals since 2021: griffinite (Al_2TiO_5), magnéliite ($Ti^{3+}{}_2Ti^{4+}{}_2O_7$), ziroite ($ZrO_2$), sassite ($Ti^{3+}{}_2Ti^{4+}O_5$), mizraite-(Ce) ($Ce(Al_{11}Mg)O_{19}$), toledoite (TiFeSi), and yeite (TiSi) [7–13]. All of them have been approved by the IMA-CNMNC (Commission on New Minerals, Nomenclature and Classification) as requested before naming and reporting the finding of a new mineral phase. Griffinite has been already published in detail [14]. Toledoite will be published in a separate paper. Reported here are another five new minerals, adding more information on the origin of reduced high-temperature minerals from the upper mantle.

Magnéliite (IMA 2021-111), $Ti^{3+}{}_2Ti^{4+}{}_2O_7$ (simply Ti_4O_7), is a new Ti-oxide mineral that corresponds to the first member of the homologous series of Ti-oxides (with Ti_nO_{2n-1}), known also as Magnéli phases [15]. The name is in honor of Arne Magnéli (1914–1996), for his pioneering work on the structural chemistry of transition-metal oxides.

Ziroite (IMA 2022-013), ZrO_2, is a new Zr-oxide mineral with the $P4_2/nmc$ zirconia(HT)-type structure. The name is derived from its composition.

Sassite (IMA 2022-014), $Ti^{3+}{}_2Ti^{4+}O_5$ (simply Ti_3O_5), is another new Ti-oxide mineral with the *Cmcm* pseudobrookite-type structure. The name is in honor of Eytan Sass (b. 1932), a geologist at the Freddy and Nadine Herrmann Institute of Earth Sciences, Hebrew University of Jerusalem. He performed the excellent mapping work on Mt. Carmel that identified the various volcanic centers.

Mizraite-(Ce) (IMA 2022-027), $Ce(Al_{11}Mg)O_{19}$, is a new Ce-rich oxide mineral belonging to the magnetoplumbite-group [16], with the $P6_3/mmc$ magnetoplumbite-type structure. The name is after the Mizra river in the Mt. Carmel region, where some corundum xenocrysts investigated in this study (including Grain 198-8) come from alluvial deposits. The tributary Mizra river flows into the Kishon River.

Yeite (IMA 2022-079) is a new alloy mineral, TiSi, with the *Pnma* FeB-type structure. The name is in honor of Danian Ye (b. 1939), a mineralogist at the Institute of Geology and Geophysics, Chinese Academy of Sciences, for his many contributions to mineralogy and crystal chemistry.

2. Materials and Methods

The corundum xenoliths hosting the new minerals as inclusions occur in the pyroclastic ejecta from small Cretaceous basaltic volcanoes on Mt. Carmel and from placer gemstone deposits found in the terraces of the Paleocene to Pleistocene proto-Kishon river; the modern Kishon River drains Mt. Carmel and the tributary Mizra river and enters the sea near Haifa in northern Israel [2]. Much of the xenolith material in the paleoterrace deposits probably also is derived from Miocene and Pliocene basalt outcroppings in the drainage area of the Kishon River. The xenoliths occur as aggregates of skeletal corundum crystals that enclose melt pockets containing reduced mineral assemblages [1,2,17,18].

All the type materials are deposited in the mineralogy collection of the Università degli Studi di Milano, Via Mangiagalli, 34-20133 Milano, Italy.

The type magnéliite in corundum Grain 767-1 from Mt. Carmel mount Corundum-SY is under the registration number MCMGPG-H2022-001.

The type ziroite in corundum Grain 479-1a from Mt. Carmel mount Corundum-18-1 is under the registration number MCMGPG-H2021-003.

The type sassite in corundum Grain 1125C1 from Mt. Carmel mount Corundum-18-1 is under the registration number MCMGPG-H2021-004.

The type mizraite-(Ce) in corundum Grain 198-8 from Mt. Carmel mount Corundum-18-1 is under the registration number MCMGPG-H2022-005.

The type yeite in corundum Grain 198c from Mt. Carmel mount Corundum-18-1 is under the registration number MCMGPG-H2021-002. This corundum grain also hosts the type griffinite (Al_2TiO_5; IMA 2021-110; [14]).

In order to characterize the composition and structure of the new minerals and associated phases, we used an electron probe microanalyzer (EPMA) and a high-resolution scanning electron microscope (SEM) with an X-ray energy dispersive spectrometer (EDS) and electron backscatter diffraction (EBSD). A ZEISS 1550VP Field-Emission SEM (ZEISS Group, Oberkochen, Germany) with an Oxford X-Max EDS was used for backscatter electron (BSE) imaging and fast elemental analysis. Quantitative WDS elemental microanalyses of the new minerals were carried out using a JEOL 8200 EPMA (JEOL Ltd., Tokyo, Japan) (15 kV and 10 nA, focused beam) and processed with the CITZAF correction procedure [19]. The focused electron beam is ~150 nm in diameter.

EBSD analyses at a submicrometer scale were performed using methods described by [20,21] for studies of micron-sized new minerals. An HKL EBSD system on the ZEISS 1550VP Field-Emission SEM was operated at 20 kV and 6 nA in focused beam mode with a 70° tilted stage and in a variable pressure mode (25 Pa). The EBSD system was calibrated using a single-crystal silicon standard. Experimental EBSD patterns allowed the collection of structural information and cell constants that were derived by matching with those of the structures of synthetic phases from the ICSD (Inorganic Crystal Structure Database).

Due to the small size of the samples, most of the physical properties (optical, hardness, fracture, cleavage, habit, density, etc.) were impossible to obtain.

3. Results

3.1. *Magnéliite*

Magnéliite occurs with alabandite, zirconolite, Ti,Al,Zr-oxide, and hibonite in one inclusion from corundum Grain 767-1 (Figure 1). Other inclusions in this corundum grain contain hibonite and osbornite. It is transparent, occurring as subhedral crystals ~4 μm in size. The Gladstone–Dale relationship [22] gives n = 2.423, obtained from the chemical composition and calculated density.

The chemical composition of magnéliite using EPMA (Table 1) shows an empirical formula (based on 7 O *pfu*) of $(Ti^{3+}_{1.66}Al_{0.13}Ti^{4+}_{0.15}Mg_{0.10}Ca_{0.01}Sc_{0.01})_{\Sigma 2.06}(Ti^{4+}_{1.93}Zr_{0.08})_{\Sigma 2.01}O_7$. The simplified formula is $(Ti^{3+},Al)_2Ti^{4+}_2O_7$. The ideal formula is $Ti^{3+}_2Ti^{4+}_2O_7$, which requires Ti_2O_3 47.36, TiO_2 52.64, total 100 wt%.

Table 1. EPMA analytical results (in wt%, n = 7) for magnéliite.

Constituent	Mean	Range	SD (σ)	Probe Standard
TiO_2 *	53.54	53.28–53.93	0.17	TiO_2
Ti_2O_3 *	38.63	38.44–38.84	0.12	TiO_2
ZrO_2	3.02	2.86–3.20	0.11	zircon
Al_2O_3	2.20	2.03–2.26	0.08	anorthite
MgO	1.36	1.26–1.51	0.09	forsterite
CaO	0.19	0.17–0.20	0.01	anorthite
Sc_2O_3	0.22	0.20–0.23	0.01	$ScPO_4$
Total	99.16			

* Total titanium has been partitioned between Ti^{3+} and Ti^{4+} for charge balance to make ideal stoichiometry.

(a)
(b)

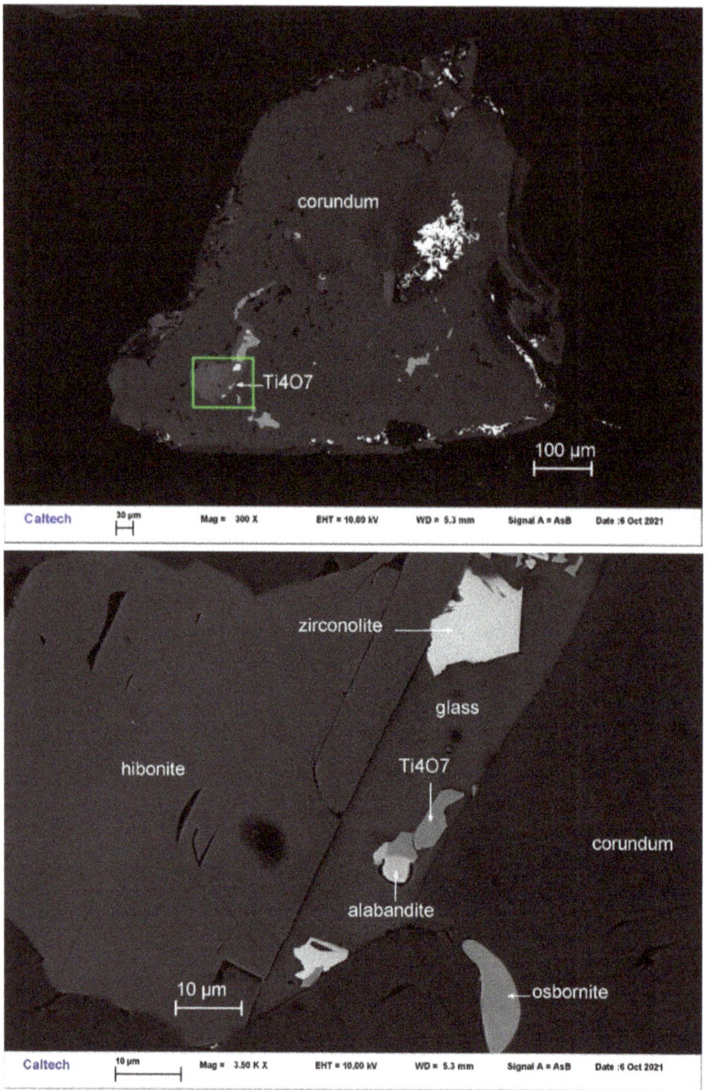

Figure 1. BSE images showing magnéliite (Ti_4O_7) in corundum Grain 767-1. The rectangular area in (**a**) is enlarged in (**b**).

The EBSD patterns can be indexed only by the $P\bar{1}$ Ti_4O_7-type structure and match the synthetic Ti_4O_7 cell from [23] (Figure 2), with a mean angular deviation of 0.32°–0.35°, revealing the following cell parameters: a = 5.60(1) Å, b = 7.13(1) Å, c = 12.47(1) Å, α = 95.1(1)°, β = 95.2(1)°, γ = 108.7(1)°, V = 466(2) Å^3, and Z = 4. The calculated density is 4.30 g·cm^{-3} using the empirical formula and the unit-cell volume estimated from the EBSD data.

Magnéliite ($Ti^{3+}{}_2Ti^{4+}{}_2O_7$) is a new Ti-oxide mineral. It belongs to the so-called Magnéli phases, i.e., a series of Ti-oxides homologous with Ti_nO_{2n-1} (with n = from 4 to 10). The first member of the series, synthetic Ti_4O_7, is well known (e.g., [23–25]). The crystal structure of magnéliite can be considered to be derived from the structure of rutile TiO_2 by crystallographic shear of the $(121)_{rutile}$ plane with a $1/2[0–11]_{rutile}$ vector every four octahedra of rutile [26]. The resulting structure has chains of edge-sharing TiO_6 octahedra truncated every four octahedra by the crystallographic shear planes (Figure 3). At room-*T*, Ti^{3+} and Ti^{4+} are disordered among the eight symmetrically independent positions, while at $T < 120$ K, Ti^{3+} and Ti^{4+} are arranged in an ordered fashion to form a Ti^{3+}-Ti^{4+} pair (bipolarons) and the material becomes a nonmagnetic insulator. Recent data by [26] show that even at room-*T* some local ordering of Ti^{3+}–Ti^{3+} and Ti^{4+}–Ti^{4+} pairs exists.

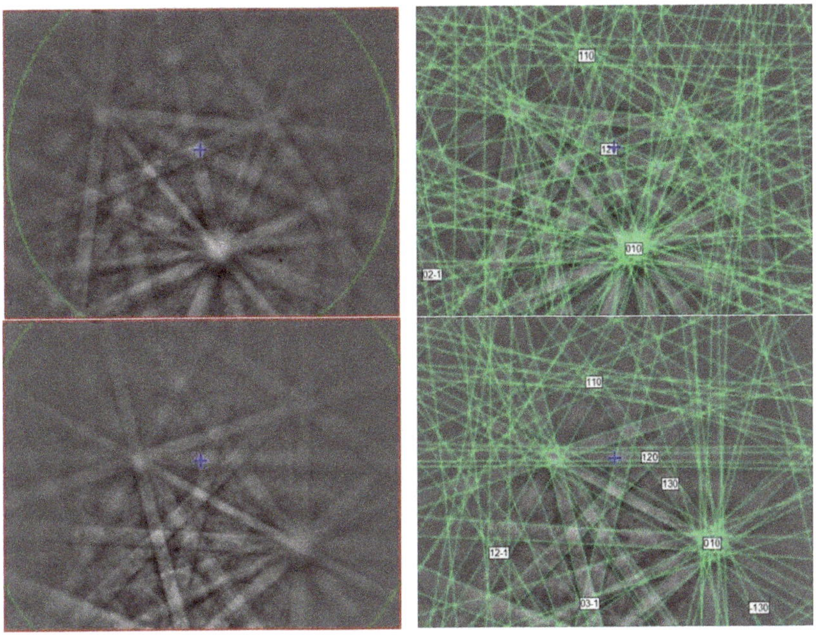

Figure 2. (**left**) EBSD patterns of the magnéliite crystal in Figure 1 at different orientations, and (**right**) the patterns indexed with the $P\overline{1}$ Ti_4O_7-type structure. Blue cross marks the pattern center.

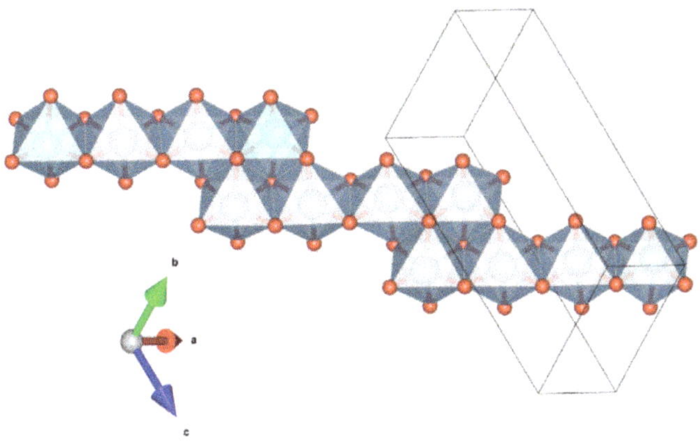

Figure 3. Detail of the structure of magnéliite (using the atom coordinates form Marezio and Dernier, 1971) projected onto (−556), showing the chains of four-member units of edge-sharing Ti-centered octahedra. Figure obtained using Vesta 3.0 [27].

3.2. *Ziroite*

Ziroite occurs with baddeleyite, Ce-rich hibonite, and Ti,Al,Zr-oxide in inclusions in corundum Grain 479-1a (Figure 4). Other inclusions in this corundum grain are MgAlspinel, fluorbritholite-(Ce), osbornite, and hapkeite (Fe_2Si). It occurs as irregular crystals ~1–4 μm in size. It is transparent and shows a brownish black tint. The Gladstone–Dale relationship gives n = 2.342.

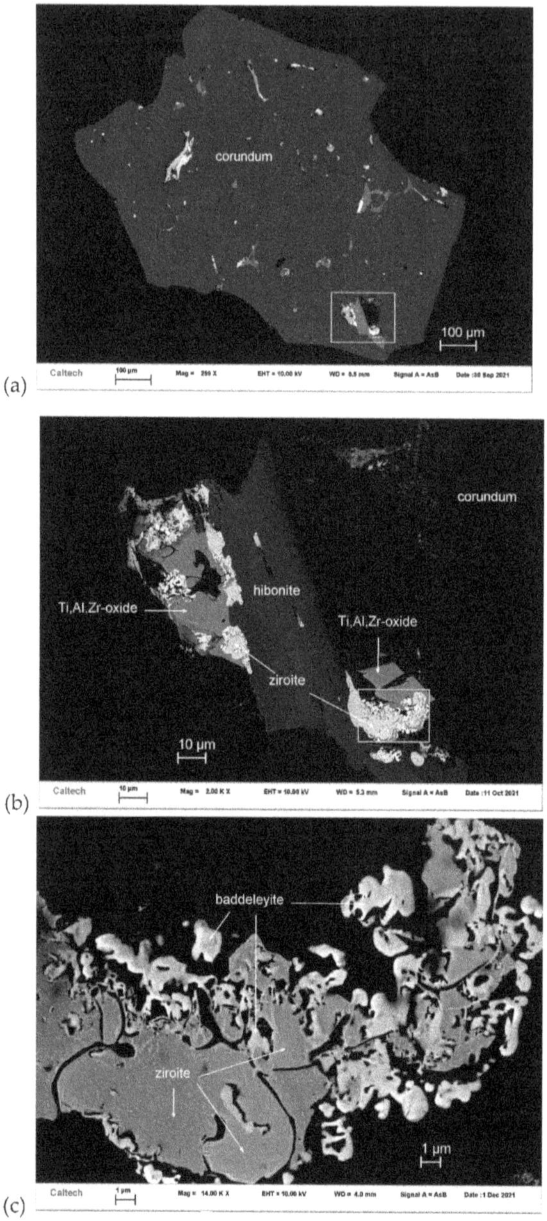

Figure 4. BSE images showing ziroite (ZrO_2) in corundum Grain 479-1a. The rectangular area in (**a**) is enlarged in (**b**). The rectangular area in (**b**) is enlarged in (**c**).

The chemical composition of ziroite (Table 2) gives rise to an empirical formula (based on 2 O *pfu*) of $(Zr_{0.72}Ti^{4+}_{0.26}Mg_{0.02}Al_{0.02}Hf_{0.01})_{\Sigma 1.03}O_2$. The simplified formula is $(Zr,Ti)O_2$. The ideal formula is ZrO_2.

Table 2. EPMA analytical results (in wt%, n = 8) for ziroite.

Constituent	Mean	Range	SD (σ)	Probe Standard
ZrO_2	78.52	76.06–79.92	1.18	zircon
TiO_2	18.21	16.68–20.47	1.18	TiO_2
HfO_2	1.43	1.35–1.54	0.08	Hf metal
MgO	0.70	0.62–0.75	0.05	forsterite
Al_2O_3	0.87	0.81–0.95	0.05	anorthite
Total	99.73			

The EBSD patterns can be indexed only by the tetragonal $P4_2/nmc$ zirconia (HT)-type and match the synthetic ZrO_2 cell values of [28] (Figure 5), with a mean angular deviation of 0.30–0.36°, revealing the following cell parameters: a = 3.60(1) Å, c = 5.18(1) Å, V = 67.1(3) Å^3, and Z = 2. The calculated density is 5.53 g·cm^{-3} using the empirical formula and the unit-cell volume estimated from the EBSD data.

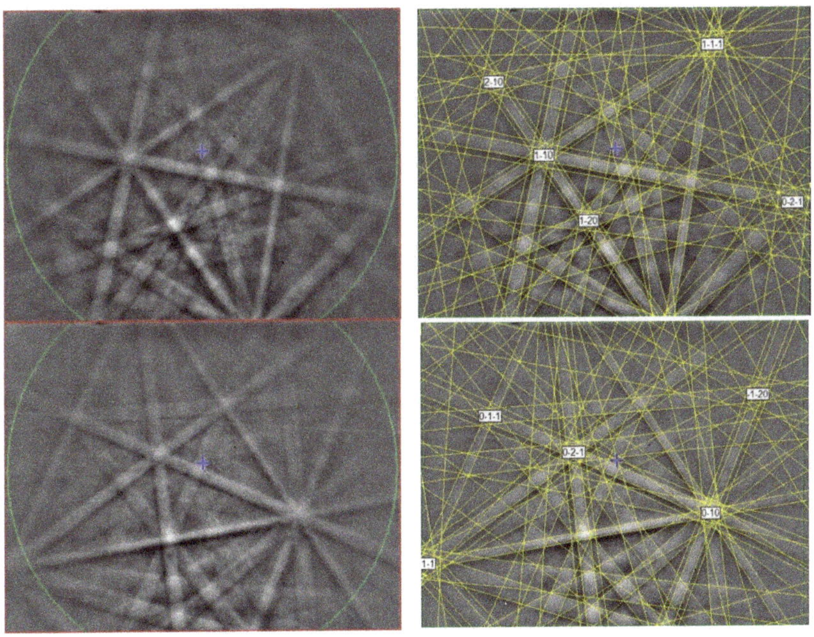

Figure 5. (**left**) EBSD patterns of the ziroite crystals in Figure 4, and (**right**) the patterns indexed with the $P4_2$/nmc zirconia(HT)-type. Blue cross marks the pattern center.

Ziroite is a tetragonal polymorph of baddeleyite (monoclinic ZrO_2). Synthetic ZrO_2 with the $P4_2/nmc$ zirconia (HT)-type structure is well known (Figure 6) (e.g., [28–30]). Reported here is the first natural occurrence of tetragonal ZrO_2.

3.3. Sassite

Sassite occurs with Ti,Al,Zr-oxide, mullite, osbornite, baddeleyite, alabandite, and Si-rich glass in melt pockets trapped in corundum Grain 1125C1 (Figure 7). The mineral occurs as subhedral-euhedral crystals ~4–16 μm in size. It is transparent with a brown color. The Gladstone–Dale relationship gives n = 2.16.

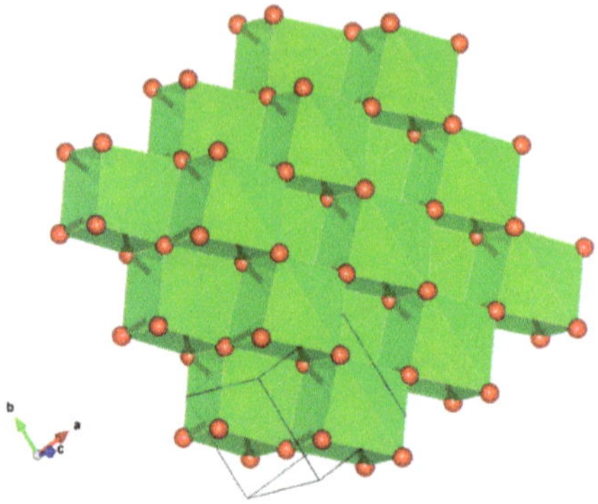

Figure 6. The structure of ziroite. Figure obtained using Vesta 3.0 [27].

Sassite (Table 3) shows an empirical formula (based on 5 O *pfu*) of $(Ti^{3+}{}_{1.35}Al_{0.49}Ti^{4+}{}_{0.08}Mg_{0.07})_{\Sigma 1.99}(Ti^{4+}{}_{0.93}Zr_{0.06}Si_{0.01})_{\Sigma 1.00}O_5$. The simplified formula is $(Ti^{3+},Al)_2Ti^{4+}O_5$. The ideal formula is $Ti^{3+}{}_2Ti^{4+}O_5$, which requires Ti_2O_3 64.29, TiO_2 35.71, total 100 wt%.

Table 3. EPMA analytical results (in wt%, n = 13) for sassite.

Constituent	Mean	Range	SD (σ)	Probe Standard
Ti_2O_3 *	45.32	44.81–45.94	0.30	TiO_2
TiO_2 *	37.32	36.90–37.83	0.24	TiO_2
Al_2O_3	11.73	10.22–13.02	0.83	anorthite
ZrO_2	3.70	3.05–5.04	0.68	zircon
MgO	1.29	0.97–1.66	0.24	forsterite
SiO_2	0.24	0.15–0.46	0.08	anorthite
MnO	0.15	0.10–0.20	0.04	Mn_2SiO_4
CaO	0.07	0.05–0.12	0.02	$ScPO_4$
Total	99.82			

* Total titanium has been partitioned between Ti^{3+} and Ti^{4+} for charge balance to achieve ideal stoichiometry.

The EBSD patterns can be indexed only by the orthorhombic *Cmcm* pseudobrookite-type structure and match the synthetic β-Ti_3O_5 cell from [31] (Figure 8), with a mean angular deviation of 0.31°–0.35°, revealing the following cell parameters: *a* = 3.80(1) Å, *b* = 9.85(1) Å, *c* = 9.99(1) Å, *V* = 374(1) Å^3, and Z = 4. The calculated density is 3.81 g·cm^{-3} using the empirical formula and the unit-cell volume estimated from the EBSD data.

Sassite ($Ti^{3+}{}_2Ti^{4+}O_5$) is a new member of the pseudobrookite group, joining pseudobrookite (Fe_2TiO_5), armalcolite [$(Mg,Fe^{2+})Ti_2O_5$], and griffinite (Al_2TiO_5; IMA 2021-110) [14]. Synthetic Ti_3O_5-pseudobrookite is well known (e.g., [31,32]). It is also known as β-Ti_3O_5 because several polymorphs of Ti_3O_5 have been described so far. α–Ti_3O_5 has monoclinic symmetry, *C*2/*m*, with *a* = 9.752(1), *b* = 3.802(1), *c* = 9.442(1) Å, and β = 91.55(1)° [33]. β-Ti_3O_5 has orthorhombic symmetry, *Cmcm*, *a* =3.798(2), *b* = 9.846(3), *c* = 9.988(4) Å [31], which corresponds to sassite. γ-Ti_3O_5 has the monoclinic V_3O_5-type structure, with *a* = 10.115, *b* = 5.074, *c* = 7.182 Å, β = 112°, and *C*2/*c* space group [34], which corresponds to the mineral kaitianite, recently described in Allende CV3 carbonaceous chondrite [35]. δ–Ti_3O_5 has a monoclinic structure and *P*2/*a* space group, with lattice parameters *a* = 9.9651(7), *b* = 5.0604(4), *c* = 7.2114(5) Å, and β = 109.3324(9)°, and is related to γ-Ti_3O_5 by decreasing temperature [36,37]. λ–Ti_3O_5 is monoclinic *C*2/*m* with *a* = 9.8357(11),

$b = 3.7935(2)$, $c = 9.9863(7)$ Å, $\beta = 90.976(6)°$, and is related to α–Ti_3O_5 by a second-order phase transition [38].

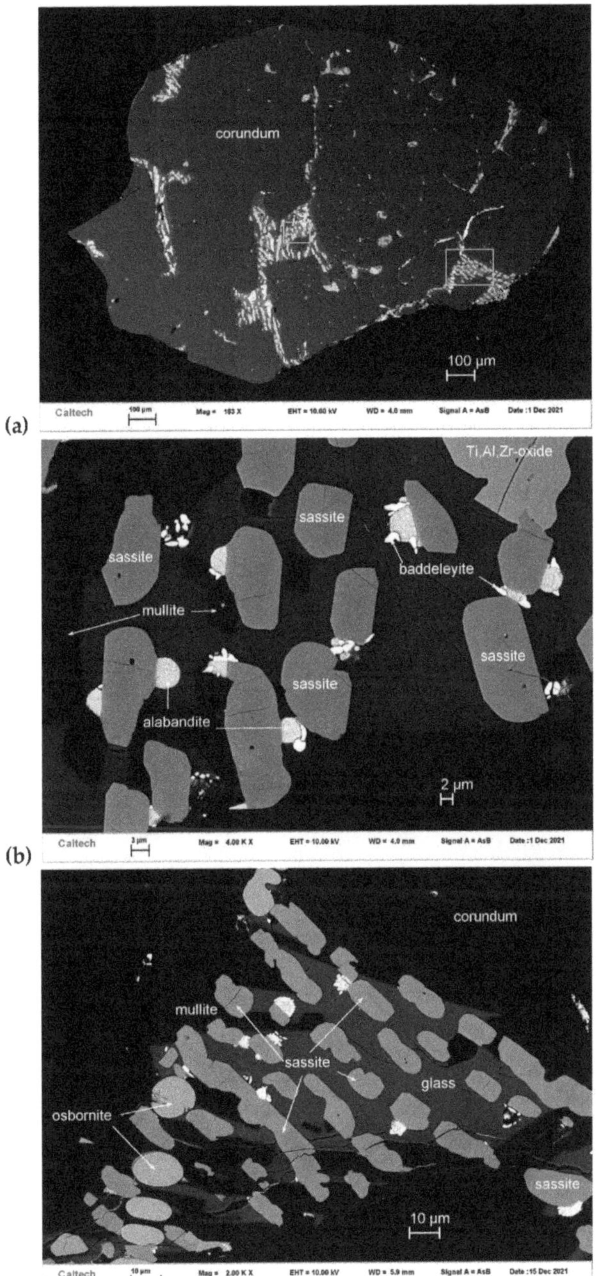

Figure 7. BSE images showing sassite ($Ti^{3+}{}_2Ti^{4+}O_5$) in corundum Grain 1125C1. The rectangular area in (**a**) is enlarged in (**b**). The rectangular area in (**b**) is enlarged in (**c**).

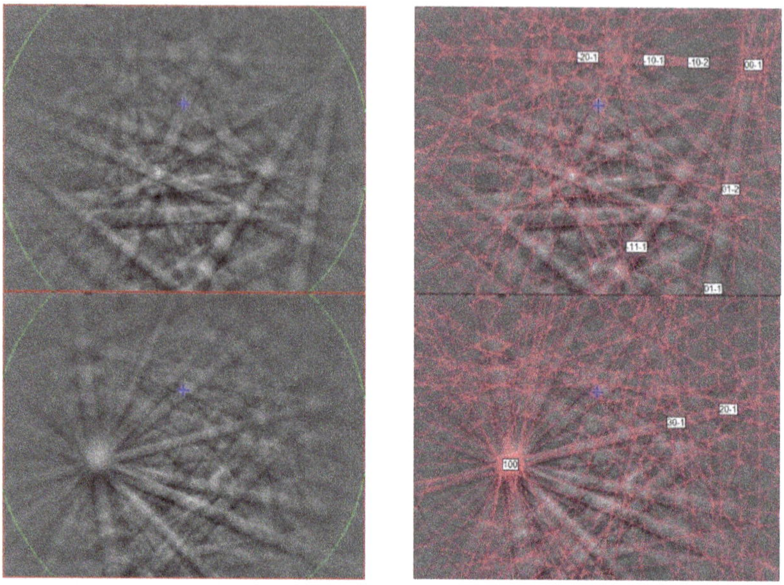

Figure 8. (**left**) EBSD patterns of two sassite crystals in Figure 7, and (**right**) the patterns indexed with the *Cmcm* pseudobrookite-type Ti_3O_5 structure. Blue cross marks the pattern center.

Kaitianite was first discovered in association with tistarite and rutile together with other refractory phases of corundum, xifengite, mullite, osbornite, and a new Ti,Al,Zr-oxide mineral in the Allende meteorite [35]; most of these minerals are also found in the melt pockets in corundum from Mt. Carmel. Ti^{3+}-rich phases are common in melt inclusions in those corundum xenocrysts, including tistarite (Ti_2O_3), magnéliite ($Ti^{3+}{}_2Ti^{4+}{}_2O_7$), sassite ($Ti^{3+}{}_2Ti^{4+}O_5$), and grossmanite. Sc-bearing sassite was identified in the SaU 290 CH3 chondrite as an ultrarefractory phase, labeled as "anosovite," among the first solids formed in the solar system [39]. While low-temperature α–Ti_3O_5 is highly ordered (three octahedrally Ti sites, with <Ti–O> = 2.015 Å, 2.033 Å, and 2.033 Å), both sassite and kaitianite show higher degrees of disorder (two crystallographically independent Ti sites, labeled Ti1 and Ti2, are octahedrally surrounded by oxygen atoms, having <Ti1–O> = 2.0271 Å and <Ti2–O> = 2.0385 Å in sassite [31] and <Ti1–O> = 2.0334 Å and <Ti2–O> = 2.0331 Å in kaitianite [34]). In sassite, the TiO_6 octahedra are linked by sharing edges and corners, building up a characteristic row extending along the *c* axis, which is joined to an adjacent row along [100] by sharing edges (Figure 9).

3.4. Mizraite-(Ce)

Mizraite-(Ce) occurs with Ce-silicate and Ti-sulfide in melt pockets between corundum and spinel within Grain 198-8 (Figure 10). Other inclusions in this corundum grain contain Ti,Al,Zr-oxide, ziroite, baddeleyite, thorianite, osbornite, zangboite ($TiFeSi_2$), wenjiite (Ti_5Si_3), and a $[(Mn,Fe,Ti,V,Cr)_4Ti_2]Si_5$ alloy. The mineral occurs as euhedral crystals < 1–14 μm in size. It is transparent with a light bluish-green color. The Gladstone–Dale relationship gives n = 1.828.

Mizraite-(Ce) (Table 4) exhibits an empirical formula (based on 19 O *pfu*) of $(Ce_{0.76}Ca_{0.10}La_{0.07}Nd_{0.01})_{\Sigma 0.94}(Al_{10.43}Mg_{0.84}Ti^{3+}{}_{0.60}Si_{0.09}Zr_{0.04})_{\Sigma 12.00}O_{19}$. The simplified formula is $(Ce,Ca,La)(Al,Mg,Ti^{3+})_{12}O_{19}$. The ideal formula is $Ce(Al_{11}Mg)O_{19}$, which requires Ce_2O_3 21.45, Al_2O_3 73.28, MgO 5.27, total 100 wt%.

Table 4. EPMA analytical results (in wt%, n = 8) for mizraite-(Ce).

Constituent	Mean	Range	SD (σ)	Probe Standard
Al_2O_3	69.75	69.41–70.07	0.23	Al_2O_3
Ce_2O_3	16.30	16.13–16.50	0.14	$CePO_4$
* Ti_2O_3	5.67	5.64–5.75	0.04	TiO_2
MgO	4.45	4.37–4.51	0.04	forsterite
La_2O_3	1.47	1.40–1.56	0.06	$LaPO_4$
SiO_2	0.72	0.69–0.79	0.03	anorthite
CaO	0.71	0.69–0.73	0.02	anorthite
ZrO_2	0.67	0.43–0.81	0.12	zircon
Nd_2O_3	0.29	0.25–0.35	0.04	$NdPO_4$
Total	100.02			

* Titanium has been assigned to be Ti^{3+} for charge balance to achieve best stoichiometry.

The EBSD patterns can be indexed only by the hexagonal $P6_3/mmc$ magnetoplumbite structure and match the Ce-bearing hibonite cell of [40] (Figure 11), with a mean angular deviation of 0.32°–0.37°, revealing the following cell parameters: a = 5.61(1) Å, c = 22.29(1) Å, V = 608(2) Å^3, and Z = 2. The calculated density is 4.16 g·cm^{-3} using the empirical formula and the unit-cell volume estimated from the EBSD data.

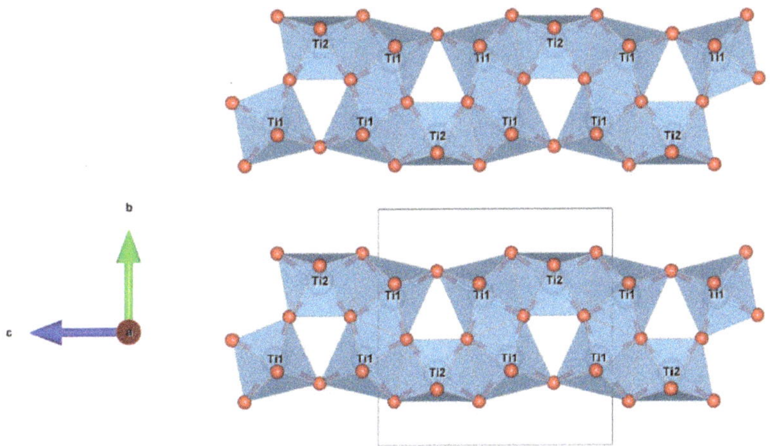

Figure 9. Detail of the structure of sassite (using atom coordinates form Onoda 1998) projected onto (010), showing the chains of TiO_6 octahedra along [001] sharing edges and vertexes. Figure obtained using Vesta 3.0 [27].

(a)

(b)

(c)

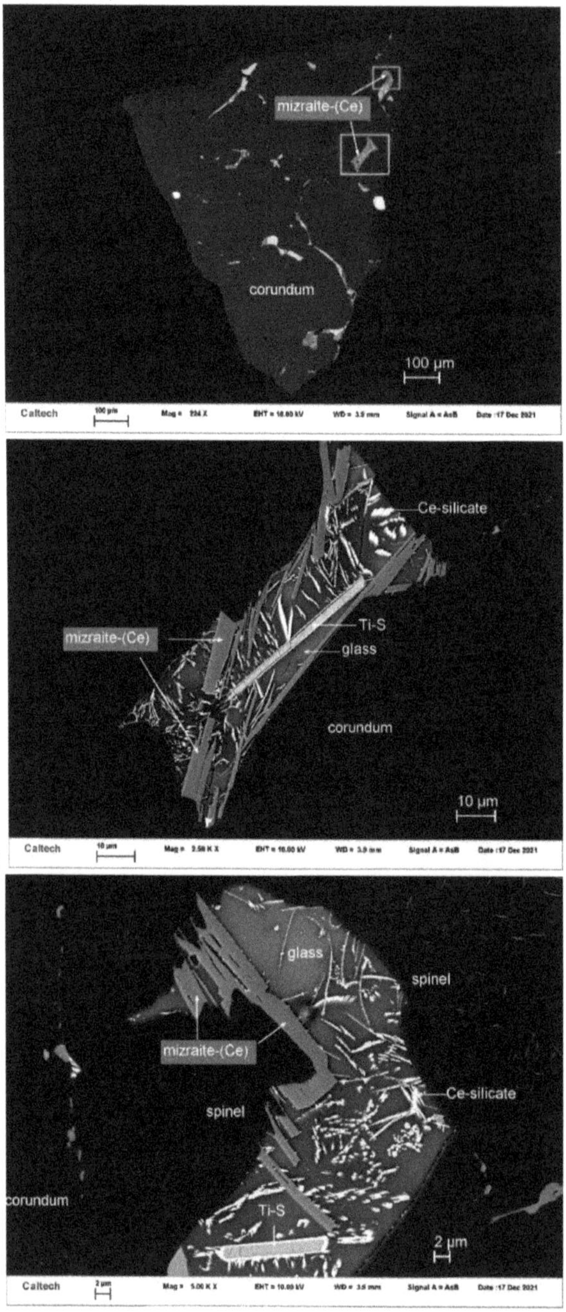

Figure 10. BSE images showing mizraite-(Ce) in corundum Grain 198-8. The rectangular areas in (**a**) are enlarged in (**b**,**c**).

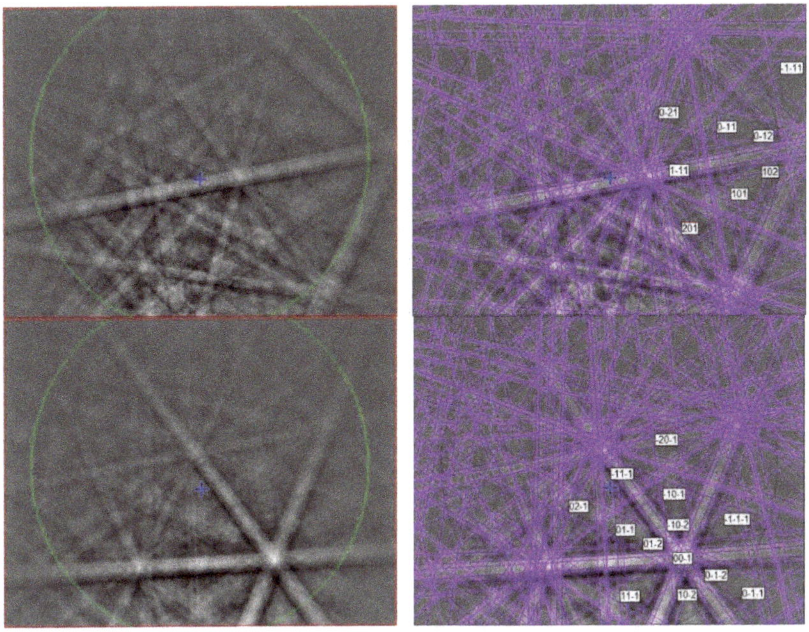

Figure 11. (**left**) EBSD patterns of two mizraite-(Ce) crystals in Figure 10, and (**right**) the patterns indexed with the $P6_3/mmc$ hibonite structure. Blue cross marks the pattern center.

Mizraite-(Ce) is the Ce-analog of hibonite, and is a new member of the magnetoplumbite group ($A[B_{12}]O_{19}$; [16]); it is the first member presenting the heterovalent substitution $A^{2+} + B^{3+} \rightarrow A^{3+} + B^{2+}$ ($Ca^{2+} + Al^{3+} \rightarrow REE^{3+} + Mg^{2+}$) as the dominant species-defining exchange. Whenever another magnetoplumbite REE-dominant mineral is described, it would represent a new subgroup along with the magnetoplumbite (*A* = Pb), hawthorneite (*A* = Ba), and hibonite (*A* = Ca) subgroups. Hibonite has a general formula of $(Ca,Ce)(Al,Ti,Mg)_{12}O_{19}$ and an ideal formula of $CaAl_{12}O_{19}$. Synthetic $Ce(Al_{11}Mg)O_{19}$ is not reported, whereas $La(Al_{11}Mg)O_{19}$, $La(Al_{11}Mn)O_{19}$, and $La(Al_{11}Ni)O_{19}$ with the hibonite structure have been synthesized [41–43]. Terrestrial hibonite often contains minor Ce and other *REE*s and has a general formula of $(Ca,Ce)(Al,Ti,Mg)_{12}O_{19}$ [40,44]. Reported here is the first natural occurrence of $Ce(Al_{11}Mg)O_{19}$, although zoned "hibonite" grains with *REE*-rich cores ($\Sigma REE > 0.6$ atoms per formula unit) have been described by [45] where kalsilite, leucite, and hibonite occur together with spinel, corundum, sphene, perovskite, Ti-phlogopite, and K-feldspar in a granulite-facies gneiss in the Punalur district in Kerala, southern India. The structure of mizraite-(Ce) has the topology of the magnetoplumbite group minerals with $Ln^{3+}(Al_{11}M^{2+})O_{19}$ stoichiometry and is made of two structural layers: the hexagonal close-packed *R*-block, containing the Ln^{3+} site, the trigonal bipyramidal *M*2 site, and the octahedral face-sharing *M*4 site; and the cubic close-packed *S*-block, containing layers of *M*5 octahedra interspaced by the *M*3 tetrahedra and the *M*1 octahedra (Figure 12). The spinel blocks contain most of the Al^{3+} in the *M*1 and *M*5 sites, and M^{2+} cations are distributed among the octahedral and tetrahedral sites. The Ln^{3+} and remaining Al^{3+} cations are localized in mirror planes, whereas *M*4 octahedra containing high-charge small cations lie on both sides of the mirror plane (Figure 12). The separation between the two Ln^{3+} sites of the same mirror plane is equal to the *a* unit cell parameter (ca. 5.6 Å), whereas between two different mirror planes it is approximately 11 Å.

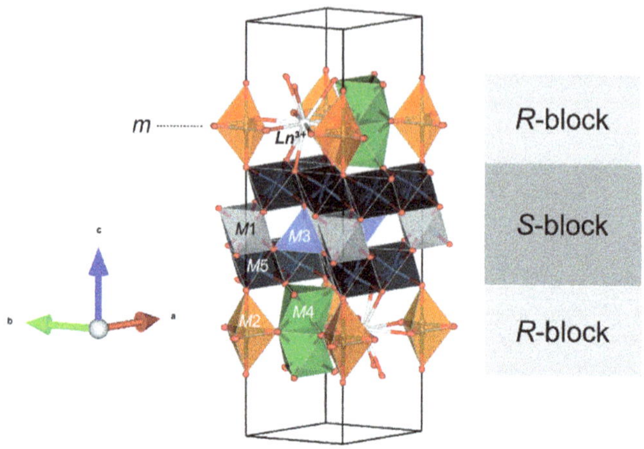

Figure 12. Detail of magnetoplumbite-type structure of mizraite-(Ce), showing the interlayering of *S*- and *R*-blocks. Ln^{3+} cations are located in the *R*-blocks along with the *M*2 and *M*4 sites. Figure obtained using Vesta 3.0 [27].

3.5. Yeite

Yeite occurs with wenjiite (Ti_5Si_3) and zhiqinite ($TiSi_2$) in Ti-Si alloy inclusions in corundum Grain 198c (Figure 13). Other inclusions in this corundum grain contain type griffinite (Al_2TiO_5) [14], rutile, baddeleyite, hibonite, osbornite, khamrabaevite, Ti,Al,Zr-oxide, zirconolite, and jingsuiite. Yeite occurs as irregular-subhedral crystals 1.2–3.5 μm in size. It is opaque and shows a black color.

The chemical composition of yeite (Table 5) gives rise to an empirical formula (based on 2 atoms *pfu*) of $(Ti_{0.995}Mn_{0.003}V_{0.001}Cr_{0.001})(Si_{0.996}P_{0.004})$. The simplified formula is TiSi. The ideal formula is TiSi, which requires Ti 63.04, Si 36.96, total 100 wt%. Associated wenjiite has an empirical formula $(Ti_{3.70}Mn_{0.43}Cr_{0.07}V_{0.02})Si_{3.68}$. Zhiqinite has an empirical formula $Ti_{0.99}Si_{2.01}$.

Table 5. EPMA analytical results (in wt%, n = 7) for yeite.

Constituent	Mean	Range	SD (σ)	Probe Standard
Ti	62.34	62.11–62.71	0.21	Ti metal
Si	36.59	36.47–36.78	0.10	Si metal
P	0.18	0.13–0.25	0.05	GaP
Mn	0.18	0.14–0.31	0.06	Mn_2SiO_4
Cr	0.08	0.00–0.13	0.04	Cr metal
V	0.06	0.00–0.17	0.06	V metal
Fe	0.00	0.00	0.00	Fe metal
Total	99.43			

The EBSD patterns of yeite can be indexed only by the orthorhombic *Pnma* FeB-type structure and match the synthetic TiSi cells of [46,47] (Figure 14), with a mean angular deviation of 0.23°–0.32°, revealing the following cell parameters: a = 6.55(1) Å, b = 3.64(1) Å, c = 4.99(1) Å, V = 119.0(4) Å^3, and Z = 4. The calculated density is 4.24 g·cm^{-3} using the empirical formula and the unit-cell volume estimated from the EBSD data.

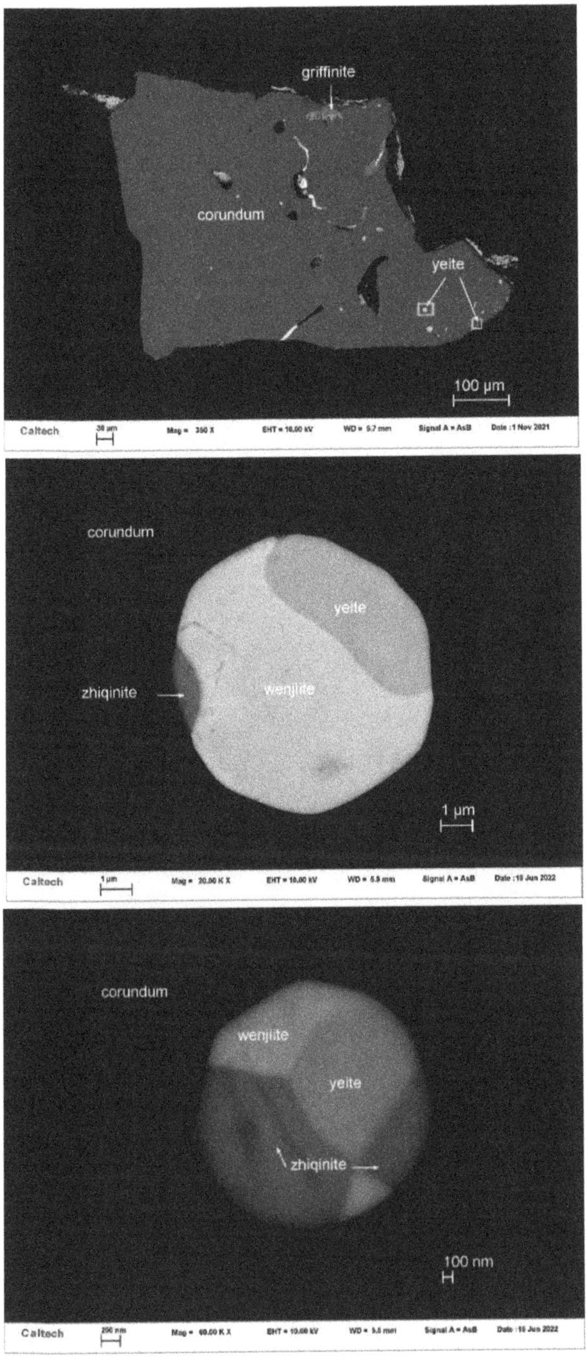

(a) (b) (c)

Figure 13. BSE images showing yeite (TiSi) with wenjiite (Ti_5Si_3) and zhiqinite ($TiSi_2$) in corundum Grain 198c. The rectangular areas in (**a**) are enlarged in (**b**,**c**).

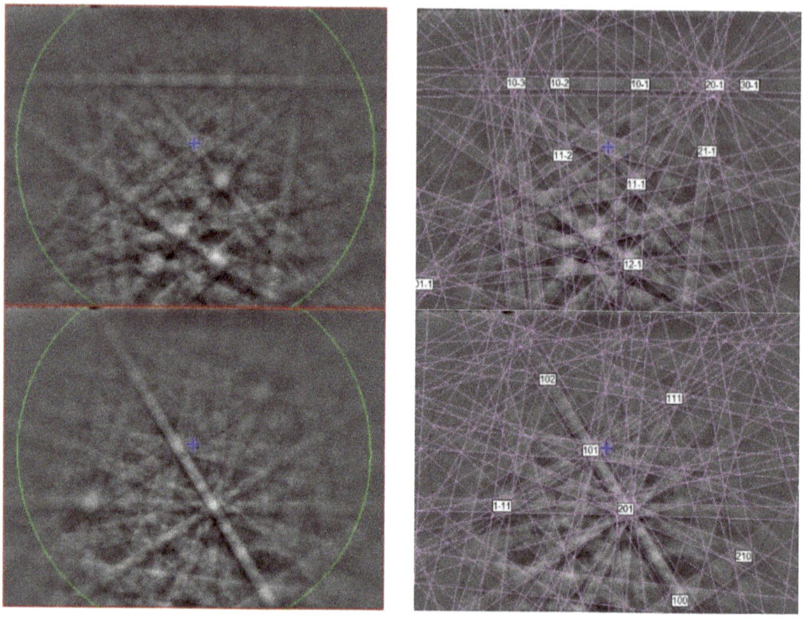

Figure 14. (**left**) EBSD patterns of yeite in Figure 13 at different orientations, and (**right**) the patterns indexed with the *Pnma* TiSi structure. Blue cross marks the pattern center.

In the structure of yeite, each Si atom is coordinated with seven Ti atoms, forming $SiTi_7$ polyhedra (capped trigonal prism, <Si-Ti> = 2.638 Å, distortion index = 0.01259) that share edges, building up a three-dimensional framework (Figure 15). The structure can be also described as $TiSi_7$ polyhedra (<Si-Ti> = 2.636 Å, distortion index = 0.01245), with an odder shape, sharing edges.

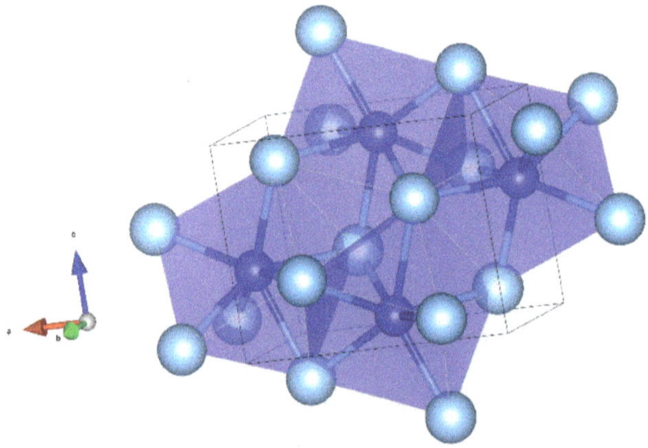

Figure 15. The structure of yeite (using the atom coordinates published by [46]). $SiTi_7$ capped triangular prims in blue. The polyhedra share edges. Figure obtained using Vesta 3.0 [27].

Yeite is natural TiSi with the *Pnma* FeB-type structure. Synthetic TiSi with the *Pnma* FeB-type structure is well known [46,47]. To our knowledge, yeite is not related to other minerals. Other Ti-Si minerals include zhiqinite with an orthorhombic *Fddd* $TiSi_2$-type structure [48], kangjinlaite ($Ti_{11}Si_{10}$) with a tetragonal *I*4/*mmm* $Ho_{11}Ge_{10}$-type structure [49], and a special wenjiite (Ti_5Si_3) with a hexagonal $P6_3/mcm$ Mn_5Si_3-type structure identified in this study.

4. Discussion

The oxide minerals described here are high-temperature phases. They crystallized from melts that were trapped in intracrystalline and interstitial voids in aggregates of corundum crystals [2]. The whole suite of corundum xenoliths is characterized by oxygen fugacity (fO_2) below the levels normally encountered in Earth's upper mantle or crust (IW to IW-9; [50]). We recognize three broad paragenetic types.

Crn-A: these are hopper to skeletal crystals showing strong zoning in Ti due to the uptake of Ti^{3+} during rapid crystal growth [51]. The composition of the trapped melts is Ca-Mg-Al silicates showing high contents of S as well as incompatible elements. Phase assemblages reflect low fO_2, with all Ti as Ti^{3+} (e.g., tistarite).

Crn-B: these are large homogeneous (unzoned) corundum crystals, which typically show Ti contents > 1 wt%. In these crystals, interstitial pockets contain small amounts of glass, which are typically high in *REE*, Zr, and other incompatible elements. In phenocrysts, Ti is present as both Ti^{3+} and Ti^{4+}.

Crn-C: these are texturally similar to *Crn-B*; however, the Ti contents in corundum are typically low (<0.5 wt% Ti). Rare glasses are rich in *LREE* and Ba. The presence of more Ti^{4+} phases (rutile, griffinite) suggests higher mean fO_2 than in *Crn-A* and *Crn-B*. Hibonite occurs in all three parageneses; in *Crn-A* and *Crn-B*, it contains high levels of Ti^{3+}, whereas in *Crn-C*, the Ti^{3+} contents are very low.

Magnéliite, sassite, and ziroite are members of a large population of Ti-Al-Zr phases, which include carmeltazite, griffinite, tistarite, rutile, "Allende-like" Ti-Zr-Al oxide [35], kaitianite [52], and many as yet undescribed minerals (Figure 16). Part of this variety is due to the presence of Ti as both Ti^{3+} and Ti^{4+}, reflecting the differences in fO_2 among the three parageneses. Individual phases may show large ranges in solid solution, reflecting substitutions of trivalent (Ti^{3+}, Al) and quadrivalent (Ti^{4+}, Zr) ions.

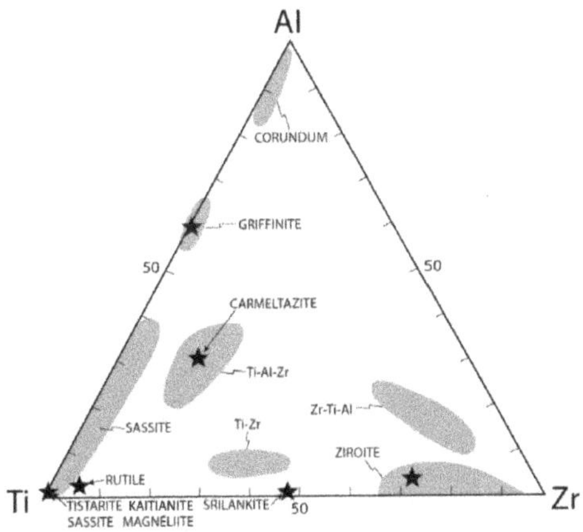

Figure 16. Ti-Al-Zr triplot showing phases from melt inclusions in corundum xenocrysts from the Mt. Carmel area, from [52].

Magnéliite shows some solid solution of both ZrO_2 and Al_2O_3 (Table 1); it has crystallized from a glass, residual after the crystallization of large hibonite crystals. The type magnéliite is associated with alabandite, which suggests that both crystallized during the ascent of the xenoliths as decreasing pressure led to lower solubility of sulfur in the melt. This assemblage and the low Ti in corundum (0.4 wt% Ti) are characteristic of paragenesis *Crn-C*.

Sassite is clearly a liquidus phase (Figure 4) together with a Ti-Al-Zr oxide and corundum; the reconstructed melt in these interstitial pockets is low in Si and Ca and very high in Ti, while the residual melt is Al, Si-rich. Sassite shows a very wide range of solid solution toward griffinite (Al_2TiO_5 [14]; Figure 16). The presence of alabandite and baddeleyite suggests quench crystallization during ascent of the xenolith, which is consistent with the quench crystallization of mullite at low P. This is a typical *Crn-B* paragenesis.

Ziroite can have a significant solid solution of TiO_2 (Table 3; Figure 16). The ability of ziroite to take up Ti can explain the coexistence of ziroite and baddeleyite (Figure 7), as the latter does not appear to take up much Ti. Like sassite, the type ziroite has crystallized from a Ca-Mg-Al-silicate glass, residual after the crystallization of hibonite and a Ti-Al-Zr oxide, in a typical *Crn-B* paragenesis.

Mizraite-(Ce) is also clearly a liquidus phase, crystallizing from a residual melt high in *LREE* and S. Its occurrence as interstitial to large exsolved spinel grains suggests that it belongs to paragenesis *Crn-B*, although the low Ti content of the adjacent corundum (0.4 wt%) is more characteristic of *Crn-C*.

The study of mixed-valence phases in paragenesis *Crn-B*, and possibly in *Crn-C*, provides new information on the interpretation of the origins of the Mt. Carmel corundum-aggregate xenoliths. While the different parageneses share many common features, it has proven difficult to establish common lines of descent between them.

The alloy phases, including yeite as described here, appear as inclusions in aggregates of corundum crystals; they represent trapped melts, melts + crystals, and subsolidus assemblages that formed from the melts on cooling, both prior to eruption and during quenching upon eruption of the host basalts [6]. The immiscible separation of these melts from the coexisting silicate melt under highly reducing conditions allowed the crystallization of Fe-free phases from the silicate melt(s). The chemistry and evolution of these melts through multiple stages of immiscibility have been described in [6]; yeite adds more detail to this picture.

Yeite occurs in spheroidal balls (Figure 13) interpreted as immiscible melts coexisting with the silicate melt from which the enclosing corundum was crystallizing. The smooth, straight, or irregular boundaries between yeite, wenjiite, and zhiqinite suggest that the original melt may have decomposed into mutually immiscible melts or crystallized into the three coexisting phases. However, examination of the phase diagram for the Ti-Si Si binary [53] suggests that the situation was more complex (Figure 17).

This binary is separated into two subsystems by a thermal divide at Ti_3Si_2; the assemblage $TiSi+TiSi_2$ appears (crystallizes) at a eutectic point (1743 K) on the Si side of the divide, while wenjiite crystallizes from melts on the Ti side of the divide from 2400 K to a eutectic ($L \rightarrow Ti + Ti_5Si_3$) at 1613 K. There is no point at which TiSi coexists with Ti_5Si_3. However, the average reconstructed composition of the melts in Figure 13 lies near several cotectics (1773–1673 K) in the Fe-Ti-Si ternary system (Figure 18) [54], making it probable that three phases may have crystallized from the melt over a very short *T* range in the high-temperature part of this ternary system. As noted by [6], the temperatures in the natural system beneath Mt. Carmel probably were lower than those in the synthetic systems due to the coexistence of a fluid phase rich in H_2, which can lower temperatures in metallic systems by up to several hundred degrees [55].

These alloy minerals thus illustrate the wide range of immiscible-melt compositions and crystallization conditions captured in the xenoliths from Mt. Carmel and give some new insights into processes in this highly reduced magmatic system. This highly reduced corundum-related assemblage is not simply a one-locality oddity; very similar associations have been reported from the Luobusa ophiolite in SE Tibet [48,49] and from many other localities in intraplate and subduction-zone tectonic settings [2,6]. These occurrences imply a significant role for mantle-derived CH_4+H_2 fluids in magmatic processes.

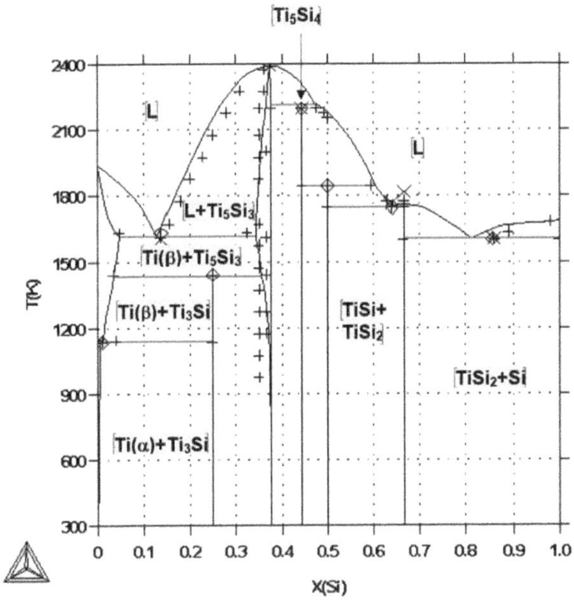

Figure 17. Calculated Ti-Si phase diagram from [53].

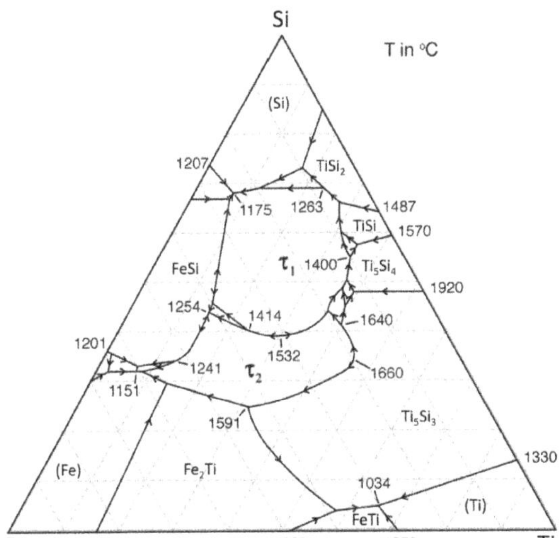

Figure 18. Liquidus projection for Fe-Si-Ti, modified from [54].

5. Conclusions

Reported here is the discovery of five new minerals, magnéliite ($Ti^{3+}{}_2Ti^{4+}{}_2O_7$), ziroite ($ZrO_2$), sassite ($Ti^{3+}{}_2Ti^{4+}O_5$), mizraite-(Ce) ($Ce(Al_{11}Mg)O_{19}$), and yeite (TiSi), in melt inclusions in corundum xenocrysts from the Mt. Carmel area, Israel. The description of their chemical composition and the crystal structures of the synthetic analogues that match the EBSD data is provided. Many physical properties cannot be obtained because of the extremely reduced dimensions of the grains (nano scale), but the data are sufficient to support their correct identification. These minerals are high-temperature oxide or alloy phases formed under extremely reduced conditions in the upper mantle, and provide new, important insights into the natural origin of super-reduced mineral assemblages.

Author Contributions: Conceptualization, C.M., F.C., L.B. and W.L.G.; Methodology, C.M.; Formal analysis, C.M., F.C., L.B. and W.L.G.; Investigation, C.M., F.C., L.B. and W.L.G.; Resources, V.T.; Data curation, C.M.; Writing—original draft, C.M.; Writing—review & editing, C.M., F.C., L.B. and W.L.G. All authors have read and agreed to the published version of the manuscript.

Funding: The research was funded by MIUR-PRIN2017, project "TEOREM deciphering geological processes using Terrestrial and Extraterrestrial ORE Minerals", prot. 2017AK8C32 (PI: Luca Bindi).

Institutional Review Board Statement: Not applicable.

Informed Consent Statement: Not applicable.

Data Availability Statement: Data are contained within the article.

Acknowledgments: SEM: EBSD and EPMA analyses were carried out at the Caltech GPS Division Analytical Facility, which is supported, in part, by NSF Grants EAR-0318518 and DMR-0080065. This is contribution 1779 from the ARC Centre of Excellence for Core to Crust Fluid Systems (www.ccfs.mq.edu.au) and 1547 from the GEMOC Key Centre (www.gemoc.mq.edu.au).

Conflicts of Interest: Vered Toledo was employed by Shefa Gems (A.T.M.) Ltd. The remaining authors declare that the research was conducted in the absence of any commercial or financial relationships that could be construed as a potential conflict of interest.

References

1. Xiong, Q.; Griffin, W.L.; Huang, J.-X.; Gain, S.E.M.; Toledo, V.; Pearson, N.J.; O'Reilly, S.Y. Super-reduced mineral assemblages in "ophiolitic" chromitites and peridotites: The view from Mt. Carmel. *Eur. J. Mineral.* **2017**, *29*, 557–570. [CrossRef]
2. Griffin, W.L.; Huang, J.-X.; Thomassot, E.; Gain, S.E.M.; Toledo, V.; O'Reilly, S.Y. Super-reducing conditions in ancient and modern volcanic systems: Sources and behaviour of carbon-rich fluids in the lithospheric mantle. *Mineral. Petrol.* **2018**, *112* (Suppl. S1), 101–114. [CrossRef]
3. Griffin, W.L.; Bindi, L.; Cámara, F.; Ma, C.; Gain, S.E.M.; Saunders, M.; Alard, O.; Huang, J.-X.; Shaw, J.; Meredith, C.; et al. Interactions of magmas and highly reduced fluids during intraplate volcanism, Mt Carmel, Israel: Implications for mantle redox states and global carbon cycles. *Gondwana Res.* **2023**, *128*, 14–54. [CrossRef]
4. Litasov, K.D.; Kagi, H.; Bekker, T.B. Enigmatic super-reduced phases in corundum from natural rocks: Possible contamination from artificial abrasive materials or metallurgical slags. *Lithos* **2019**, *340–341*, 181–190. [CrossRef]
5. Griffin, W.L.; Toledo, V.; O'Reilly, S.Y. Discussion of "Enigmatic super-reduced phases in corundum from natural rocks: Possible contamination from artificial abrasive materials or metallurgical slags" by Litasov et al. (Lithos, 340–341, p.181–190). *Lithos* **2019**, *348–349*, 105122. [CrossRef]
6. Griffin, W.L.; Gain, S.E.M.; Saunders, M.; Huang, J.-X.; Alard, O.; Toledo, V.; O'Reilly, S.Y. Immiscible metallic melts in the upper mantle beneath Mount Carmel, Israel: Silicides, phosphides and carbides. *Am. Mineral.* **2022**, *107*, 532–549. [CrossRef]
7. Ma, C.; Bindi, L.; Cámara, F.; Toledo, V. Griffinite, IMA 2021–110, in CNMNC Newsletter 66. *Mineral. Mag.* **2022**, *86*, 361–362.
8. Ma, C.; Griffin, W.L.; Bindi, L.; Cámara, F.; Toledo, V. Magnéliite, IMA 2021-111, in CNMNC Newsletter 66. *Mineral. Mag.* **2022**, *86*, 362.
9. Ma, C.; Griffin, W.L.; Bindi, L.; Cámara, F.; Toledo, V. Ziroite, IMA 2022-013, in CNMNC Newsletter 68. *Mineral. Mag.* **2022**, *86*, 854–855.
10. Ma, C.; Griffin, W.L.; Bindi, L.; Cámara, F.; Toledo, V. Sassite, IMA 2022-014, in CNMNC Newsletter 68. *Mineral. Mag.* **2022**, *86*, 855.
11. Ma, C.; Griffin, W.L.; Bindi, L.; Cámara, F.; Toledo, V. Mizraite-(Ce), IMA 2022-027, in CNMNC Newsletter 68. *Mineral. Mag.* **2022**, *86*, 857.
12. Ma, C.; Griffin, W.L.; Bindi, L.; Cámara, F. Toledoite, IMA 2022-036, in CNMNC Newsletter 66. *Mineral. Mag.* **2022**, *86*, 858.
13. Ma, C.; Griffin, W.L.; Bindi, L.; Cámara, F.; Toledo, V. Yeite, IMA 2022-079, in CNMNC Newsletter 70. *Mineral. Mag.* **2022**, *87*, 165.
14. Ma, C.; Cámara, F.; Toledo, V.; Bindi, L. Griffinite, Al_2TiO_5: A new oxide mineral from inclusions in corundum xenocrysts from the Mount Carmel area, Israel. *Crystals* **2023**, *13*, 1427. [CrossRef]
15. Andersson, S.; Magnéli, A. Diskrete Titanoxydphasen im Zusammensetzungsbereich $TiO_{1,75}$-$TiO_{1,90}$. *Naturwissenschaften* **1956**, *43*, 495–496. [CrossRef]
16. Holtstam, D.; Hålenius, U. Nomenclature of the magnetoplumbite group. *Mineral. Mag.* **2020**, *84*, 376–380. [CrossRef]
17. Griffin, W.L.; Gain, S.E.M.; Adams, D.T.; Huang, J.-X.; Saunders, M.; Toledo, V.; Pearson, N.J.; O'Reilly, S.Y. First terrestrial occurrence of tistarite (Ti_2O_3): Ultra-low oxygen fugacity in the upper mantle beneath Mt Carmel, Israel. *Geology* **2016**, *44*, 815–818. [CrossRef]

18. Griffin, W.L.; Gain, S.E.M.; Huang, J.-X.; Saunders, M.; Shaw, J.; Toledo, V.; O'Reilly, S.Y. A terrestrial magmatic hibonite-grossite-vanadium assemblage: Desilication and extreme reduction in a volcanic plumbing system, Mt Carmel, Israel. *Am. Mineral.* **2019**, *104*, 207–217. [CrossRef]
19. Armstrong, J.T. CITZAF: A package of correction programs for the quantitative electron microbeam X-ray analysis of thick polished materials, thin films, and particles. *Microbeam Anal.* **1995**, *4*, 177–200.
20. Ma, C.; Rossman, G.R. Barioperovskite, $BaTiO_3$, a new mineral from the Benitoite Mine, California. *Am. Mineral.* **2008**, *93*, 154–157. [CrossRef]
21. Ma, C.; Rossman, G.R. Tistarite, Ti_2O_3, a new refractory mineral from the Allende meteorite. *Am. Mineral.* **2009**, *94*, 841–844. [CrossRef]
22. Mandarino, J.A. The Gladstone-Dale relationship. I. Derivation of new constants. *Can. Mineral.* **1976**, *14*, 498–502.
23. Marezio, M.; Dernier, P.D. The crystal structure of Ti_4O_7, a member of the homologous series Ti_nO_{2n-1}. *J. Solid State Chem.* **1971**, *3*, 340–348. [CrossRef]
24. Marezio, M.; McWhan, D.B.; Dernier, P.D.; Remeika, J.P. Structural aspect of the metal-insulator transition in Ti_4O_7. *J. Solid State Chem.* **1973**, *6*, 213–221. [CrossRef]
25. le Page, Y.; Marezio, M. Structural chemistry of magnéli phases tin O(2n − 1) (4< n< 9) IV. Superstructure in Ti_4O_7 at 140 K. *J. Solid State Chem.* **1984**, *53*, 13–21.
26. Zhang, Q.; Liu, W.; Zhou, Y.; Li, J.; Sun, T.; Liu, Q.; Ma, Y.; Wang, J.; Li, J.; Zhao, R.; et al. Andersson-Magnéli phases Ti_nO_{2n-1}: Recent progress inspired by Swedish scientists. *Z. Anorg. Allg. Chem.* **2021**, *647*, 126–133. [CrossRef]
27. Momma, K.; Izumi, F. VESTA 3 for three-dimensional visualization of crystal, volumetric and morphology data. *J. Appl. Crystallogr.* **2011**, *44*, 1272–1276. [CrossRef]
28. Bondars, B.; Heidemane, G.; Grabis, J.; Laschke, K.; Boysen, H.; Schneider, J.; Frey, F. Powder diffraction investigations of plasma sprayed zirconia. *J. Mater. Sci.* **1995**, *30*, 1621–1625. [CrossRef]
29. Howard, C.J.; Kisi, E.H.; Roberts, R.B.; Hill, R.J. Neutron diffraction studies of phase transformations between tetragonal and orthorhombic zirconia in magnesia-partiallystabilized zirconia. *J. Am. Ceram. Soc.* **1990**, *73*, 2828–2833. [CrossRef]
30. Bouvier, P.; Djurado, E.; Lucazeau, G.; Le Bihan, T. High-pressure structural evolution of undoped tetragonal nanocrystalline zirconia. *Phys. Rev. B Condens. Matter Mater. Phys.* **2000**, *62*, 8731–8737. [CrossRef]
31. Onoda, M. Phase transitions of Ti_3O_5. *J. Solid State Chem.* **1998**, *136*, 67–73. [CrossRef]
32. Rusakov, A.A.; Zhdanov, G.S. The crystal structure and the chemical formula of titanic oxide Ti_3O_5 (Anosovite). *Dokl. Akad. Nauk SSSR* **1951**, *77*, 411–414.
33. Åsbrink, S.; Magnéli, A. Crystal structure studies on trititanium pentoxide, Ti_3O_5. *Acta Cryst.* **1959**, *12*, 575–581. [CrossRef]
34. Hong, S.-H.; Åsbrink, S. The structure of γ-Ti_3O_5 at 297 K. *Acta Cryst.* **1982**, *38*, 2570–2576. [CrossRef]
35. Ma, C.; Beckett, J.R. Kaitianite, $Ti^{3+}{}_2Ti^{4+}O_5$, a new titanium oxide mineral from Allende. *Meteorit. Planet. Sci.* **2021**, *56*, 96–107. [CrossRef]
36. Åsbrink, G.; Åsbrink, S.; Magnéli, A.; Okinaka, H.; Kosuge, K.; Kachi, S. A Ti_3O_5 modification of V_3O_5-type structure. *Acta Chem. Scand.* **1971**, *25*, 3889–3890. [CrossRef]
37. Åsbrink, S.; Pietraszko, A. Bond diffractometer measurements on Ti_3O_5 in the temperature region 90 to 720 K. *Phys. Status Solidi A* **1991**, *128*, K77–K81. [CrossRef]
38. Ohkoshi, S.; Tsunobuchi, Y.; Matsuda, T.; Hashimoto, K.; Namai, A.; Hakoe, F.; Tokoro, H. Synthesis of a metal oxide with a room-temperature photoreversible phase transition. *Nat. Chem.* **2010**, *2*, 539–545. [CrossRef]
39. Zhang, A.C.; Ma, C.; Sakamoto, N.; Wang, R.C.; Hsu, W.B.; Yurimoto, H. Mineralogical anatomy and implications of a Ti–Sc-rich ultrarefractory inclusion from Sayh al Uhaymir 290 CH3 chondrite. *Geochim. Cosmochim. Acta* **2015**, *163*, 27–39. [CrossRef]
40. Bermanec, V.; Holtstam, D.; Sturman, D.; Criddle, A.J.; Back, M.E.; Scavnicar, S. Nezilovite, a new member of the magnetoplumbite group, and the crystal chemistry of mognetoplumbite and hibonite. *Can. Mineral.* **1996**, *34*, 1287–1297.
41. Efremov, V.A.; Chernaya, N.G.; Trunov, V.K.; Pisarenko, V.F. Crystal structure of lanthanum magnesium hexaaluminate. *Kristallografiya* **1988**, *33*, 38–42.
42. Gasperin, M.; Saine, M.C.; Kahn, A.; Laville, F.; Lejus, A.M. Influence of M^{2+} ions substitution on the structure of lanthanum hexaaluminates with magnetoplumbite structure. *J. Solid State Chem.* **1984**, *54*, 61–69. [CrossRef]
43. Laville, F.; Perrin, M.; Lejus, A.M.; Gasperin, M.; Moncorge, R.; Vivien, D. Synthesis, crystal growth, structural determination, and optical absorption spectroscopy of the magnetoplumbite type compound $LaNiAl_{11}O_{19}$. *J. Solid State Chem.* **1986**, *65*, 301–308. [CrossRef]
44. Curien, H.; Guillemin, C.; Orcel, J.; Sternberg, M. La hibonite, nouvelle espèce minérale. *Comptes Rendus Hebd. Des Séances De L'académie Des Sci.* **1956**, *242*, 2845–2847.
45. Sandiford, M.; Santosh, M. A granulite facies kalsilite-leucite-hibonite association from Punalur, Southern India. *Mineral. Petrol.* **1991**, *43*, 225–236. [CrossRef]
46. Brukl, C.E.; Nowotny, H.; Schob, O.; Benesovsky, F. Die Kristallstrukturen von TiSi, $Ti(Al,Si)_2$ und $Mo(Al,Si)_2$. *Monatshefte Fuer Chem.* **1961**, *92*, 781–788. [CrossRef]
47. Svechnikov, V.N.; Kocherzhinskii, Y.A.; Yupko, L.M.; Kulik, O.G.; Shishkin, E.A. Phase diagram of the titanium-silicon system. *Dokl. Chem. Technol.* **1970**, *193*, 82–85.

48. Xiong, F.; Xu, X.; Mugnaioli, E.; Gemmi, M.; Wirth, R.; Grew, E.S.; Robinson, P.T.; Yang, J. Two new minerals, badengzhuite, TiP, and zhiqinite, $TiSi_2$, from the Cr-11 chromitite orebody, Luobusa ophiolite, Tibet, China: Is this evidence for super-reduced mantle-derived fluids? *Eur. J. Mineral.* **2020**, *32*, 557–574. [CrossRef]
49. Xiong, F.; Xu, X.; Mugnaioli, E.; Gemmi, M.; Wirth, R.; Yang, J.; Grew, E.S. Wenjiite, $Ti_{10}(Si,P,\square)_7$, and kangjinlaite, $Ti_{11}(Si,P)_{10}$, new minerals in the ternary Ti-P-Si system from the Luobusa ophiolite, Tibet, China. *Am. Mineral.* **2023**, *108*, 197–210. [CrossRef]
50. Griffin, W.L.; Gain, S.E.M.; Cámara, F.; Bindi, L.; Shaw, J.; Alard, O.; Saunders, M.; Huang, J.-X.; Toledo, V.; O'Reilly, S.Y. Extreme reduction: Mantle-derived oxide xenoliths from a hydrogen-rich environment. *Lithos* **2020**, *358–359*, 105404. [CrossRef]
51. Oliviera, B.B.; Griffin, W.L.; Gain, S.E.M.; Saunders, M.; Shaw, J.; Toledo, V.; Afonso, J.C.; O'Reilly, S.Y. Ti^{3+} in corundum: Tracing crystal growth in a highly reduced magma. *Sci. Rep.* **2021**, *11*, 2439. [CrossRef] [PubMed]
52. Ma, C.; Cámara, F.; Bindi, L.; Toledo, V.; Griffin, W.L. First terrestrial occurrence of kaitianite ($Ti^{3+}{}_2Ti^{4+}O_5$) from the upper mantle beneath Mount Carmel, Israel. *Minerals* **2023**, *13*, 1097. [CrossRef]
53. Fiore, M.; Beneduce, F.; Azevado, C.R.d.F. Simplification of the thermodynamic description of the Ti-Si system. *Tecnol. Em Metal. Mater. Mineração* **2016**, *13*, 91–97. [CrossRef]
54. Weitzer, F.; Schuster, J.C.; Naka, M.; Stein, F.; Palm, M. On the reaction scheme and liquidus surface in the ternary system Fe-Si-Ti. *Intermetallics* **2008**, *16*, 273–282. [CrossRef]
55. Fukai, Y. The Metal-Hydrogen System: Basic Bulk Properties. In *Springer Series in Materials Science*, 2nd ed.; Springer: Berlin/Heidelberg, Germany, 2005; Volume 21, 500p.

MDPI AG
Grosspeteranlage 5
4052 Basel
Switzerland
Tel.: +41 61 683 77 34

Materials Editorial Office
E-mail: materials@mdpi.com
www.mdpi.com/journal/materials

www.ingramcontent.com/pod-product-compliance
Lightning Source LLC
Chambersburg PA
CBHW040812250926
51850CB00007BA/1150
* 9 7 8 3 7 2 5 8 6 1 4 3 9 *